INVENTORY AND SUPPLY CHAIN MANAGEMENT WITH FORECAST UPDATES

Recent titles in the **INTERNATIONAL SERIES IN OPERATIONS RESEARCH & MANAGEMENT SCIENCE**
Frederick S. Hillier, Series Editor, *Stanford University*

Zhu/ *QUANTITATIVE MODELS FOR PERFORMANCE EVALUATION AND BENCHMARKING*
Ehrgott & Gandibleux/ *MULTIPLE CRITERIA OPTIMIZATION: State of the Art Annotated Bibliographical Surveys*
Bienstock/ *Potential Function Methods for Approx. Solving Linear Programming Problems*
Matsatsinis & Siskos/ *INTELLIGENT SUPPORT SYSTEMS FOR MARKETING DECISIONS*
Alpern & Gal/ *THE THEORY OF SEARCH GAMES AND RENDEZVOUS*
Hall/*HANDBOOK OF TRANSPORTATION SCIENCE - 2nd Ed.*
Glover & Kochenberger/ *HANDBOOK OF METAHEURISTICS*
Graves & Ringuest/ *MODELS AND METHODS FOR PROJECT SELECTION: Concepts from Management Science, Finance and Information Technology*
Hassin & Haviv/ *TO QUEUE OR NOT TO QUEUE: Equilibrium Behavior in Queueing Systems*
Gershwin et al/ *ANALYSIS & MODELING OF MANUFACTURING SYSTEMS*
Maros/ *COMPUTATIONAL TECHNIQUES OF THE SIMPLEX METHOD*
Harrison, Lee & Neale/ *THE PRACTICE OF SUPPLY CHAIN MANAGEMENT: Where Theory and Application Converge*
Shanthikumar, Yao & Zijm/ *STOCHASTIC MODELING AND OPTIMIZATION OF MANUFACTURING SYSTEMS AND SUPPLY CHAINS*
Nabrzyski, Schopf & Węglarz/ *GRID RESOURCE MANAGEMENT: State of the Art and Future Trends*
Thissen & Herder/ *CRITICAL INFRASTRUCTURES: State of the Art in Research and Application*
Carlsson, Fedrizzi, & Fullér/ *FUZZY LOGIC IN MANAGEMENT*
Soyer, Mazzuchi & Singpurwalla/ *MATHEMATICAL RELIABILITY: An Expository Perspective*
Chakravarty & Eliashberg/ *MANAGING BUSINESS INTERFACES: Marketing, Engineering, and Manufacturing Perspectives*
Talluri & van Ryzin/ *THE THEORY AND PRACTICE OF REVENUE MANAGEMENT*
Kavadias & Loch/*PROJECT SELECTION UNDER UNCERTAINTY: Dynamically Allocating Resources to Maximize Value*
Brandeau, Sainfort & Pierskalla/ *OPERATIONS RESEARCH AND HEALTH CARE: A Handbook of Methods and Applications*
Cooper, Seiford & Zhu/ *HANDBOOK OF DATA ENVELOPMENT ANALYSIS: Models and Methods*
Luenberger/ *LINEAR AND NONLINEAR PROGRAMMING, 2nd Ed.*
Sherbrooke/ *OPTIMAL INVENTORY MODELING OF SYSTEMS: Multi-Echelon Techniques, Second Edition*
Chu, Leung, Hui & Cheung/ *4th PARTY CYBER LOGISTICS FOR AIR CARGO*
Simchi-Levi, Wu & Shen/ *HANDBOOK OF QUANTITATIVE SUPPLY CHAIN ANALYSIS: Modeling in the E-Business Era*
Gass & Assad/ *AN ANNOTATED TIMELINE OF OPERATIONS RESEARCH: An Informal History*
Greenberg/ *TUTORIALS ON EMERGING METHODOLOGIES AND APPLICATIONS IN OPERATIONS RESEARCH*
Weber/ *UNCERTAINTY IN THE ELECTRIC POWER INDUSTRY: Methods and Models for Decision Support*
Figueira, Greco & Ehrgott/ *MULTIPLE CRITERIA DECISION ANALYSIS: State of the Art Surveys*
Reveliotis/ *REAL-TIME MANAGEMENT OF RESOURCE ALLOCATIONS SYSTEMS: A Discrete Event Systems Approach*
Kall & Mayer/ *STOCHASTIC LINEAR PROGRAMMING: Models, Theory, and Computation*

*** A list of the early publications in the series is at the end of the book ***

INVENTORY AND SUPPLY CHAIN MANAGEMENT WITH FORECAST UPDATES

SURESH P. SETHI
School of Management, University of Texas at Dallas, Richardson, TX, USA

HOUMIN YAN
The Chinese University of Hong Kong, Hong Kong, China

HANQIN ZHANG
Academy of Mathematics and Systems Science, Academia Sinica, China

Springer

Suresh P. Sethi
University of Texas @ Dallas
Richardson, TX, USA

Houmin Yan
The Chinese University of Hong Kong
China

Hanqin Zhang
Academia Sinica
China

Library of Congress Cataloging-in-Publication Data

A C.I.P. Catalogue record for this book is available
from the Library of Congress.

ISBN-10: 1-4020-8123-5 e-ISBN: 0-387-25663-6
ISBN-13: 978-1-4020-8123-1

Printed in the United States of America.

9 8 7 6 5 4 3 2 1 SPIN 11051428

springeronline.com

Contents

List of Figures

List of Tables

Preface

Supply chain management research has attracted a great deal of attention over the last ten years. This research covers an enormous territory involving multiple disciplines. It is carried out in the academia as well as by practitioners. A number of interesting topics that are examined are coordination of supply chains, supply chain design and re-engineering, competition of supply chain players, information dynamics, and contracts and incentive design.

From cottage industries and corner stores to today's search-engines in internet commerce, obtaining information and sourcing merchandise have been a major issue. Over the last 20 years, modern information technology has greatly changed the landscape of acquisition and distribution of both product and demand information. Companies have recognized the importance of learning about their customers needs and obtaining advance information. In addition, the progress in manufacturing technology, logistics services, and globalization makes it possible for companies to satisfy their customers from sources with different prices and lead times. Therefore, investigating ways to effectively distribute and obtain information, and to efficiently make use of different sources of production and transportation have been and are important foci of supply chain research.

With a careful analysis of real data collected from industry, we demonstrate the dynamics of information in the forecasting process. Our approach considers the forecasting process as one analogous to peeling away the layers of an onion—that is, the information at any given time has a number of sources of uncertainties that are resolved one by one in successive periods. We study the problem of supply chain decision making with such an information-updating process. The models considered in this book are inventory decisions with multiple delivery modes, supply-contract design and evaluation, and a two-player competitive supply chain. We formulate mathematical description of real problems, develop approaches for analysis of these models, and gain insights into better supply chain management. Much attention is given to characterization of the solutions—that is, inventory decisions prior and subsequent to information updates and the impact of the quality of information on these decisions.

Mathematical tools employed in this book involve dynamic programming and game theory. This book is written for students, researchers, and practitioners in the areas of Operations Management and Industrial Engineering. It can also be used by those working in the areas of Operations Research and Applied Mathematics.

The models and applications of supply chain decision making with information updates presented in this book are in their early stages of development. There have been a series of advances, but there is still much to be done. Therefore, many of the models addressed in the book could be further extended to capture more realism.

We wish to thank Qi Feng, Xiang-Hua Gan, Art Hsu, Hong-Yan Huang, Ke Liu, Ruihua Liu, Si-Tong Tan, and Hua Xiang, who have worked with us in the area of inventory and supply chain decision making with information updates. For their careful reading of the manuscript and able assistance at various stages in the writing of this book, we also want to thank our students Yumei Hou, Hui Li, Lijun Ma, Jun Wu, Jiankui Yang, and Haibo Yu. In addition we express our appreciation to Barbara Gordon and Joyce Xu for their assistance in the preparation of the various drafts of the manuscript.

This book is supported in part by the faculty research grants from School of Management, University of Texas at Dallas, the RGC (Hong Kong) Competitive Earmarked Research Grants, a Distinguished Young Investigator Grant from the National Natural Sciences Foundation of China, and a grant from the Hundred Talents Program of the Chinese Academy of Sciences. We are grateful for their support.

Richardson, TX, USA, January, 2005 — Suresh P. Sethi
Hong Kong, China, January, 2005 — Houmin Yan
Beijing, China, January, 2005 — Hanqin Zhang

To the memory of my brothers Champalal, Pannalal, and Sripal
To the memory of my sister Mohini
To my brothers Mahipal, Laxmipal, and Shantipal
To my sisters Mena, Kamalsri, and Sulochana — *Suresh P. Sethi*

To my wife Joyce and sons Ron and Justin — *Houmin Yan*

To my parents Shu-Ping Zhang and Yu-Jie Song — *Hanqin Zhang*

Notation

This book is divided into eight chapters. In any given chapter, say Chapter 2, sections are numbered consecutively as 2.1, 2.2, 2.3, 2.4, and so on. Subsections and sub-subsections are also numbered consecutively as 2.4.1, 2.4.2, ... and 2.4.3.1, 2.4.3.2, ..., respectively. Similarly, mathematical expressions such as equations, inequalities, and conditions, are numbered consecutively as (2.1), (2.2), (2.3), ... Figures, tables and propositions are numbered consecutively as Figure 2.1, Figure 2.2, ..., Table 2.1, Table 2.2, ..., and Proposition 2.1, Proposition 2.2,.... The same numbering scheme is used for theorems, lemmas, corollaries, definitions, remarks, and examples.

We provide clarification of some frequently-used terms in this book. The terms "surplus", "inventory/shortage", and "inventory/backlog" are used interchangeably. The terms "control", "policy", and "decision" are used interchangeably.

We make use of the following notation in this book:

$w.p.1$	with probability one
i.i.d.	independent, identically distributed
$\Longrightarrow$	denotes "implies"
$\Phi(x)$	$= \frac{1}{\sqrt{2\pi}} \int_{-\infty}^{x} e^{-\frac{t^2}{2}} dt$
$\Phi^{-1}(\cdot)$	the inverse function of $\Phi(\cdot)$
$\delta(x)$	$= \begin{cases} 1, & x > 0 \\ 0, & x \leq 0 \end{cases}$
I_D	the indicator function of a set D
$\emptyset$	the empty set
$\square$	end of a proof

$(\Omega, \mathcal{F}, \mathsf{P})$	the probability space
$\mathsf{P}(\xi \in \cdot)$	the probability distribution of a random variable ξ
$\mathsf{E}\xi$	the expectation of a random variable ξ
$\mathsf{Var}\xi$	the variance of a random variable ξ
$(a_1, \ldots, a_l) > 0$	means $a_1 > 0, \ldots, a_l > 0$
$(a_1, \ldots, a_l) \geq 0$	means $a_1 \geq 0, \ldots, a_l \geq 0$
$a \geq b$	means $a - b \geq 0$ for any vectors a and b
$a_1 \wedge \cdots \wedge a_l$	$= \min\{a_1, \ldots, a_l\}$ for any real numbers a_i, $i = 1, \ldots, l$
$a_1 \vee \cdots \vee a_l$	$= \max\{a_1, \ldots, a_l\}$ for any real numbers a_i, $i = 1, \ldots, l$
x^+	$= \max\{x, 0\}$ for a real number x
x^-	$= \max\{-x, 0\}$ for a real number x
$\lfloor x \rfloor$	the largest integer smaller than or equal to a real number x

Chapter 1

INVENTORY AND SUPPLY CHAIN MODELS WITH FORECAST UPDATES

1.1. Introduction

Most global companies deal with customers that have different degrees of demand variability and forecasting ability. Companies with superior forecasting abilities can afford to procure or produce a large fraction of their demand by making use of slow production modes and inexpensive logistics services, paying a premium for faster production and logistics services only when demand surges unexpectedly. Companies with irregular demands and inferior forecasting ability have to pay dearly for using fast production modes to respond to unexpected surges in demand.

Companies have recognized the importance of managing a portfolio of customers with different needs and have recognized the value of learning about customer demands in advance. As observed by Fisher, Hammond, Obermeyer, and Raman [22] in the case of the apparel industry, regrouping forecasting efforts from all sources (such as firm orders received, preseasonal sale information, and the point-of-sales data) have been remarkably effective in obtaining demand information in advance. Effective use of early demand information has been a major initiative in many industries, such as the apparel industry (Fisher, Hammond, Obermeyer, and Raman [22]; Iyer and Bergen [37]), the toy industry (Barnes-Schuster, Bassok, and Anupindi [5]), and the computer and electronics industry (Tsay and Lovejoy [67]; Brown and Lee [9]; Yan, Liu, and Hsu [72]).

In addition, the advances in manufacturing technology, logistics services, and globalization make it possible for companies to satisfy their customer needs from sources with different prices and lead times. The advance demand information improves their understanding of customer demand. On the one hand, the ability to provide a better forecasting increases as the delivery date approaches. On the other hand, the costs of products and logistics services increase as a shorter lead time is required. Therefore, it is critical for companies to use advance demand information, different manufacturing technologies, and logistics services to strike a balance between the quality of demand information and the costs of production and logistics services.

In the last decade or so, supply chain management has attracted a great deal of attention from people in academia and industry. Research in supply chain management covers an enormous territory, involves multiple disciplines, and employs both quantitative and qualitative tools. A wide range of topics have been explored, and a great diversity of details of those topics have been examined. Managerial introductions to supply chain management can be found in Copacino [17], and Handfield and Nichols [32]. Simchi-Levi, Kaminsky, and Simchi-Levi [62], and Tayur, Ganeshan, and Magazine [65] provide more technical, model-based treatments of supply chain management.

In the literature, we see that the focus is primarily on methods for coordination and improving system efficiencies, supply chain re-engineering by delaying product differentiation, information dynamics and its impact on supply chain performance, competition among supply chain players, and the design of supply and purchase contracts and incentives. In these models, it is generally agreed that information is critical in supply chain decisions and therefore that it is important to explore advance demand information. However, the problem of including demand-information updates in supply chain decisions remains largely an open research area.

In the last four to five years, we have been modeling the problem of dynamic supply chain decision making with information updating. Our model includes inventory decisions with multiple sources and delivery modes, supply contracts design, and a competitive supply chain model. We mathematically formulate real problems into tractable models, develop approaches for their analysis, and present insights into better supply chain management. In this volume, we provide a unified treatment of the above models, summarize our major results, present a critique of the existing results, and point out potential research directions.

1.2. Aims of the Book

Customer demands, supply conditions, sales, and raw-material prices are the fundamental pieces of information that companies need to plan their operations in stocking, production, and distribution. Market uncertainties, information disparity and distortion, globalization, and shortening lead times make supply chain planning a challenging venture.

Modern technologies for obtaining advance information and manufacturing logistics provide companies with the means and tools that they need to deal with the above challenges. However, the classical stochastic inventory models, which provide information about when to buy and how many units to purchase, do not take advance information and multiple alternatives in manufacturing logistics into consideration. Therefore, our first task is to model the process of obtaining advance information and its impact on inventory decisions.

We model the forecast-updating process as one analogous to peeling away the layers of an onion (Sethi, Yan, and Zhang [61]). That is, the information in any given period (hidden in the core of an onion) has a number of sources of uncertainties (hidden in the layers of the onion), and these uncertainties are resolved successively in periods leading to the period in which the demand materializes (peeling the layers one by one to get to the core). With such an information-updating process, it is possible for us to consider the optimal inventory decisions with new information.

It is common that, for the same sets of goods, companies provide their customers with a choice between different lead times or delivery alternatives. For examples, Hewlett-Packard's MODO boxes are assembled in its Singapore factory, but the factory allows HP's distribution centers in Roseville (California), Grenoble, Guadalajara, and Singapore to choose between ocean and air shipments (Beyer and Ward [7]). These differences in lead times and delivery alternatives may result in different charging schemes. It is generally assumed that the faster delivery modes are more expensive than the slower ones. So a combination of multiple delivery modes and information updating will doubtlessly present an efficient approach to coordinating the supply chain and mitigating the distortion of demand and price information. To take advantage of multiple delivery modes, we likened the forecast-updating process to that of peeling an onion. In a model with multiple delivery modes, we explore the form of an optimal policy.

A supply contract is an agreement between a buyer and a supplier that stipulates the terms of the purchase in an environment of incomplete information and possible reaction alternatives. Different forms of contracts have received a great deal of attention recently from practitioners and researchers. Research in this area focuses mainly on contract management and incentive design. The former tackles an optimization problem, while the latter addresses an issue of supply chain coordination. However, the main incentive to having both the buyer and the supplier to get some form of contract is that the contract provides the buyer with an option to revise its decision with incomplete information in addition to some degree of certainty to the supplier in allocating its capacity. Therefore, it is critical for both parties to understand the potential of new information before designing and executing a contract. This brings us to the second task—information updating and contract design, execution, and management.

For contract management, we pursue research in two directions: quantity-flexibility contract models and competitive models. Quantity-flexibility contracts allow the buyer in a supply chain to postpone some of his purchases to a later date and at a favorable price after an improved forecast of customer demand becomes available. Thus the contract provides the buyer with a cushion against demand uncertainty. The supplier, on the other hand, benefits from having a smoother production schedule as a result. Here we focus mainly on quantity-flexibility contracts that involve one demand-forecast update in each period and a spot market with or without a fixed exercise price. With regard to competitive models, there is a body of research work on the supply-contract context that investigates channel performance through the competitive-behavior study of the supply chain. Information updating adds another dimension to both speculative and reactive decisions. Therefore, our last task is to investigate the competitive behavior of supply chain players with respect to the impact of information updating.

Before closing this section, we want to point out that for some forms of contracts, it takes months from the signing to the execution of the contract. Therefore, it is problematic to consider such criteria as expected profit maximization or expected cost minimization particularly when profit and cost variances and the uncertainty in information are large. This brings us to the risk analysis of supply contracts with information updating. In the fields of economics and finance, agents are often assumed to be risk-averse, and they maximize a concave utility of wealth (von Neumann and Morgenstern [68]). A simple operational approach to dealing with risk aversion is that of mean-variance analysis (Markowitz [50]). There have been a few attempts in the inventory and supply chain management literature to deal with risk aversion. Lau and Lau [42] study a single-supplier, single-retailer supply chain, where both the retailer and the supplier use objective functions that increase with the expected profit and decrease with the variance of profit. Note that while they consider aversion to risk, their objective function is not a von Neumann-Morgenstern-type utility in general. Chen and Federgruen [14] revisit a number of basic inventory models using the mean-variance approach. They conclude that for risk-averse decision makers, the optimal order quantity is less than the one that corresponds to maximizing the expected profit. Gan, Sethi and Yan [26, 27] consider supply chains with risk-averse agents. They provide a general definition of coordination for such supply chains. They obtain coordinating contracts explicitly in a number of cases. In a case with utility-maximizing agents, they also show that the contract yields a Nash bargaining solution. Buzacott, Yan and Zhang [10] study a class of commitment and option supply contracts in the mean-variance framework with demand-information updating. It is shown that a mean-variance tradeoff analysis with advance reservation can be carried out efficiently. Moreover,

Yan, Yano and Zhang [73] consider multiperiod inventory models in which the risk aversion is measured by a probability constraint to a target performance index. They prove that the optimal policies are threshold-control type and not base-stock type. Further work on this topic is currently in progress.

To summarize, we present in this book our research results in inventory and supply chain management that involve information updates. The topics span from the stochastic dynamic inventory models with different delivery modes, contracts with exercise prices, quantity-flexibility contracts accompanied with spot-market purchase decisions, to competitive supply chains. The rest of this chapter reviews the related literature and highlights our modeling approaches and main results.

1.3. Information Dynamics in Supply Chains

Sourcing and obtaining information have been major ventures since the earliest form of trading and commerce. Over the past 20 years, modern information technology has greatly improved the efficiency of obtaining and distributing information. Examples of these technologies include continuous-replenishment programs (CRPs) based on electronic-data-exchange technology (at Procter & Gamble) and vendor-managed-inventory (VMI) systems based on point-of-sale data technology (at Wal-Mart). Massive investments in information technology have been made by manufacturers, distributors, and retailers with the hope of achieving supply chain coordination. Investigating ways to effectively distribute and use information in a supply chain have been a centerpiece in supply chain management research.

The information we refer to is primarily about demand and price. Demand information has a direct impact on production scheduling, inventory control, and delivery plans of individual members in the supply chain. At the same time, price information affects the buyers' allocation of their purchasing quantities, which in turn affects the demand. Since demand information is a key factor in supply chain management, we review various demand models and ways that demand affects supply chain management. The key objective for supply chain management is to better match supply with demand to reduce the costs of inventory and stockout. Researchers have found that disparities in supply and demand result partially from distorted demand and price information. On the normative side, the combination of sell-through data, inventory-status information, order coordination, and simplified pricing schemes can help mitigate information distortion. To overcome this shortcoming, the information-updating and information-sharing processes deserve thorough investigation. Many companies have embarked on initiatives that enable more demand information sharing between their downstream customers and their upstream suppliers. Research on the effects of information on supply chain management has focused on three issues—information distortion, information sharing, and information updating. To better understand the importance of obtaining advance information and making use of information updates, we believe that it is necessary to review the literature on information distortion and on information sharing and to recall recent initiatives and practices from various industries.

1.3.1 Information Distortion in Supply Chains

It is commonly agreed that meeting customer demand is the primary goal of a supply chain. Therefore, information about customer demand should be the basis for decision making by a supply chain manager. However, the orders at the upstream of a supply chain have been observed to exhibit a higher level of variability than those at the downstream, which is nearer to the customer. The phenomenon of information distortion, popularly known as the *bullwhip effect*, is one of the early finds in the study of the information dynamics of a supply chain. If companies make their supply chain decisions based on their orders instead of on customer demand, the bullwhip effect leads companies to make inaccurate demand forecasts, acquire excessive inventory, and be less efficient in capacity utilization. Lee, Padmanabhan, and Whang [43, 44] systematically investigate the cause of the information distortion within a supply chain. They conclude that demand-signal processing, rational games, order batching, and price variation are the major causes. Remedies for these causes are also provided.

Following the work of Lee, Padmanabhan, and Whang [43, 44], there is a large body of work that explores the causes of the bullwhip effect as well as methods for controlling its impact. Metters [51] establishes an empirical lower bound of detrimental effect that the bullwhip effect may have. His results indicate that reduction of the bullwhip effect can improve profitability in a dramatic fashion. Chen, Dreaner, Ryan, and Simchi-Levi [13] identify the causes and quantify the increase in variability due to demand forecasting and lead times. They further extend their results to consider the impact of centralized demand information on the bullwhip effect. Methods for reducing the impact of the bullwhip effect are also proposed. These methods include reducing the variability that is inherent in the customer demand process, reducing lead times, and establishing strategic partnerships.

1.3.2 Information Sharing in Supply Chains

By reducing lead times (information delays), multiple data entries, and the bullwhip effect, information technology has had a substantial impact on supply chains. Many industries have embarked on information sharing efforts to improve the efficiency of their supply chains. Scanners collect sales data at the point of sale, and electronic data interchange (EDI) allows these data to be transmitted and broadcasted immediately to individuals in the supply chain. The application of information technologies, especially in the grocery industry, has substantially helped better match supply with demand to reduce production and delivery lead times, and the costs of inventory and stockout. Sharing information among parties in a supply chain has been viewed as a major strategy for countering difficulties such as inaccurate demand forecasts, low capacity utilization, excessive inventory, double marginalization, and poor customer service. For example, letting the supplier have access to retailers' sale data can help ameliorate the detrimental effects of demand distortion. The benefits of information sharing in the supply chain also motivate industrial application programs like vendor-managed-inventory (VMI), continuous-replenishment programs (CRPs), and quick-response programs (QRPs).

In coordinating supply chain models with information sharing, some members of the supply chain are happy with improved information, while others believe that its benefit does not justify its cost (see Takac [64]). Thus, while information is beneficial in general, it is interesting to quantify the value of information sharing between members of a supply chain. Bourland, Powell, and Pyke [8], Cachon and Fisher [11], Gavirneni [28], Gavirneni, Kapuscinski, and Tayur [29], Lee, So, and Tang [46], Li and Zhang [47], Moinzadeh [53], and Simchi-Levi and Zhao [63] are some works dealing with the value of information sharing in a supply chain. Bourland, Powell, and Pyke [8] examine the case in which the review period of the manufacturer is not synchronized with the retailer. Similarly, Cachon and Fisher [11] show analytically how the manufacturer can benefit from using information about the retailer's inventory levels when the retailers use a batch-ordering policy. Also studied in [8, 11] is the value of resolving a part of uncertainty by obtaining some information about the retailer's demand. Gavirneni [28], and Gavirneni, Kapuscinski, and Tayur [29] consider two cases of information sharing between manufacturer and retailer. In the first case, the manufacturer obtains information from the retailer about the parameters of the underlying demand and the cost of the (s, S) ordering policy adopted by the retailer. In the second case, the manufacturer obtains additional information from the retailer about the period-to-period inventory level. Under various types of demand distributions, they compare the optimal costs associated with these two cases. Conditions under which gaining information about the retailer's inventory is beneficial are also explored. Li and Zhang [47] study the relationships among demand variability, inventory management, and information sharing in a supply chain consisting of one retailer as well as multiple retailers. The retailers have private information about their customer demands and may share it with the supplier. They prove that the strategic reactions of the retailers change the values of information as well as the supplier's inventory decisions. Moinzadeh [53] considers a supply chain model consisting of a single product, one supplier, and multiple retailers. The supplier has online information about the demand and inventory activities of the product at each retailer and uses this information when making ordering decisions. Numerical work is carried out to identify the parameter setting under which information sharing is most beneficial. Simchi-Levi and Zhao [63] consider a single-product, periodic-review, two-stage production-inventory system with a single capacitated supplier and a single retailer facing independent demand and using an order-up-to inventory policy. For this supply chain model, they solve the problems that arise when information sharing provides significant cost savings and address how the supplier can use this information most effectively in make-to-stock production systems.

Lee, So, and Tang [46] use a serially correlated demand model to explore the value of information sharing in a two-stage supply chain. They also examine the impact of the correlation coefficient and the lead times on expected inventory reduction.

1.3.3 Information Updates in Supply Chains

Related research has been carried out in the area of inventory management with demand-information updates. It is possible to classify this line of research into the following three categories.

The first category is to use time series to update the demand forecast. This approach is very powerful when there is a significant intertemporal correlation among the demands of consecutive periods (see Johnson and Thompson [39], and Lovejoy [49]). They model the demand process as an integrated autoregressive moving-average process and show the optimality of myopic policies under certain conditions. Recently, Aviv [3] has formulated the underlying demand process of a supply chain in a linear state-space framework. As a result, the demand realization during each period can be written as a linear function of a state vector that evolves as a vector autoregressive time series. Employing the Kalman filter technique, the minimum mean-square error forecast of future demands at each location of the supply chain can be obtained, and an adaptive inventory order policy can be given.

The second category is concerned with forecast updates. This approach is developed by Hausmann [33], Sethi and Sorger [60], Graves, Meal, Dasu, and Qiu [30], Heath and Jackson [35], Donohue [18], Yan, Liu, and Hsu [72], Gurnani and Tang [31], Barnes-Schuster, Bassok, and Anupindi [5], Huang, Sethi, and Yan [36], and Gallego and Özer [25]. Hausmann [33] models the evolution of the forecast as a quasi-Markovian process and provides optimal decision rules for sequential decision problems. Sethi and Sorger [60] formulate a fairly general model that allows for unrestricted forecast updates at some forecast cost. They also provide an optimality framework for the usual practice of rolling-horizon decision making. They develop dynamic programming equations to determine optimal rolling horizons, optimal forecast decisions, and optimal production plans. While their model represents a significant conceptual advance, the computation of optimal decisions suffers from the curse of dimensionality. Graves, Meal, Dasu, and Qiu [30] and Heath and Jackson [35] use a martingale to model the forecast evolution. They analyze economic safety-stock levels for a multi-product, multi-facility production system. Yan, Liu, and Hsu [72] obtain the optimal order quantity in a single-period, two-stage model with dual supply modes and demand-information updates. For uniformly distributed demand forecasts, they show further that an optimal solution can be myopic, if some regularity conditions are satisfied. Donohue [18] considers a risk-sharing supply contract between a buyer and a supplier. She discusses pricing issues when the demand-information update is perfect. For a bivariate normal demand, Gurnani and Tang [31] provide an explicit solution in the cases of worthless and perfect information updates. Barnes-Schuster, Bassok, and Anupindi [5] consider a single-period, two-stage model with updating information arriving at the beginning of the second stage. They provide structural properties of the objective functions of the buyer and the supplier. The issue of channel coordination is also discussed. Huang, Sethi, and Yan [36] consider a single-period, two-stage supply contract model with both fixed and variable costs and demand-information updates. The information updates can vary from being worthless to being perfect. For a uniformly distributed demand forecast, they are able to provide an explicit solution. The explicit nature of the solution leads to important insights into a better supply-contract management. Gallego and Özer [25] model the forecast evolution as a supermartingale and prove the optimality of a state-dependent (s, S) policy.

The third and last category is Bayesian analysis. Bayesian models are first introduced in the inventory literature by Dvoretzky, Kiefer, and Wolfowitz [19]. In this framework, the demand distribution is chosen from a family of distributions whose parameters are not specified with certainty. Bayes's rule defines a procedure to update this distribution as new information becomes available. Scarf [58] characterizes an adaptive optimal order policy, which depends on the past history, for the case of exponential family of distributions. Azoury [4], and Lariviere and Porteus [41] extend the work of Scarf [58] to other classes of distributions. Eppen and Iyer [21] analyze a quick-response program in a fashion-buying problem by using Bayes's rule to update demand distributions.

Here we emphasize two approaches with the Bayesian analysis framework. One models the demand process as a normal distribution with a known variance. The other employs a normal distribution with an unknown variance to investigate the dynamics of demand updating. We elaborate them in what follows.

Iyer and Bergen [37] analyze a quick-response system in the fashion industry by using the Bayesian method to update demand distribution. In a quick-response system, demand is modeled as a normal distribution with unknown mean and known variance. Then there are two folds of demand uncertainty: one arises from the demand itself; the other results from the uncertain mean. With a Bayesian updating mechanism, they show there is a decrease of demand variance as information updating is introduced. Many papers in the literature assume that demand variance is known. Then the decrease of demand variance through updating demand information before the selling season is reasonable and practicable. Therefore, a lot of literature further explores the value of information updating in supply chain performance, especially the use of a dual mode of supply to improve the efficiency of supply contract. However, if the variance of demand itself is also uncertain, we are interested in what the demand-uncertainty structure is.

Based on the analysis of the data obtained from an electronic company, we make an interesting observation. The company uses a rolling horizon method to update its forecast. The data provides us an opportunity to observe the evolution of the forecasting process and the forecast-error process. We observe that the forecast error decreases as more demand information comes in. However, when compared with the initial forecast, the updated forecast exhibits a larger variance. We provide our analysis and interpretation of this observation in Chapter 2.

1.4. Inventory and Supply Chains with Multiple Delivery Modes

Starting with Fukuda [24], several researchers have investigated inventory problems with limited or no information updating on the ordering costs and demands. Most studies focus on two delivery modes with different costs and lead times separated by one review period. For two delivery modes, Fukuda [24] shows that the optimal policy is similar to those of the dynamic inventory problem with a single-procurement mode—that is, a base-stock type of inventory-control policy with a stock order-up-to level for procurement mode. Under a similar framework, Hausmann, Lee, and Zhang [34] study an inventory system with two procurement modes for a stationary demand. They derive an explicit formula for the optimal order quantities, assuming linear inventory holding and shortage costs. Whittemore and Saunders [71] consider air and surface delivery modes with lead times of τ and $(\tau+1)$ review periods, where τ is any positive integer. They allow for fixed and variable ordering costs associated with ordering placements. Rosenshine and Obee [57] examine a standing-order inventory system, where a regular order of constant size is received every period and an emergency order of fixed size may be placed once per period and arrives immediately.

Chiang and Gutierrez [15] analyze a different periodic-review inventory system with a faster supply channel and a slower supply channel, using both the dynamic programming approach and the approach of minimizing the average cost per unit of time. They allow lead times to be shorter than a review period. At each review epoch, the manager must decide whether to place a regular order or an emergency order. In a sequel paper, Chiang and Gutierrez [16] consider a problem where multiple emergency orders can be placed at any time within a review period, including the time of the regular order. Scheller-Wolf and Tayur [59] study a periodic-review nonstationary Markovian dual-source production inventory model with stochastic demand and holding and penalty costs (all state-dependent). It is shown that under certain ordering cost and demand conditions, there exists an optimal policy indexed by the state of the Markov chain. However, for the general case, an optimal policy is not easy to be constructed.

For three or more delivery modes or for two modes separated by more than one review period, the problem becomes much more complex. To our knowledge, Fukuda [24] and Zhang [74] are the only papers that address the three-procurement-mode problem. Fukuda [24] investigates a three-procurement-mode problem under the assumption that orders can be placed only in every other period. He shows that, under this assumption, the problem is equivalent to a two-procurement-mode problem. Zhang [74] extends Fukuda's work to three procurement modes with infinite horizon and discounted cost. Assuming that the difference between the lead times is one period and that the inventory-holding and shortage costs are linear, she analyzes two cases and obtains the structure of the optimal order policy. In the first case, explicit formulas to calculate the optimal order-up-to levels are derived. In the second case, she discusses some structural properties and proposes a newsvendor-based heuristic policy.

The models investigated in Chapters 3, 4, and 5 of this book, as has already been mentioned, consider both advance demand information and multiple supply sources. Chapter 3 is concerned with the case of two consecutive delivery modes without set-up costs for each supply source. We show that state-dependent (dependent on the observed information) base-stock policies are optimal for finite-horizon problems as well as for discounted infinite-horizon problems. Such policies are defined by a pair of numbers—one for the fast mode and the other for the slow mode. These numbers are known as the base-stock levels. Chapter 4 is related to the case of two consecutive delivery modes: with a set-up cost for each supply source, the (s, S)-type policies can be proved to be optimal. Chapter 5 is devoted to the case of three consecutive delivery modes (fast, medium, and slow) without set-up costs for each supply source.

It is shown that in all cases, there is a base-stock policy for fast and medium modes that is optimal. Furthermore, the optimal policy for the slow mode is not a base-stock policy in general. At the same time, we also investigate why the base-stock policy is or is not optimal in different situations.

1.5. Supply Contracts

It is well documented that imperfect demand information influences the buyer's decision about order quantity and the manufacturer's decision about production plan, especially when production lead time can be significantly large. To facilitate the tradeoff between production lead time and imperfect demand information, various forms of supply contracts exist in industries. A contract provides flexibility either in absolute order size or in combination of different products or provides a so-called downside risk for buyers. In the last few years, supply contracts have attracted much attention.

Bassok and Anupindi [6] analyze a single-product periodic-review inventory system with a minimum-quantity contract, such that the cumulative purchase over a multiple periods must exceed a minimum quantity to qualify for a price-discount schedule. Bassok and Anupindi [6] are able to demonstrate that the optimal inventory policy for the buyer is an order-up-to type and that the order-up-to level can be determined by a newsvendor model. Anupindi and Bossok [2] further extend their previous work to the case of multiple products. For the case of multiple products, the supply contract requires that the total purchase over different products exceeds a minimum dollar amount to obtain the price discount. Tsay [66] studies incentives, causes of inefficiency, and possible ways of performance improvement over a quantity-flexibility contract between a buyer and a supplier. In particular, Tsay [66] investigates the quantity revision in responding to demand-information revisions, where the information is the location parameter of the demand distribution.

Similar to the structure of quantity-flexibility contracts, a form of minimum commitment or take-or-pay provision has been used in many long-term natural-resources and energy-supply contracts (Tsay [66]). A take-or-pay contract is an agreement between a buyer and a supplier. A take-or-pay contract often specifies a minimum volume that the buyer must purchase (take) and a maximum volume that the buyer can obtain (pay) over the contract period. Brown and Lee [9] note that the problem of capacity-reservation agreements in the semiconductor industry has a similar structure. Brown and Lee [9] examine how much capacity should be reserved (take) and how much capacity should be reserved for the future (pay). In a general case of a minimum-commitment contract, Anupindi and Akella [1], Moinzadeh and Nahmias [54], Bassok and Anupindi [6], and Anupindi and Bassok [2] study the optimal order policy for finite horizon problems.

In a buy-back contract, the supplier specifies his selling price and promises to take the unsold goods back at a predetermined price. Therefore, the buy-back contract establishes the responsibility for unsold inventory. One can make an analogy between a buy-back contract and a quantity-flexibility contract, in that both structures lay out ground rules to compensate the buyer for a decision that was made prior to the demand realization. However, a subtle difference exists such that the buy-back takes effect after demand is observed, whereas the execution takes place when demand uncertainty may still remain.

An analytical treatment of a buy-back contract was first carried out by Pasternack [55]. His model deals with one supplier and one retailer in a supply chain. The result shows that if a setting can be manipulated to look like a newsvendor problem, it can be successfully decentralized through a system of linear prices. Pasternack determines that coordination of the channel can be achieved by a buy-back contract that allows a full return at a partial refund and that the efficient prices can be set in a way that guarantees Pareto improvement. Kandel [40] covers much of the same ground as Pasternack [55]. In particular, he emphasizes the incentive for a supplier to implement a consignment policy. He also notes that if the demand distribution depends on the retail price, coordination cannot be achieved through buy-backs unless the supplier can impose resale price maintenance.

Gurnani and Tang [31] and Yan, Liu, and Hsu [72] study the effect of information updates on the decision making of the buyer in a dual-mode supply chain. More specifically, Yan, Liu, and Hsu [72] study how an updated forecast affects a buyer's commitment with a supplier, and Gurnani and Tang [31] assume that there is an uncertain unit-purchasing cost faced by the buyer at the second stage—namely, a high one and a low one. They investigate the impact of uncertain cost and forecast updating in a supply chain from the perspective of the buyer.

In Chapter 6, we develop a model that analyzes quantity-flexibility contracts in a setting with single or multiple periods involving one demand-forecast update in each period and a spot market. We obtain the optimal order quantity at the beginning of a period and order quantities on contract and from the spot market at the prevailing price after the forecast revision and before the demand materialization. The amount that can be purchased on contract is bounded by a given flexibility limit. We discuss the impact of the forecast quality and the level of flexibility on the optimal decisions and managerial insights behind the results.

In Chapter 7, we study a supply contract with a fixed exercise price. The purchase contract provides the buyer with an opportunity to adjust an initial commitment based on an updated demand forecast obtained at a later stage. An adjustment, if any, incurs a fixed as well as a variable cost. We formulate the buyer's problem as a dynamic programming problem. We derive explicit optimal solutions for a class of demand distributions including uniform distributions. In addition, we obtain the critical value of the fixed contract-exercise cost, below (or above) which the buyer would (or would not) sign the contract. Our results lead to valuable insights into better supply chain management.

1.6. Competitive Supply Chains

Competitive study is another body of research that investigates the efficiency of supply chain management. In this book, Chapter 8 is concerned with the pricing issue and the value-of-information issue based on game theory.

The behaviors of the decision makers are locally rational and are often inefficient from a global point of view. The attention of some researchers has turned to mechanisms for improving the efficiencies of the entire supply chain. Contractual arrangements and information sharing fall mainly into this area. It is understood that no single agent has control over the entire supply chain. Therefore, no agent has the power to optimize the entire supply chain. It is also reasonable to assume that each agent will attempt to optimize his own preference, knowing that all of the other agents will do the same.

The methodological tool employed in this field is game theory. The modeling of a game can be either static or dynamic, with or without complete information, in settings of supply chain management. With game theory, the behavior of players can be determined when they seek to maximize their own welfare. The key issues include whether there exists a Nash equilibrium, the uniqueness of the equilibrium, and whether the optimal policies belong to the set of Nash equilibria. The most interesting part is finding whether competitive and optimal behavior coincide, assessing which party would benefit, and examining cases where the supply chain coordination is a matter of interest.

In a single-period setting, Lippman and McCardle [48] extend the standard newsboy problem to a competitive setting, where the random demand is split between two or more firms. Suppliers compete with others to maximize their own profits. The authors examine the effect of competition on industry inventory and the relation between equilibrium inventory levels and the splitting rule.

A number of papers provide more detailed models of supply chain inventory management with information updates and collaborative decision making within two independent parties. Recent examples include Tsay [66], Cachon and Zipkin [12], Barnes-Schuster, Bassok, and Anupindi [5], and Donohue [18]. Of these four, the last one is the most relevant to our model, as described below. For those papers considering supply-contract issues in inventory management with prior demand information, see Tsay [66] for a detailed review. Tsay [66] investigates quantity-flexibility contracts in a multiparty supply chain: the buyer purchases no less than a certain percentage below the forecast, whereas the supplier delivers up to a certain percentage above. He focuses on the implications of quantity-flexibility contracts for the behavior and performance of both parties and for the supply chain as a whole.

Cachon and Zipkin [12] analyze channel competition and cooperation in a supply chain with one supplier and one retailer. In a one-period setting, the Nash equilibrium of the game, between the supplier and the retailer, is derived through choosing their individual order quantity to their own objectives. The optimal solution is derived if the objective is to minimize total supply chain costs. They emphasize the contracting issues in realizing the value of cooperation. They also provide a Stackelberg model in the same setting, which is different from ours mainly in that we consider a two-stage problem with information updating within a period.

Barnes-Schuster, Bassok, and Anupindi [5] provide a two-period correlated-demand model for analysis of the role of options in a buyer-supplier system. In the first period, while the buyer decides profit-maximizing order quantities for both periods, as well as the options that would be exercised partially or totally in the second period, the supplier makes decisions on the profit-maximizing production quantity. In the second period, the buyer chooses to exercise quantity options based on the observed demand in a previous period. The authors give a numerical evaluation of the value of options and coordination as a function of demand correlation and the service level offered.

Donohue [18] investigates a supply-contract problem in which a manufacturer and a buyer are involved in a two-stage problem. She designs a centralized system where the manufacturer decides the production quantities in both periods and faces the demand in the market directly, which means only one player in the channel. With this centralized system as a benchmark, the decentralized system includes the two players in the two-stage problem. The contract pricing scheme is fixed—that is, (w_1, w_2, b) where w_i is the wholesale price in stage i and b is the return price for excess product at the end of the season. For the issue of supply-contract pricing, Emmons and Gilbert [20], Monahan [52], Lee and Rosenblatt [45], and Rosenblatt and Lee [56] investigate supply contracts with quantity-discount schemes. In innovative works from a marketing perspective,

Jeuland and Shugan [38] and Weng [69, 70] consider the impact of pricing in channel coordination.

Chapter 8 focuses on a problem that can be stated as follows: the production lead time of the manufacturer requires a buyer to make purchase decisions without accurate demand information. The buyer is aware that improved demand information will be available at a later time. A purchase contract that allows the buyer to modify its initial order quantity before a specific date with both fixed and variable penalties provides volume flexibility to the buyer and brings additional income to the manufacturer (supplier). To the buyer, the problem is how to make initial orders and how to react to the demand information obtained in the later stage to minimize total cost. To the supplier, the problem is how to design the contract to maximize profit.

References

[1] R. Anupindi and R. Akella. An inventory model with commitments. Working Paper, Northwestern University, Evanston, IL, 1993.

[2] R. Anupindi and Y. Bassok. Approximations for multiproduct contracts with stochastic demands and business volume discounts: Single-supplier case. *IIE Transactions*, 30:723–734, 1998.

[3] Y. Aviv. A time-series framework for supply-chain inventory management. *Operations Research*, 51:210–227, 2003.

[4] K.S. Azoury. Bayes solution to dynamic inventory models under unknown demand distribution. *Management Science*, 31:1150–1160, 1985.

[5] D. Barnes-Schuster, Y. Bassok, and R. Anupindi. Coordination and flexibility in supply contracts with options. *Manufacturing and Service Operations Management*, 4:171–207, 2002.

[6] Y. Bassok and R. Anupindi. Analysis of supply contracts with total minimum commitment. *IIE Transactions*, 29:373–381, 1997.

[7] D. Beyer and J. Ward. Network server supply chain at HP: A case study. In *Suuply Chain Structures: Coordination, Information and Optimization*, S. Song and D.D. Yao (editors), pp. 257–282, Kluwer Academic Publishers, Boston, 2001.

[8] K.E. Bourland, S. Powell, and D. Pyke. Exploring timely demand information to reduce inventories. *European Journal of Operational Research*, 92:239–253, 1996.

[9] A.O. Brown and H.L. Lee. Optimal pay-to-delay capacity reservation with application to the semiconductor industry. Working Paper, Stanford University, Stanford, CA, 1997.

[10] J. Buzacott, H. Yan, and H. Zhang. Risk analysis of commitment-option contracts with forecast updates. Working Paper, York University, Toronto, Canada, 2002.

[11] G.P. Cachon and M. Fisher. Supply chain inventory management and the value of shared information. *Management Science*, 46:1032–1048, 2000.

[12] G.P. Cachon and P.H. Zipkin. Competitive and cooperative inventory policies in a two-stage supply chain. *Management Science*, 45:936–953, 1999.

[13] F. Chen, Z. Dreaner, J.K. Ryan, and D. Simchi-Levi. The bullwhip effect: managerial insights on the impact of forecasting and information on variability in supply chain. In *Quantitative Models for Supply Chain Management*, S. Tayur, R. Ganeshan, and M. Magazine (editors), pp. 301–336, Kluwer Academic Publishers, Boston, 1999.

[14] F. Chen and A. Federgruen. Mean-variance analysis of basic inventory models. Working Paper, Graduate School of Business, Columbia University, New York, 2000.

[15] C. Chiang and G.J. Gutierrez. A periodic review inventory system with two supply modes. *European Journal of Operations Research*, 94:527–547, 1996.

[16] C. Chiang and G.J. Gutierrez. Optimal control policies for a periodic-review inventory system with emergency orders. *Naval Research Logistics*, 45:187–204, 1998.

[17] W.C. Copacino. *Supply Chain Management: The Basics and Beyond.* St. Lucie Press/APICS Series on Resource Management, Hall Creek, VA, 1997.

[18] K.L. Donohue. Efficient supply contracts for fashion goods with forecast updating and two production modes. *Management Science*, 46:1397–1411, 2000.

[19] A. Dvoretzky, J. Kiefer, and J. Wolfowitz. The inventory problem ii. Case of unknown distributions of demand. *Econometrica*, 20:450–466, 1952.

[20] H. Emmons and S. Gilbert. The role of returns policies in pricing and inventory decisions for catalogue goods. *Management Science*, 44:276–283, 1998.

[21] G.D. Eppen and A.V. Iyer. Improved fashion buying with Bayesian updates. *Operations Research*, 45:805–819, 1997.

[22] M. Fisher, J.H. Hammond, W.R. Obermeyer, and A. Raman. Making supply meet demand in an uncertain world. *Harvard Business Review*, May-June, 83–93, 1994.

[23] M. Fisher and A. Raman. Reducing the cost of demand uncertainty through accurate response to early sales. *Operations Research*, 44:87–99, 1996.

[24] Y. Fukuda. Optimal policies for the inventory problem with negotiable leadtime. *Management Science*, 10:690–708, 1964.

[25] G. Gallego and Ö. Özer. Integrating replenishment decisions with advance demand information. *Management Science*, 47:1344–1360, 2001.

[26] X.H. Gan, S.P. Sethi, and H. Yan. Coordination of supply chain with risk-averse agents. *Production and Operations Management*, 13:135–149, 2004.

[27] X.H. Gan, S.P. Sethi, and H. Yan. Channel coordination with a risk-neutral supplier and a downside-risk-averse retailer. *Production and Operations Management*, 14, 1, 2005, to appear.

[28] S. Gavirneni. Information flows in capacitated supply chains with fixed ordering costs. *Management Science*, 48:644–651, 2002.

[29] S. Gavirneni, R. Kapuscinski, and S. Tayur. Value of information in capacitated supply chains. *Management Science*, 45:16–24, 1999.

[30] S. Graves, H.C. Meal, S. Dasu, and Y. Qiu. Two-stage production planning in a dynamic enviorenment. *Multi-Stage Production Planning and Control: Lecture Notes in Economic and Mathematics Systems*, Vol. 266, S. Axsäter, C. Schneeweiss, and E. Silver (editors), pp. 9–43. Springer-Verlag, Berlin, 1986.

[31] H. Gurnani and C.S. Tang. Note: optimal ordering decisions with uncertain cost and demand forecast updating. *Management Science*, 45:1456–1462, 1999.

[32] R.B. Handfield and E.L. Nichols. *Introduction to Supply Chain Management.* Prentice Hall, New York, 1998.

[33] W.H. Hausman. Sequential decision problems: A model to exploit existing forecast. *Management Science*, 16:B93–B111, 1969.

[34] W.H. Hausman, H.L. Lee, and V.L. Zhang. Optimal ordering for an inventory system with dual lead times. Working Paper, Stanford University, Stanford, CA, 1993.

[35] D. Heath and P. Jackson. Modeling the evolution of demand forecast with application to safety stock analysis in production/distribution systems. *IIE Transactions*, 26:17–30, 1994.

[36] H.Y. Huang, S.P. Sethi, and H. Yan. Purchase contract management with demand forecast updates. *IIE Transactions*, to appear.

[37] A.V. Iyer and M.E. Bergen. Quick response in manufacturer-retailer channels. *Management Science*, 43:559–570, 1997.

[38] A. Jeuland and S. Shugan. Managing channel profits. *Marketing Science*, 2:239–272, 1983.

[39] O. Johnson and H. Thompson. Optimality of myopic inventory policies for certain dependent demand processes. *Management Science*, 21:1303–1307, 1975.

[40] E. Kandel. The right to return. *Journal of Law and Economics*, 39:329–356, 1996.

[41] M.A. Lariviere and E.L. Porteus. Stalking information: Bayesian inventory management with unobserved lost sales. *Management Science*, 45:346–363, 1999.

[42] S.H. Lau and A.H. Lau. Manufacturer's pricing strategy and return policy for a single-period commodity. *European Journal of Operational Research*, 116:291–304, 1999.

[43] H.L. Lee, V. Padmanabhan, and S. Whang. The bullwhip effect in supply chains. *Sloan Management Review*, 38:93–102, 1997.

[44] H.L. Lee, V. Padmanabhan, and S. Whang. Information distortion in a supply chain: The Bullwhip effect. *Management Science*, 43:546–558, 1997.

[45] H.L. Lee and M.J. Rosenblatt. A generalized quantity discount pricing model to increase suppliers' profits. *Management Science*, 32:1177–1185, 1986.

[46] H.L. Lee, K.C. So, and C.S. Tang. The value of information sharing in a two-level supply chain. *Management Science*, 46:626–643, 2000.

[47] L. Li and H.T. Zhang. Competition, inventory, demand variability, and information sharing in a supply chain. Working Paper, Yale School of Organization and Management, Yale University, New Haven, CT, 2004.

[48] S. Lippman and K. McCardle. The competitive newsboy. *Operations Research*, 45:54–65, 1997.

[49] W.S. Lovejoy. Myopic policies for some inventory models with uncertain-demand distributions. *Management Science*, 36:724–738, 1990.

[50] H.M. Markowitz. *Portfolio Selection: Efficient Diversification of Investments.* Yale University Press, New Haven, CT, 1959.

[51] R. Metters. Quantifying the bullwhip effect in supply chains. *Journal of Operations Management*, 15:89–100, 1997.

[52] G.P. Monahan. A quantity discount pricing model to increase vendor profits. *Management Science*, 30:720–726, 1984.

[53] K. Moinzadeh. A multi-echelon inventory system with information exchange. *Management Science*, 48:414–426, 2002.

[54] K. Moinzadeh and S. Nahmias. Adjustment strategies for a fixed delivery contract. *Operations Research*, 48:408–423, 2000.

[55] B.A. Pasternack. Optimal pricing and return policies for perishable commodities. *Marketing Science*, 4:166–176, 1985.

[56] M.J. Rosenblatt and H.L. Lee. Improving profitability with quantity discounts under fixed demand. *IIE Transactions*, 17:388–395, 1985.

[57] M. Rosenshine and D. Obee. Analysis of a standing order inventory system with emergency orders. *Operations Research*, 24:1143–1155, 1976.

[58] H. Scarf. Bayes solution of the statistical inventory problem. *Annals of Mathematical Statistics*, 30:490–508, 1959.

[59] A. Scheller-Wolf and S. Tayur. A Markovian dual-source production-inventory model with order bands. Working Paper, Carnegie Mellon University, Pittsburgh, PA, 1998.

[60] S.P. Sethi and G. Sorger. A theory of rolling horizon decision making. *Annals of Operations Research*, 29:387–416, 1991.

[61] S.P. Sethi, H. Yan, and H. Zhang. Peeling layers of an onion: Inventory model with multiple delivery modes and forecast updates. *Journal of Optimization Theory and Applications*, 108:253–281, 2001.

[62] D. Simchi-Levi, P. Kaminsky, and E. Simchi-Levi. *Designing and Managing the Supply Chain: Concepts, Strategies and Case Studies.* Irwin McGraw Hill, Boston, 1998.

[63] D. Simchi-Levi and Y. Zhao. The value of information sharing in a two-stage supply chain with production capacity constraints. *Naval Research Logistics*, to appear.

[64] P.F. Takac. Electronic data interchange (EDI): An avenue to better performance and the improvement of trading relationships. *International Journal of Computer Applications in Technology*, 5:22–36, 1992.

[65] S. Tayur, R. Ganeshan, and M. Magazine. *Quantitative Models for Supply Chain Management.* Kluwer Academic Publishers, Boston, 1999.

[66] A. Tsay. The quantity flexibility contract and supplier-customer incentives. *Management Science*, 45:1339–1358, 1999.

[67] A. Tsay and W.S. Lovejoy. Quantity flexibility contracts and supply chain performance. *Manufacturing and Service Operations Management*, 1:89–111, 1999.

[68] J. von Neumann and O. Morgenstern. *Theory of Games and Economic Behavior.* Princeton University Press, Princeton, NJ, 1953.

[69] Z.K. Weng. Channel coordination and quantity discount. *Management Science*, 41:1509–1522, 1995.

[70] Z.K. Weng. Pricing and ordering strategies in manufacturing and distribution alliances. *IIE Transactions*, 29:681–692, 1997.

[71] A.S. Whittemore and S.C. Saunders. Optimal inventory under stochastic demand with two supply options. *SIAM Journal of Applied Mathematics*, 32:293–305, 1977.

[72] H. Yan, K. Liu, and A. Hsu. Optimal ordering in a dual-supplier system with demand forecast updates. *Production and Operations Management*, 12:30–45, 2003.

[73] H. Yan, C. Yano, and H. Zhang. Risk-aversion multi-period inventory models. Working Paper, The Chinese University of Hong Kong, Hong Kong, 2003.

[74] V.L. Zhang. Ordering policies for an inventory system with three supply modes. *Naval Research Logistics*, 43:691–708, 1996.

Chapter 2

EXAMPLES FROM INDUSTRY

2.1. Introduction

Forecasting demand is a management function that is a key contributor to corporate success. Thus, a thorough understanding of the demand variations and forecast errors is essential. The research in this area has seen a renewed interest in the study of supply chains. Multistage forecasting mechanisms allow better accuracy of demand forecasts. This research work is based on our consulting work with a Hong Kong electronics manufacturer. Our motivation is to explore the dynamics of demand uncertainty and what are its main drivers.

Consider the case of a single-manufacturer single-retailer supply chain in which the retailer observes customer demand and places orders with the manufacturer. To determine how much to order from the manufacturer, the retailer must forecast customer demand. Generally, the retailer uses its historical customer-demand data and standard forecasting techniques to perform the forecasting.

As reviewed in Section 1.3, forecasting analysis can go in two directions. One school assumes that the variance of the market demand is known; the other assumes it to be unknown. The former is popular in the literature. But in many applications, the variance of demand is unknown. For real data collected from the electronics company, we have observed the following interesting and surprising phenomenon: as the forecasting horizon decreases, the variances of forecasts increase, and the variances of forecasting errors decrease.

The closer that the forecasting horizon is to the end, the larger the fluctuation in demand displays is, and the more accurate the forecast is. This seems to be counterintuitive. It certainly deserves closer investigation. Focusing on this problem, this chapter examines the real data and attempts to provide an explanation. Understanding the phenomena will lead to the design of optimal supply chains and various other management practices. The chapter can be summarized as follows:

1. We examine the dynamics of a multistage forecasting process. Using the Bayesian decision theory, we model the forecasting process as a stochastic process and we observe that the forecast changes with the variation of demand. At each stage, the forecast estimates the demand. After Bayesian updating, the forecast fluctuates more than the initial forecast, but more accurately corresponds to the end demand.

2. For a single node in a supply chain, we examine four sources that contribute to variances in demand forecasts—price promotion, lot sizing, new-product introduction, and make-to-stock policy.

In this chapter, Section 2.2 presents data from industry. The analysis of the data is carried out with statistical tools. In Section 2.3, we adopt the Bayesian decision theory to investigate the dynamics of multistage-demand forecasting. We prove that under the multistage-demand-forecasting structure, forecast variances and precisions both increase, which supports our observation. Section 2.4 concerns the operational factors that cause the demand forecasts to fluctuate and approach real demand over stages—price promotion, lot sizing, new-product introduction, and pre-confirmed orders. In Section 2.5, the managerial implications developed in this chapter for the design of a supply chain are described, and the chapter is concluded.

2.2. Industry Observations

A major security-system manufacturing company produces and distributes security systems for military, residential, commercial, and industrial applications. It has a design center in California, a manufacturing center in Asia, and three regional distribution centers in San Francisco, Amsterdam, and Singapore. The company sources components and subassemblies around the world. The management objectives are to improve the response time to meet market demand, to reduce inventory, and to shorten lead time (including the time for manufacturing and distribution). In the security-system market, customers expect to have the required device or system within one month. Therefore, given long lead times in procurement and production, the manufacturing operation relies largely on forecasts.

From a practical point of view, forecasts are never accurate, and the company updates its demand forecasts until the real demand is realized. When too little raw material is ordered, the company has to pay a higher price to secure them or use air shipment to expedite them (if these options are feasible). When too many raw materials and subassemblies are ordered, the company has to keep them in inventory. These materials often become obsolete. These updates in forecasting also make it difficult for the company to allocate its production capacity efficiently.

A key component in security systems is the microcontroller, which makes up 30% to 40% of the total materials cost. A microcontroller is a central processing unit (CPU) chip with a built-in memory and interface circuits. The read-only memory (ROM) contains permanent data (program code). See Spasov [5] for a discussion of related concepts about microcontrollers and their technology. The company can order microcontrollers with user-supplied data requirements. If user-supplied data is provided, the semiconductor manufacturing includes a process known as custom photo masking in the wafer-fabrication process. Alternatively, the company can purchase microcontrollers with a programmable ROM such as one-time-programmable (OTP) read-only memory or erasable programmable read-only memory (EPROM). The company inputs the data into these programmable microcontrollers after the chips are received. To order custom-masked chips, the users are required to provide the data (program code) prior to manufacturing, and a significant lead time is required. On the other hand, since programmable ROMs are generic, these microcontrollers can be produced with a considerably shorter lead time. However, the OTP chips are about twice as expensive as custom-masked chips and EPROM chips are even more expensive. The company must decide how to order both custom-masked and OTP chips.

The company uses a half-year rolling window for demand forecasting. These forecasts are made and updated monthly by the regional offices. The headquarter coordinates the forecasts and passes them to its logistics and manufacturing functions. Procurement decisions are made based on the demand forecast and the lead time required by its vendors. The company divides the raw materials into two classes: critical and regular. The components that have fewer sources, and have a higher value content, and require a longer lead time are classified into their critical materials. Microcontrollers are a typical example.

In what follows, we first analyze the demand-forecast data. We assume that the forecast data are arranged in a rolling K-stage horizon, where the first $(K-1)$ updates are forecasts, and the last one represents the realized demand. The major security-system manufacturing company (see Yan [7]) provides us with two years of data for seven products. Based on these data, using the Bayesian theory, we establish the demand forecast.

For the seven products investigated by us, the logistics and manufacturing functions of the company receive a monthly demand update. In the six-month rolling horizon ($K = 6$), the first five updates are forecasts, the last one is the real demand. We obtain the data from February 1996 to September 1997. Our purpose is to investigate the dynamics of forecast and forecast error and the possible managerial decisions and implications.

We depict the demand and forecast in the scatter charts shown in Figures 2.1 and 2.2, where the x-axis is the actual demand and y-axis is the forecasted demand. In both figures, a star represents actual demand. Therefore, the demand data are scattered on the 45 degree line. In Figure 2.1, a circle represents the forecast when the forecasting horizon is five months. When the circle is above the 45 degree line, it indicates that the forecast is higher than the real demand; on the other hand, when below the 45 degree line, it indicates that the forecast is lower than the real demand. Similarly, in Figure 2.2, a black dot represents the relationship between the forecast and actual demand.

It is easily observed that the later forecasts are much closer to actual demand. We study forecast errors and their variances, which are given in Table 2.1. Note that the numbers listed are scaled from the real data for the sake of confidentiality. In the table, errors and standard deviations of the five-month, three-month, and one-month forecasts with respect to the real demand are tabulated. We also calculate the percentage of the standard deviation changes in percentage with respect to the standard deviations of the five-month forecast. As we have discussed earlier, the variance (or the standard deviation) is used to measure the accuracy of the demand forecast. From Table 2.1, comparing the latest and the earliest forecasts, the latter improves remarkably for most of the products. The phenomenon of decreasing forecasting errors can be further demonstrated by the indicator of mean absolute deviation (MAD). We provide the MAD values in Table 2.2. Further, for all seven products, we notice that the forecast variance increases as the forecast horizon decreases. In addition, the variance of demand is larger than the variance of its each stage forecast.

Denote S_i^2 and S_{ei}^2 as sample variance of i-month forecast and forecast error, where $i = 1, \cdots, 5$. We study forecast variances and forecast errors. Table 2.3 provides values of F-statistic F_0, which is defined as $F_0 = S_i^2/S_j^2$, where $i > j$, and $i, j = 1, 3, 5$. Again, let the significance level α be 0.1. We find that $F_{0.1}(19, 19) = 0.549$, and with exception of 236UL, other products' F_0 values in the second hypothesis are all less than $F_{0.1}(19, 19)$. We reject H_0 and conclude that the variance of a one-month forecast is obviously greater than that of a five-month forecast (refer to Table 2.3). Second, we study the variances of forecast errors. Table 2.4 contains values of F-statistic F_0, which is defined as $F_0 = S_{ei}^2/S_{ej}^2$, where $i < j$, and $i, j = 1, 3, 5$.

The hypotheses tests that are carried out above check whether our observations on forecast variances and forecast errors are significant. In conclusion, our observations and analysis reveal that forecasts become more accurate as the forecast horizon becomes shorter. In addition, demand fluctuates much more than initially thought.

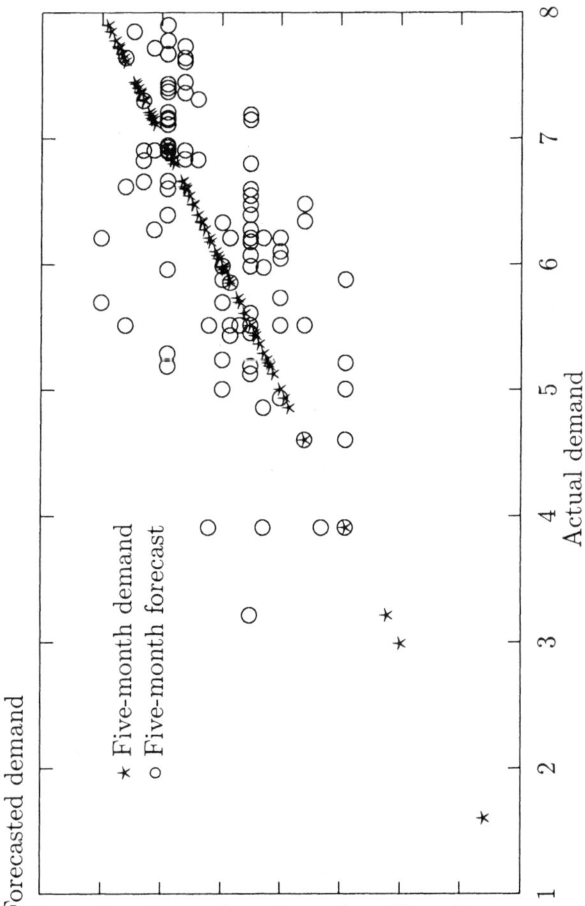

Figure 2.1. A scatter chart of five-month forecasts and actual five-month demands

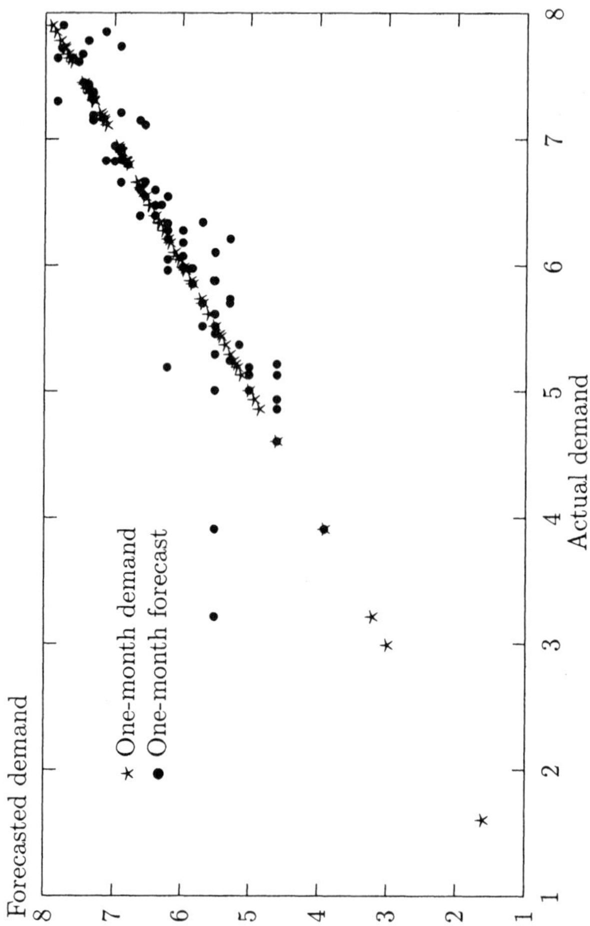

Figure 2.2. A scatter chart of one-month forecasts and actual one-month demands

Products	Five-Month Forecast		Three-Month Forecast			One-Month Forecast		
	Average	St. Dev.	Average	St. Dev.	Changes %	Average	St. Dev.	Changes %
Product 1	-21.5	64.7	-20.3	62.1	4.0	-30.2	60.3	6.8
Product 2	-52.0	208.6	22.5	250.9	20.3	-28.8	78.0	62.0
Product 3	33.0	1211.0	5.0	819.0	32.4	-139.8	366.3	69.8
Product 4	30.0	275.8	42.5	266.9	3.2	12.5	72.9	73.6
Product 5	-53.5	234.9	-73.5	139.9	40.4	-50.8	70.6	69.9
Product 6	-162.0	487.0	-90.0	528.0	8.4	-42.0	201.4	58.6
Product 7	86.0	962.0	-24.0	737.0	23.4	-47.5	443.9	53.9

Table 2.1. Forecast errors and improvements

Forecasts	Seven Products						
	2316I	2316UL	236UL	238UL	2316TL	236TL	238TL
Five months	106.00	280.00	1720.50	441.00	399.00	778.00	1632.50
Three months	96.50	284.00	1090.50	390.00	240.00	744.00	1062.50
One month	68.50	72.50	414.50	60.00	123.60	258.00	487.00

Table 2.2. Mean absolute deviation: An indicator of forecast improvements

Hypothesis	Seven Products						
	2316I	2316UL	236UL	238UL	2316TL	236TL	238TL
$H_0 : \sigma_3^2 \geq \sigma_1^2$ $H_1 : \sigma_3^2 < \sigma_1^2$	1.067280	1.793970	1.600610	0.912340	0.748418	0.677037	0.674127
$H_0 : \sigma_5^2 \geq \sigma_1^2$ $H_1 : \sigma_5^2 < \sigma_1^2$	0.457407	0.273383	1.615550	0.317203	0.539510	0.422045	0.269064
$H_0 : \sigma_5^2 \geq \sigma_3^2$ $H_1 : \sigma_5^2 < \sigma_3^2$	0.428573	0.152390	1.009340	0.347680	0.720867	0.623370	0.399129

Table 2.3. Tests of hypotheses on forecast variances

Hypothesis	Seven Products						
	2316I	2316UL	236UL	238UL	2316TL	236TL	238TL
$H_0 : \sigma_{e1}^2 \geq \sigma_{e3}^2$ $H_1 : \sigma_{e1}^2 < \sigma_{e3}^2$	0.942869	0.096647	0.200363	0.074603	0.254668	0.158790	0.362773
$H_0 : \sigma_{e1}^2 \geq \sigma_{e5}^2$ $H_1 : \sigma_{e1}^2 < \sigma_{e5}^2$	0.868613	0.139817	0.091492	0.079963	0.090332	0.171026	0.212922
$H_0 : \sigma_{e3}^2 \geq \sigma_{e5}^2$ $H_1 : \sigma_{e3}^2 < \sigma_{e5}^2$	0.921244	1.446681	0.457382	0.936502	0.354707	1.175466	0.586928

Table 2.4. Tests of hypotheses on the variances of forecast errors

2.3. Multistage Forecasts

For a multistage forecast, in each cycle the demand forecast is updated several times. At the beginning of the forecasting cycle, the mean and the variance of the demand are unknown. As the demand information is updated, the demand forecast evolves until the end of the cycle. A multistage forecasting process can be described as a stochastic process. For a fixed time, the forecast is a random variable, which estimates demand according to existing information. With the fixed sample point, demand is a function of time. To catch up with the changing demand, the forecast experiences fluctuation. In what follows, we first analyze the dynamics of the forecast updating. Then we analyze the uncertain factors that cause fluctuations in forecasts.

2.3.1 Dynamics of Forecast Updates

We denote the densities of a univariate normal and an inverse Gamma distribution by $f_N(\cdot)$ and $f_{G^{-1}}(\cdot)$, respectively, with

$$f_N(y|\mu,\sigma^2) = \frac{1}{\sqrt{2\pi}\sigma}\exp\left(-\frac{(y-\mu)^2}{2\sigma^2}\right), \quad y \in R,$$

and

$$f_{G^{-1}}(y|\alpha,\beta) = \frac{1}{\Gamma(\alpha)\beta^\alpha y^{(\alpha+1)}}\exp\left(-\frac{1}{\beta y}\right), \quad y \geq 0,$$

where α, $\beta > 0$ (see Berger [1], pp. 559–561). Assume that the demand X follows a normal distribution with mean θ and variance V. Because of the uncertainty inherently present in the market, both the mean and the variance of the demand are unknown. To estimate them, forecast is performed. Under the Baysian framework, as information about the demand is accumulated, the forecast undergoes Bayesian updates.

Using the Bayesian decision theory (Chapter 4 in Berger [1] and Chapter 17 in Pratt, Raiffa, and Schlaifer [4]), we assume that the joint distribution of (θ, V) is a normal inverted-Gamma distribution with density

$$f_{N,G^{-1}}(\theta,V) = f_N(\theta|\mu,\tau_1 V)\cdot f_{G^{-1}}(V|\alpha_1,\beta_1), \tag{2.1}$$

where $\tau_1 = 1/n$ and n is the sample size. The data about uncertain demand collected are used to form d, an estimate of the demand. This d can be used to generate a posterior distribution of the demand. Then the prior joint distribution of θ and V in (2.1) is updated to obtain the posterior joint distribution of $(\theta, V|d)$, which is also a normal inverted-Gamma distribution with density

$$f_{N,G^{-1}|d}(\theta,V|d) = f_N(\theta|\mu(d),\tau_2 V)\cdot f_{G^{-1}}(V|\alpha_2,\beta_2), \tag{2.2}$$

where the parameters in the prior and posterior distributions satisfy

$$\begin{aligned}\mu(d) &= \frac{\mu + \tau_1 d}{\tau_1 + 1}, \\ \frac{1}{\tau_2} &= \frac{1}{\tau_1} + 1, \\ \alpha_2 &= \alpha_1, \\ \frac{1}{\beta_2} &= \frac{1}{\beta_1} + \frac{(d-\mu)^2}{2(1+\tau_1)}. \end{aligned} \tag{2.3}$$

To illustrate that the update decreases the uncertainty of demand, we need to investigate the forecast dynamics further. The marginal distributions of variance and precision are derived as follows.

2.3.2 Marginal Distribution of the Variance of Demand

It is straightforward to show that the marginal distribution of variance V is an inverted-Gamma distribution (see Berger [1], p. 288) with density

$$f_{G^{-1}}(V) = f_{G^{-1}}(V|\alpha_1, \beta_1). \tag{2.4}$$

In addition, given the forecast update, the conditional distribution of V follows an inverted-Gamma distribution with parameters (α_2, β_2)—that is,

$$f_{G^{-1}}(V|d) = f_{G^{-1}}(V|\alpha_2, \beta_2).$$

It follows that the mean of V and the mean of V conditioned on d are

$$\mathsf{E}[V] = \frac{1}{\beta_1(\alpha_1 - 1)} \quad \text{and} \quad \mathsf{E}[V|d] = \frac{1}{\beta_2(\alpha_2 - 1)},$$

respectively. By virtue of (2.3), we have $\alpha_2 = \alpha_1$ and $\beta_2 \leq \beta_1$. Thus,

$$\mathsf{E}[V|d] \geq \mathsf{E}[V].$$

That is, the conditional mean of the variance is larger than the mean of the variance.

2.3.3 Forecast Precision

The forecast precision, denoted by $J(X)$, is defined as the reciprocal of the variance of X—that is, $J(X) = 1/V$. Note that the precision of X is a random variable that represents the amount of information about X contained in the distribution of X (see Pratt, Raiffa, and Schlaifer [4]). When the forecast is carried out, $J(X)$ needs to be updated as well.

Since the marginal distribution of V is inverted-Gamma, $1/V$ follows a Gamma distribution (see Berger [1]), whose density can be expressed as

$$f_G\left(\frac{1}{V}\right) = f_G\left(\frac{1}{V}\bigg|\alpha_1, \beta_1\right).$$

Similarly, $1/V$ conditioned on d follows a Gamma distribution with parameters (α_2, β_2)—that is,

$$f_{G|d}\left(\frac{1}{V}\bigg|d\right) = f_G\left(\frac{1}{V}\bigg|\alpha_2, \beta_2\right).$$

Therefore, we obtain both unconditional and conditional variances of $1/V$ as

$$\mathsf{Var}\left[\frac{1}{V}\right] = \alpha_1\beta_1^2, \ \mathsf{Var}\left[\frac{1}{V}\bigg|d\right] = \alpha_2\beta_2^2.$$

Again, by using (2.3), $\alpha_2 = \alpha_1$, and $\beta_2 < \beta_1$, we have

$$\mathsf{Var}\left[\frac{1}{V}\right] > \mathsf{Var}\left[\frac{1}{V}\bigg|d\right].$$

In terms of $J(X)$, $\mathsf{Var}[J(X)] > \mathsf{Var}[J(X)|d]$. Therefore, the variance of the precision of X decreases as the forecast of the demand is updated. That is, the estimation of precision becomes increasingly accurate, and $1/V$ conditioned on d contains more information about uncertain demand than $1/V$. We summarize the above results in the following proposition.

PROPOSITION 2.1 *In a multistage forecast, as the forecast horizon is shortened, the forecast variance and the forecast precision both increase.*

2.4. Operational Factors Affecting Forecasting Process

For a global supply chain scenario, a firm must redesign its own operations and coordinate with its upstream and downstream partners. The uncertainty that exists longitudinally within a single firm and the uncertainty that exists in the whole supply chain (known as a bullwhip phenomenon) increase the variance of demand information. Here we examine the uncertain factors that exist longitudinally within a single node in the supply chain.

2.4.1 Price Promotion

Let d^H denote the forecast when the price is high, and d^L denote the forecast when the price is low. In Lee, Padmanabhan, and Whang [2, 3], it is proved that faced with the price variations, the retailer's optimal inventory policy is as follows. At a low price get as close as possible to the stock level d^L, and at a high price bring the stock level to d^H, where $d^H < d^L$. It has been pointed out that price promotion gives rise to the distortion of demand. To proceed, we give an exact expression of the variance of demand forecast that is produced by a price promotion, in connection with the stage in which the price promotion takes place and the depth $(d^H - d^L)$ of the price promotion.

To examine the forecast variance, we assume the probability that the downstream location meets the price promotion conducted by an upstream location is ζ at stage i. At the ith stage, the expectation and variance of demand forecast d_i can be calculated as follows:

$$\begin{aligned} \mathsf{E}[d_i] &= d^L\zeta + d^H \cdot (1-\zeta), &(2.5)\\ \mathsf{Var}[d_i] &= \mathsf{E}[d_i^2] - \mathsf{E}^2[d_i] \\ &= \left[\left(d^L\right)^2\zeta + \left(d^H\right)^2(1-\zeta)\right] - \left[d^L\zeta + d^H(1-\zeta)\right]^2 \\ &= \left[\zeta - \zeta^2\right]\left(d^L\right)^2 - 2\zeta(1-\zeta)d^H d^L \\ &\quad + \left[(1-\zeta) - (1-\zeta)^2\right]\left(d^H\right)^2 \\ &= \zeta(1-\zeta)\left(d^H - d^L\right)^2. &(2.6) \end{aligned}$$

It is observed that the variance is proportional to ζ and that the depth of the price promotion is $\left(d^H - d^L\right)$.

Assume that the forecasting cycles of downstream and upstream locations are K_l and K_u stages, respectively. For both $K_u \geq K_l$ and $K_u < K_l$, we could divide ζ into two parts. The first part is that promotion will happen in the forecasting cycle, and the second is that the promotion will happen at the ith forecasting stage. Thus,

$$\begin{aligned} \zeta &= \begin{cases} \dfrac{K_l}{K_u}\dfrac{1}{K_l}, & \text{if } K_u \geq K_l, \\ \left(1 - \dfrac{k}{K_u}\right)\dfrac{m}{K_l} + \dfrac{k}{K_u}\dfrac{m+1}{K_l}, & \text{if } K_u < K_l, \end{cases} \\ &= \frac{1}{K_u}, &(2.7) \end{aligned}$$

where we use the fact that for $K_u < K_l$,

$$K_l = mK_u + k, \quad 0 \leq k \leq K_u.$$

Consequently, using (2.6) and (2.7), we get

$$\mathsf{Var}[d_i] = \frac{1}{K_u}\left(1 - \frac{1}{K_u}\right) \cdot (d^H - d^L)^2.$$

Observe that for $0 \le \zeta \le 1$, the function $\zeta(1-\zeta)$ first increases in ζ, attains its maximum at $\zeta = 0.5$, and decreases in ζ thereafter. On the other hand, as $\zeta = 1/K_u$, when $K_u = 1$, we have $\zeta = 1$; when $K_u \ge 2$, we have $\zeta \le 0.5$. Hence, for $K_u = 1$, $\zeta(1-\zeta) = 0$; for $K_u \ge 2$, $\zeta(1-\zeta)$ is decreasing in K_u. For $K_u \ge 2$, it is obvious that $\mathsf{Var}[d_i]$ is a monotone decreasing function of K_u and its upper bound is $(1/4)(d^H - d^L)^2$, which is reached when $K_u = 2$. Decreasing $1/K_u$, which means decreasing the frequency of price adjustment, could lead to a decrease in the variance. For $K_u = 1$, which means that the cycle of price promotion is one unit of time, the variance caused by price promotion will disappear.

PROPOSITION 2.2 *The variance of a forecast that is caused by price fluctuation is proportional to the product of ζ and the depth $d^H - d^L$ of price promotion. The variance increases with the decrease of the price-promotion cycle of an upstream location K_u. When $K_u = 2$, the variance will reach its maximum. On the other hand, an everyday-low-pricing (EDLP) strategy (that is, $K_u = 1$) can eliminate the effect of a distorted demand forecast caused by a price promotion.*

Decreasing the frequency and depth of a price promotion at an upstream location could mitigate the demand variance caused by price promotion. Lee, Padmanabhan, and Whang [2] noted: "One way to control the bullwhip effect due to price fluctuation is to reduce the frequency and depth of the manufacturer's trade promotions". Undoubtedly, the expression tells us that the best way to eliminate the variance caused by price promotion is to use an EDLP strategy.

2.4.2 Lot Sizing

Economies of scale in the order quantity (lot size) occur whenever a fixed set-up cost is incurred for each order that is not depending on the lot size. Suppose that $\widetilde{N}$ retailers give orders to the manufacturer. Retailer j will give an order ξ_j in a forecasting cycle, where $\{\xi_j\}$ is a sequence of independent and identically distributed (i.i.d.) random variables satisfying

$$\xi_j = n_j Q + l_j, \quad l_j \in [0, \ Q-1].$$

Here, Q is the lot size. Without loss of generality, we assume that l_j is a uniformly distributed random variable within the interval $[0, Q-1]$ and that

$$\mathsf{P}\{l_j = l\} = \frac{1}{Q}, \quad l \in [0, \ Q-1].$$

At the beginning of the forecasting cycle, the orders are aggregated to form the forecast denoted by

$$d_b = \sum_{j=1}^{\tilde{N}} \xi_j = \sum_{j=1}^{\tilde{N}} (n_j Q + l_j).$$

At the end of the cycle, the aggregated order is lot-sized to the last-stage demand forecast represented by

$$d_e = \sum_{j=1}^{\tilde{N}} s_j,$$

where s_j is an integer-valued function controlled by $\varepsilon \in [0, 1]$, s_j is lot-sized to $n_j Q$ if $0 \le l_j \le \varepsilon Q$, and to $(n_j + 1)Q$ if $\varepsilon Q < l_j \le Q$, with

$$s_j = \begin{cases} 0 & \text{for} \quad 0 \le l_j \le \varepsilon Q, \\ Q & \text{for} \quad \varepsilon Q < l_j \le Q - 1. \end{cases}$$

Therefore,

$$d_e = \sum_{j=1}^{\tilde{N}} (n_j Q + s_j).$$

Then we can get the variance of d_e conditioned on d_b—that is,

$$\begin{aligned} \mathsf{Var}[d_e | d_b] &\equiv \mathsf{E}[(d_e - d_b)^2] \\ &= \sum_{j=1}^{\tilde{N}} \mathsf{E}[(s_j - l_j)^2]. \end{aligned}$$

Note that

$$\begin{aligned} \mathsf{E}[(s_j - l_j)^2] &= \sum_{\ell=0}^{\lfloor \varepsilon Q \rfloor \wedge (Q-1)} \ell^2 \cdot \frac{1}{Q} + \sum_{\ell=(\lfloor \varepsilon Q \rfloor + 1) \wedge (Q-1)}^{Q-1} (Q - \ell)^2 \cdot \frac{1}{Q} \\ &= \frac{1}{Q} \cdot \left(\sum_{\ell=0}^{\lfloor \varepsilon Q \rfloor \wedge (Q-1)} \ell^2 + \sum_{\ell=(\lfloor \varepsilon Q \rfloor + 1) \wedge (Q-1)}^{Q-1} (Q - \ell)^2 \right). \end{aligned}$$

Therefore,

$$\mathsf{Var}[d_e | d_b] = \frac{\tilde{N}}{Q} \cdot \left(\sum_{\ell=0}^{\lfloor \varepsilon Q \rfloor \wedge (Q-1)} \ell^2 + \sum_{\ell=(\lfloor \varepsilon Q \rfloor + 1) \wedge (Q-1)}^{Q-1} (Q - \ell)^2 \right).$$

The range of variance is the interval

$$\left[\frac{\widetilde{N}}{Q}\cdot\left(\sum_{\ell=0}^{\lfloor Q/2\rfloor}\ell^2+\sum_{\ell=\lfloor Q/2\rfloor+1}^{Q-1}(Q-\ell)^2\right),\ \frac{\widetilde{N}}{Q}\sum_{\ell=0}^{Q-1}\ell^2\right],$$

obtained as ε ranges over $[0,\ 1]$.

When $\varepsilon = 1$ or 0, the variance reaches its maximum,

$$\frac{\widetilde{N}}{Q}\sum_{\ell=0}^{Q-1}\ell^2,$$

which equals approximately $(1/3)\widetilde{N}Q^2$. As a result if we use a policy that lotsizes any order ξ_j to n_jQ or to $(n_j+1)Q$, respectively, no matter what the size is, the variance caused by price promotion reaches its maximum. When $\varepsilon = 1/2$, the variance reaches its minimum,

$$\frac{\widetilde{N}}{Q}\cdot\left(\sum_{\ell=0}^{\lfloor Q/2\rfloor}\ell^2+\sum_{\ell=\lfloor Q/2\rfloor+1}^{Q-1}(Q-\ell)^2\right),$$

which equals approximately $(1/12)\widetilde{N}Q^2$. It is apparent that decreasing lot size Q or setting $\varepsilon = 1/2$ could alleviate the variance caused by lot sizing.

PROPOSITION 2.3 *The lot-sizing policy controlled by ε, as defined above, will cause fluctuation of demand forecast. When $\varepsilon = 1$ or 0, the variance reaches its maximum, $(\widetilde{N}/Q)\sum_{\ell=0}^{Q-1}\ell^2$. When $\varepsilon = 1/2$, the variance reaches its minimum,*

$$\frac{\widetilde{N}}{Q}\cdot\left(\sum_{\ell=0}^{\lfloor Q/2\rfloor}\ell^2+\sum_{\ell=\lfloor Q/2\rfloor+1}^{Q-1}(Q-\ell)^2\right).$$

Moreover, the variance is proportional to the number of retailers $\widetilde{N}$ and the square of lot size Q.

2.4.3 New-Product Launch

The number of introductions of new products has exploded in recent years. Frequent introductions of new products reduce the average lifetime of products. With shortened life cycles, many products are either at the beginning or at the end of their lives. When a retailer knows that a new product will be launched in the near future, it has no reason to order or to keep the old product in stock with full capacity. Consequently, demand forecast becomes more difficult and dynamically changing over time.

To demonstrate how the new-product launch influences the dynamic of demand forecasting, we first investigate the accuracy of forecasting for new-product launches. Then we prove that improved forecasts for new-product introductions increase the variance of demand forecasting.

2.4.3.1 Forecasting for Introduction of a New Product

Denote the development cycle of a new product by r, which is assumed to be normally distributed with mean η and variance σ^2 or with a normal density $f_N(r|\eta, \sigma^2)$. Under the Bayesian framework, the mean of the development cycle η is modeled as a normal distribution with mean μ and variance τ^2 or with a normal density $f_N(\eta|\mu, \tau^2)$. This implies (see Berger [1]) that r also follows a normal distribution with mean μ and variance $(\sigma^2 + \tau^2)$ or with a normal density $f_N(r|\mu, \sigma^2 + \tau^2)$.

$f_{N|\eta}(r|\eta) = f_N(r|\eta, \sigma^2)$ is the conditional density of r given η. $f_N(r) = f_N(r|\mu, \sigma^2 + \tau^2)$ is the unconditional density of r. This is the so-called normal process with an unknown mean and a known variance. The technique we use in Section 2.3 is a normal process with an unknown mean and an unknown variance (see Pratt, Raiffa, and Schlaifer [4], Chapters 16 and 17).

To account for the effect of information updating about r, consider a stochastic process $\{r_i,\ i = 1, \cdots, M\}$. Beginning from the first stage, r_1 is obtained and used to update the estimate of μ. Then the forecast of r_2 is produced. The round is repeated again and again until a terminal time M. The distribution of r_i depends on the observed values of $r_1, \cdots, r_{i-1}$, since these observations affect the estimation of η and the forecast of r_i.

Given r_1, the posterior distribution of η (see Pratt, Raiffa, and Schlaifer [4], p. 383) is

$$f_{N|r_1}(\eta|r_1) = f_N\left(\eta \middle| \mu(r_1), \frac{1}{\rho_{r_1}}\right),$$

where

$$\mu(r_1) = \frac{\mu/\tau^2 + r_1/\sigma^2}{1/\tau^2 + 1/\sigma^2}$$

and

$$\rho_{r_1} = \frac{1}{\sigma^2} + \frac{1}{\tau^2}.$$

Similarly,

$$f_{N|r_i}(\eta|r_i) = f_N\left(\eta \middle| \mu(r_i), \frac{1}{\rho_{r_i}}\right),$$

where

$$\mu(r_i) = \frac{\rho_{r_{i-1}} \cdot \mu(r_{i-1}) + r_i/\sigma^2}{\rho_{r_i}}, \tag{2.8}$$

$$\rho_{r_i} = \frac{i}{\sigma^2} + \frac{1}{\tau^2}. \tag{2.9}$$

After observing $r_1, \cdots, r_{i-1}$, the distribution of r_i is given by

$$f_N(r_{i+1}) = f_N\left(r_{i+1}\middle|\mu(r_i), \sigma^2 + \frac{1}{\rho_{r_i}}\right). \tag{2.10}$$

By (2.9), it is straightforward that $\rho_{r_{i+1}} > \rho_{r_i}$. Thus,

$$\mathsf{Var}[r_{i+1}] < \mathsf{Var}[r_i]. \tag{2.11}$$

With information gathering, the forecast variance decreases. This illustrates that more accumulated information results in more accurate forecasting about the introduction of new products. This relationship will be used in the following section to demonstrate the effect of new-product launches on demand forecast.

2.4.3.2 Variances of Demand Forecasts due to New-Product Launches

We denote by R_f the cycle of forecast. Starting from time zero, we anticipate that there will be a new-product launch in the market. Thus, it is reasonable to assume

$$R_f \leq \mu \leq 2R_f. \tag{2.12}$$

If some information is available about the new-product launch, the forecasts at this stage are relevant to the existing product as it approaches the end of its life. Then the company will defer some orders to the next stage. We assume that αd is deferred to the next stage, where α is the fraction of the order that will be deferred to next stage because the new product is anticipated to be introduced into market. Consequently, the forecast is reduced to d_N, which equals $(1-\alpha)d$ and follows the same distribution as d. The variance caused by the information about new-product launch at the ith stage is

$$\mathsf{Var}[d_N|d_i] = \mathsf{E}[(d_N - d_i)^2] = \alpha^2 d^2 p_i,$$

where p_i is the ith estimate of the probability that there is promotion at the next stage. As r_i follows normal distribution (see (2.10)), p_i can be calculated as the sum of probability that r_i falls within $[R_f,\ 2R_f]$. Here, the left-tail cumulative function of a standard normal distribution, which is denoted as $F_{N^*}(\cdot)$, is used to compute $\mathsf{Var}[d_N|d_i]$ as follows:

$$\begin{aligned}\mathsf{Var}[d_N|d_i] &= \alpha^2 d^2 \cdot \mathsf{P}(r_i \in [R_f, 2R_f]|r_{i-1}) \\ &= \alpha^2 d^2 \left\{ F_{N^*}\left(\frac{2R_f - \mu(r_{i-1})}{\sqrt{\mathsf{Var}(r_i)}}\right) - F_{N^*}\left(\frac{R_f - \mu(r_{i-1})}{\sqrt{\mathsf{Var}(r_i)}}\right)\right\}.\end{aligned}$$

In view of (2.11), it is noted that the distribution of r_i is more and more concentrated around $\mu(r_i)$ over time. As the length of the summation range is fixed, the value of the expression

$$F_{N^*}\left(\frac{2R_f - \mu(r_{i-1})}{\sqrt{\mathsf{Var}(r_i)}}\right) - F_{N^*}\left(\frac{R_f - \mu(r_{i-1})}{\sqrt{\mathsf{Var}(r_i)}}\right)$$

increases as the forecasting procedure approaches the end of the cycle. Hence,

$$\mathsf{E}[\mathsf{Var}[d_N|d_i]] > \mathsf{E}[\mathsf{Var}[d_N|d_{i-1}]].$$

That is, as information about the new product accumulates, the variance of demand forecast increases.

PROPOSITION 2.4 *In the forecasting window, as information about the new-product launch accumulates, the increasing accuracy of the forecasting about new-product launch causes increasing variance of demand forecast.*

2.4.4 Pre-confirmed Orders

A make-to-stock policy is a generally employed inventory management procedure that implies that there is limitation on the capacity of the inventory. Under such a situation, if a pre-confirmed order is accepted by the firm, which should arrive in the next stage, it will definitely have an influence on the demand forecast.

Assume that the make-to-stock level is S and that the order confirmed in advance is S_p. We assume that the pre-confirmed order S_p is normally distributed with mean μ_s and variance σ_s—that is,

$$f_N(S_p) = f_N(S_p|\mu_s, \sigma_s).$$

Due to the limitations of the capacity of the inventory, the real demand forecast reduces to $S - S_p$. That is, the pre-confirmed order will result in a large variation of demand forecast. We can show that the variance caused by the pre-confirmed order is

$$\mathsf{Var}[S - S_p|S] = \mathsf{Var}[S_p] = \sigma_s^2 > 0.$$

PROPOSITION 2.5 *Under a make-to-stock policy, an order is confirmed in advance, and the variance of demand forecast is proportional to the variance of the uncertain pre-confirmed order.*

2.5. Concluding Remarks

The volatile demand may cause inaccurate forecasts, which is one of the main sources of backlogs and markdowns. As companies attempt to produce at a fast pace, they frequently require more advanced forecasting techniques to meet demand at the lowest possible cost. Based on the analysis provided here, companies should understand the dynamics of demand first and investigate the main factors in their own operation that cause the fluctuation. To cope with variance in the variety and volume of demand, companies should redesign their planning processes so that they incorporate fluctuation into their manufacturing planning.

The requirement for flexibility in manufacturing arises from volatile customer demand. To eliminate the effects of variance that exist in demand forecasts, many strategies have been proposed, such as mass customization, accurate response, quick response, and postponement. To eliminate the effects of fluctuation in the volume of demand, we recommend that the planning processes be reengineered.

Since late forecast involves more information about uncertain demand, it is more accurate. It appears that a quick-response strategy is reasonable. Production should be triggered as late as possible. A late forecast involve a larger variance, especially for those products requiring long production cycles. The large variance results in large quantities of either stockout or markdowns. On the other hand, when production is planned according to a late forecast, there will be production gaps. This is not beneficial for smooth production planning. To limit production capacity and to cope with increasing variance, the planning processes should be reengineered. Based on this idea, at a late stage, unpredictable factors become more certain. Hence, an accurate response approach is reasonable and practicable.

This chapter gives a mathematical explanation for the fluctuations and variations of the demand forecast. Moreover, we show that to cope with the uncertainty in demand, the forecasting, planning, and production process should be reengineered. More specifically, we provide an illustration of the phenomena. The uncertainty factors existing in practical operations also are examined.

Our research in supply chain decision making with forecast revision is motivated in part by our collaboration with industry. In this chapter, we present a real-life example to highlight some of the opportunities in this area. The example is the procurement of microcontrollers by a major security manufacturer. With data obtained from the company, we demonstrate the nature of forecast revision and the urgent needs for decision-making methodologies. In the following chapters, given the precise relationships among the data, the demand, and the price, we investigate how information affects the optimal policy and the minimum (maximum) cost (profit) in the supply chains.

2.6. Notes

This chapter is based on Xiang and Yan [6] and Yan and Zhang [8].

References

[1] J. Berger. *Statistics Decision Theory and Bayesian Analysis*. Springer-Verlag, New York, 1985.

[2] H.L. Lee, V. Padmanabhan, and S. Whang. The bullwhip effect in supply chains. *Sloan Management Review*, 38:93–102, 1997.

[3] H.L. Lee, V. Padmanabhan, and S. Whang. Information distortion in the supply chain: The bullwhip effect. *Management Science*, 43:546–558, 1997.

[4] J. Pratt, H. Raiffa, and S. Schlaifer. *Introduction to Statistical Decision Theory*. MIT Press, Cambridge, MA, 1995.

[5] P. Spasov. *Microcontroller Technology*. Prentice-Hall, Englewood Cliffs, NJ, 1993.

[6] H. Xiang and H. Yan. Fluctuation and variation in dynamic forecasts. Working Paper, The Chinese University of Hong Kong, Hong Kong, 2002.

[7] H. Yan. Report on the analysis of forecast data for C&K security systems. *Consulting Report*, The Chinese University of Hong Kong, Hong Kong, 1998.

[8] H. Yan and H. Zhang. Information revision and decision making in supply chain management. In *Stochastic Modelings and Optimization*, D.D. Yao, H. Zhang, and X. Zhou (editors), pp. 429–457, Springer-Verlag, New York, 2003.

multiple layers of an onion. The simplified version examined in this chapter could be termed a two-layered-onion model.

It is common that for the same sets of goods, companies provide their customers with choices of different lead times and delivery methods. These differences in lead times and delivery alternatives may result in different charging schemes. It is generally assumed that the faster delivery modes are more expensive than the slower ones.

The models investigated in this chapter consider both advanced demand information and multiple supply sources. We show that the state-dependent base-stock policy is optimal for finite-horizon problems as well as for discounted infinite-horizon problems. This policy is defined by a pair of numbers—one for the fast mode and the other for the slow mode. These numbers are known as the *base-stock levels.*

The remainder of this chapter is organized as follows. In Section 3.2, we provide the required notation and the model formulation. Dynamic programming equations for the problem are developed in Section 3.3. In Section 3.4, we obtain the characterization of the optimal policy for the finite-horizon problem. Section 3.5 is devoted to extending the optimality results to the infinite-horizon case. In Section 3.6, we give an example to illustrate the results obtained in Sections 3.3 and 3.4 and develop more insights on the demand-information updates. The chapter is concluded in Sections 3.7 and 3.8. A technical appendix is provided in Section 3.9.

3.2. Notation and Model Formulation

Consider a discrete-time, single-product, periodic-review inventory system. The dynamics of the system contain the material flows and the information flows. The inbound material flows come from two supply sources (fast and slow), and the outbound material flows to customers. After they are ordered at the beginning of a period, materials from the fast and slow sources arrive at the end of the current period and at the end of the next period, respectively. The information flows include the initial demand forecast, regular forecast updates, and the realized customer demand. At the beginning of each period, the forecast of the demand, which will materialize at the end of the period, is updated. When the realized customer demand occurs at the end of the period, the customer is satisfied if there is sufficient available inventory, and the excess is carried over to the next period. Otherwise, the customer demand is partially satisfied, and the unsatisfied demand is fully backlogged.

The decision variables are the quantities ordered from the fast and slow sources at the beginning of each period. The decisions are made based on the current inventory position and the current (updated) demand information, where the inventory position at the beginning of the period is on-hand inventory plus

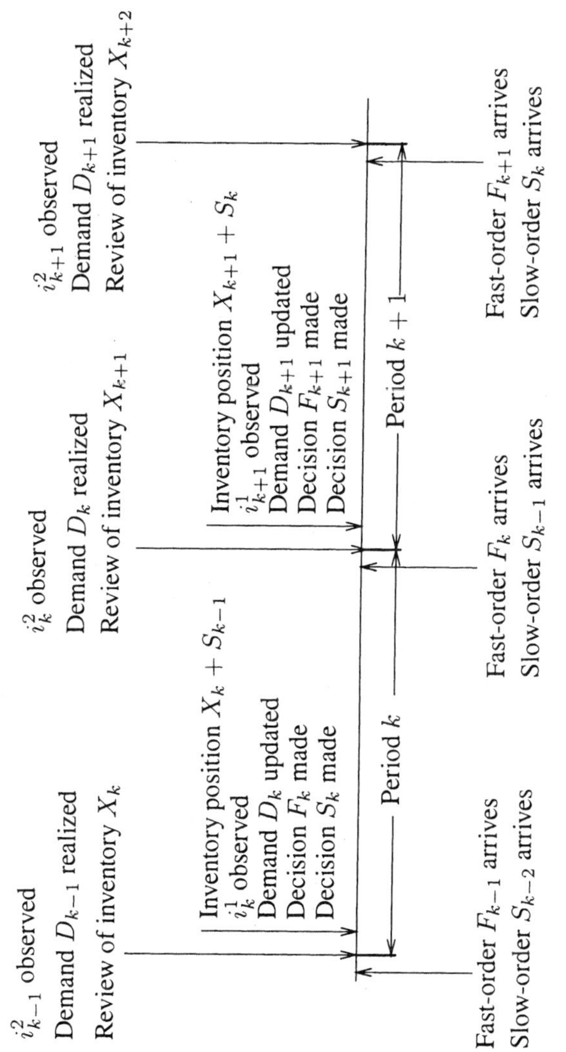

Figure 3.1. A time line of a periodic-reviews inventory system

the amount already on order to be delivered at the end of the period. A time line of the system dynamics and the ordering decisions is illustrated in Figure 3.1.

We introduce the following notation to formulate our model:

$$
\begin{aligned}
\langle 1, N\rangle &= \{1, 2, ..., N\}, \text{ the time horizon;} \\
F_k &= \text{the nonnegative fast-order quantity in period } k,\ 1 \le k \le N; \\
S_k &= \text{the nonnegative slow-order quantity in period } k, \\
&\quad\ 1 \le k \le N-1; \\
C_k^f(u) &= \text{the cost of fast-order } u \ge 0 \text{ units in period } k; \\
C_k^s(u) &= \text{the cost of slow-order } u \ge 0 \text{ units in period } k; \\
I_k^1 &= \text{the first determinant (a random variable) of the demand in} \\
&\quad\ \text{period } k \text{ observed at the beginning of period } k; \\
I_k^2 &= \text{the second determinant (a random variable) of the demand} \\
&\quad\ \text{in period } k \text{ observed at the end of period } k; \\
v_k &= \text{the third determinant (a constant) of the demand in} \\
&\quad\ \text{period } k; \\
D_k &= \text{the demand in period } k \text{ modeled as a function} \\
&\quad\ g_k(I_k^1, I_k^2, v_k); \\
X_k &= \text{the inventory level at the beginning of period } k; \\
Y_k &= X_k + S_{k-1} = \text{the inventory position at the beginning of} \\
&\quad\ \text{period } k; \\
X_{N+1} &= \text{the inventory level at the end of the last period } N; \\
H_k(x) &= \text{the inventory holding or backlog cost when } X_k = x; \\
H_{N+1}(x) &= \text{the holding cost when } X_{N+1} = x \ge 0 \text{ or penalty cost} \\
&\quad\ \text{when } X_{N+1} = x < 0.
\end{aligned}
$$

For notational convenience, let $I_1^1 = i_1^1$ be a deterministic constant. We impose the following assumptions on I_k^1 and I_k^2:

$$\{(I_k^1, I_k^2), 1 \le k \le N\} \quad \text{is a sequence of independent random vectors.} \tag{3.1}$$

Let us define $\mathcal{F}_{k+1}$, $k \ge 1$, to be the sigma algebra or σ-field generated by the random variables I_ℓ^1, I_ℓ^2, $1 \le \ell \le k$, and I_{k+1}^1—that is,

$$\mathcal{F}_{k+1} = \sigma\{(I_\ell^1, I_\ell^2), 1 \le \ell \le k, I_{k+1}^1\}, \quad 1 \le k \le N-1. \tag{3.2}$$

Let $\mathcal{F}_0 = \mathcal{F}_1 = \{\emptyset, \Omega\}$ and $\mathcal{F}_{N+1} = \mathcal{F}$. It is clear that demand D_k is an $\mathcal{F}_{k+1}$-measurable random variable. We assume further that

$$\mathsf{E}[D_k] = \mathsf{E}[g_k(I_k^1, I_k^2, v_k)] < \infty, \ 1 \le k \le N. \tag{3.3}$$

We also suppose that for each k, the cost functions

$$C_k^f(u) \text{ and } C_k^s(u) \text{ are increasing, nonnegative and convex} \tag{3.4}$$

and that the inventory-cost functions $H_k(x)$ satisfies

$$\begin{cases} H_k(x) \text{ is convex and} \\ |H_k(x) - H_k(\hat{x})| \le c_H \cdot |x - \hat{x}|, \ 1 \le k \le N+1, \end{cases} \tag{3.5}$$

for some $c_H > 0$. Furthermore, we assume that

$$C_k^f(t) + \mathsf{E}[H_{k+1}(t - g_k(I_k^1, I_k^2, v_k))] \to \infty \text{ as } t \to \infty, \tag{3.6}$$

$$C_k^s(t) + \mathsf{E}[H_{k+1}(t - g_k(I_k^1, I_k^2, v_k))] \to \infty \text{ as } t \to \infty. \tag{3.7}$$

Throughout this chapter, we assume that (3.1) and (3.3)–(3.7) hold.

The inventory-balance equations are defined as

$$\begin{aligned} X_{k+1} &= X_k + F_k + S_{k-1} - D_k \\ &= X_k + F_k + S_{k-1} - g_k(I_k^1, I_k^2, v_k), \ 1 \le k \le N, \end{aligned} \tag{3.8}$$

where S_0 is an outstanding slow order to be delivered at the end of period 1, and

$$X_1 = x_1, \text{ initial inventory level.} \tag{3.9}$$

Furthermore, the decision F_k is adapted to the σ-field $\mathcal{F}_k$, and the decision S_{k-1} is adapted to the σ-field $\mathcal{F}_{k-1}$. From (3.8), therefore, we can see that X_k is an $\mathcal{F}_k$-measurable random variable and X_{k+1} is an $\mathcal{F}_{k+1}$-measurable random variable, since $D_k = g_k(I_k^1, I_k^2, v_k)$ is $\mathcal{F}_{k+1}$-measurable.

Before going further, let us explain the dynamics (3.8) in words. At the beginning of period k, we can observe the value x_k of the on-hand inventory X_k and the value i_k^1 of the random variable I_k^1. This provides us with the updated forecast $g_k(i_k^1, I_k^2, v_k)$ for the kth-period demand D_k. We also know the inventory position $Y_k = X_k + S_{k-1}$, where S_{k-1} is the amount to be delivered at the end of period k as a result of the slow-order decision made in period $(k-1)$. Given these and the future demand forecasts D_j, $k+1 \le j \le N$, we can decide on the slow-order S_k and the fast-order F_k. Since F_k is to be delivered at the end of the period, the total quantity available to meet the kth-period demand D_k is $X_k + S_{k-1} + F_k$. At the end of period k, the value

i_k^2 of the random variable I_k^2 is observed, which is tantamount to the demand $D_k = g_k(i_k^1, i_k^2, v_k)$ to be observed. The difference of $X_k + S_{k-1} + F_k$ and D_k is the on-hand inventory X_{k+1} at the beginning of period $(k+1)$. This last statement represents a sample path of the dynamics (3.8). For the last period N, on-hand inventory X_N and the value i_N^1 of the random variable I_N^1 become available. Since any ordering from the slow source would not be delivered before the end of the problem horizon, it is obvious in view of the ordering cost given in (3.4) that

$$S_N = 0. \tag{3.10}$$

The objective is to choose a sequence of orders from the fast and the slow sources over time to minimize the total expected value of all the costs incurred during the interval $\langle 1, N \rangle$. Thus, the objective function is

$$\begin{aligned} &J_1(x_1, s_0, i_1^1, (\boldsymbol{F}, \boldsymbol{S})) \\ &\quad = H_1(x_1) + \mathsf{E}\left[\sum_{\ell=1}^{N} \Big(C_\ell^f(F_\ell) + C_\ell^s(S_\ell) + H_{\ell+1}(X_{\ell+1})\Big)\right], \end{aligned} \tag{3.11}$$

where

$$(\boldsymbol{F}, \boldsymbol{S}) = ((F_1, ..., F_N), (S_1, ..., S_N)) \tag{3.12}$$

is a sequence of *history-dependent* or *nonanticipative admissible decisions*—that is, (F_k, S_k) is a positive real-valued function of the history of the demand information up to period $(k-1)$, which is given by $\{(I_\ell^1, I_\ell^2), 0 \le \ell \le k-1\}$ and I_k^1; F_N is a positive real-valued function of the history of the demand information up to period $(N-1)$, which is given by $\{(I_\ell^1, I_\ell^2), 0 \le \ell \le N-1\}$ and I_N^1; and s_0 is an outstanding slow order to be delivered at the end of period 1 and has the same meaning as S_0 given by (3.8) with $k = 1$.

Finally, we define the value function for the problem over $\langle 1, N \rangle$ with the initial inventory level x_1 to be

$$V_1(x_1, s_0, i_1^1) = \inf_{(\boldsymbol{F}, \boldsymbol{S}) \in \mathcal{A}_1} \left\{ J_1(x_1, s_0, i_1^1, (\boldsymbol{F}, \boldsymbol{S})) \right\}, \tag{3.13}$$

where $\mathcal{A}_1$ denotes the class of all history-dependent admissible decisions for the problem over $\langle 1, N \rangle$. Note that the existence of an optimal policy is not required to define the value function. Of course, once the existence is established, the "inf" in (3.13) can be replaced by "min".

Note that we use a general form $g_k(I_k^1, I_k^2, v_k)$ to represent the process of demand-information updates and demand realizations. This representation covers some specific models in the literature. For example, let $g_k(I_k^1, I_k^2, v_k)$ equal

$v_k + I_k^1 + I_k^2$, where I_k^1 and I_k^2 are nonnegative random variables. If we let I_k^1 be a deterministic constant, our system reduces to the Scheller-Wolf and Tayur [12] model of two sources of deliveries without demand-information updates. If there is only one source for ordering, our system reduces to the advanced demand-information model of Gallego and Özer [7].

In addition, we would like to point out the role of v_k when $g_k(I_k^1, I_k^2, v_k)$ takes some special forms. If $g_k(I_k^1, I_k^2, v_k)$ is a multiplicative form—that is, $g_k(I_k^1, I_k^2, v_k) = v_k I_k^1 I_k^2$, where I_k^1 and I_k^2 are random variables with values in interval $[0, 1]$—then v_k is the upper bound of the demand in period k. On the other hand, if $g_k(I_k^1, I_k^2, v_k) = v_k + I_k^1 + I_k^2$, where I_k^1 and I_k^2 are nonnegative random variables, then v_k is the lower bound of the period-k demand and I_k^1 and I_k^2 are components of demands observed one period apart. In particular, v_k could represent a contracted periodic demand and I_k^1 could represent the firm orders received at the beginning of period k to be filled at the end of period k. Finally, I_k^2 represents the demand that arrives at the end of period k to be immediately filled (subject to a backlog situation if the inventory position $X_k + S_{k-1}$ is not sufficient to cover $v_k + I_k^1 + I_k^2$).

3.3. Dynamic Programming and Optimal Nonanticipative Policy

In this section, we use dynamic programming to study the problem. We verify that the cost of the nonanticipative policy obtained from the solution of the dynamic programming equations equals the value function of the problem over $\langle 1, N \rangle$. First, we define the problem over $\langle n, N \rangle$. Let

$$J_n\left(x_n, s_{n-1}, i_n^1, (\boldsymbol{F}, \boldsymbol{S})\right)$$
$$= H_n(x_n) + \mathsf{E}\left[\sum_{\ell=n}^{N}\left(C_\ell^f(F_\ell) + C_\ell^s(S_\ell) + H_{\ell+1}(X_{\ell+1})\right)\right], \tag{3.14}$$

where

$$(\boldsymbol{F}, \boldsymbol{S}) = ((F_n, ..., F_N), (S_n, ..., S_N))$$

is a *history-dependent* or *nonanticipative admissible decision* for the problem defined over periods $\langle n, N \rangle$. That is, given x_n, s_{n-1}, and i_n^1 as constants, (F_n, S_n) is a vector of nonnegative constants, (F_k, S_k) $(n < k < N)$ are positive real-valued functions of the history of the demand information from period n to period k, given by $\{(I_{\ell+1}^1, I_\ell^2), n \leq \ell \leq k-1\}, k = n+1, ..., N-1$, and F_N is a nonnegative real-valued function of the history of the demand information up from period n to period $(N-1)$, given by $\{(I_\ell^1, I_\ell^2), n \leq \ell \leq N-1\}$ and I_N^1. Here s_{n-1} has the same meaning as s_0 and is an outstanding slow order to be delivered at the end of period n. Define the value function associated with the problem over periods $\langle n, N\rangle$ as follows:

$$V_n\left(x_n, s_{n-1}, i_n^1\right) = \inf_{(\boldsymbol{F},\boldsymbol{S})\in\mathcal{A}_n} \left\{ J_n\left(x_n, s_{n-1}, i_n^1, (\boldsymbol{F}, \boldsymbol{S})\right) \right\}, \tag{3.15}$$

where $\mathcal{A}_n$ denotes the class of all history-dependent admissible decisions for the problem over $\langle n, N\rangle$. We have the following theorem on the property of the value function $V_n(x_n, s_{n-1}, i_n^1)$.

THEOREM 3.1 *Assume that* (3.1) *and* (3.3)–(3.7) *hold. Then the value function* $V_1(x_1, s_0, i_1^1)$ *is convex and Lipschitz continuous in* x_1 *on* $(-\infty, +\infty)$*, and the value functions* $V_n(x_n, s_{n-1}, i_n^1)$ $(2 \le n \le N)$ *are convex and Lipschitz continuous in* (x_n, s_{n-1}) *on* $(-\infty, +\infty) \times [0, +\infty)$.

Proof To prove the convexity, it suffices to show that for any $\theta \in [0,1]$, $(\tilde{x}_n, \tilde{s}_{n-1}) \in (-\infty, +\infty)\times[0, +\infty)$, and $(\hat{x}_n, \hat{s}_{n-1}) \in (-\infty, +\infty)\times[0, +\infty)$, we have

$$\begin{aligned} & V_n\left(\theta\tilde{x}_n + (1-\theta)\hat{x}_n, \theta\tilde{s}_{n-1} + (1-\theta)\hat{s}_{n-1}, i_n^1\right) \\ & \le \theta\cdot V_n\left(\tilde{x}_n, \tilde{s}_{n-1}, i_n^1\right) + (1-\theta)\cdot V_n\left(\hat{x}_n, \hat{s}_{n-1}, i_n^1\right). \end{aligned} \tag{3.16}$$

We note that for any two admissible decisions $(\tilde{\boldsymbol{F}}, \tilde{\boldsymbol{S}})$ and $(\hat{\boldsymbol{F}}, \hat{\boldsymbol{S}})$ of the problem over $\langle n, N\rangle$, the convex combination $(\theta\tilde{\boldsymbol{F}} + (1-\theta)\hat{\boldsymbol{F}}, \theta\tilde{\boldsymbol{S}} + (1-\theta)\hat{\boldsymbol{S}})$ is also an admissible decision. It follows from (3.4), (3.5), and (3.14) that

$$\begin{aligned} & J_n\Big(\theta\tilde{x}_n + (1-\theta)\hat{x}_n, \theta\tilde{s}_{n-1} + (1-\theta)\hat{s}_{n-1}, i_n^1, \\ & \qquad \left(\theta\tilde{\boldsymbol{F}} + (1-\theta)\hat{\boldsymbol{F}}, \theta\tilde{\boldsymbol{S}} + (1-\theta)\hat{\boldsymbol{S}}\right)\Big) \\ & \le \theta\cdot J_n\left(\tilde{x}_n, \tilde{s}_{n-1}, i_n^1, \left(\tilde{\boldsymbol{F}}, \tilde{\boldsymbol{S}}\right)\right) \\ & \quad +(1-\theta)\cdot J_n\left(\hat{x}_n, \hat{s}_{n-1}, i_n^1, \left(\hat{\boldsymbol{F}}, \hat{\boldsymbol{S}}\right)\right), \end{aligned} \tag{3.17}$$

which, in turn, implies (3.16).

The proof for $V_1(x_1, s_0, i_1^1)$ can be similarly established. Here we give the proof of the Lipschitz continuity for $V_n(x_n, s_{n-1}, i_n^1)$. To prove the Lipschitz continuity, it is sufficient to show that there exists a constant $L > 0$ such that for any $(x_n, s_{n-1}), (\hat{x}_n, \hat{s}_{n-1}) \in (-\infty, +\infty) \times [0, +\infty)$,

$$\begin{aligned} & \left|J_n(x_n, s_{n-1}, i_n^1, (\boldsymbol{F}, \boldsymbol{S})) - J_n(\hat{x}_n, \hat{s}_{n-1}, i_n^1, (\boldsymbol{F}, \boldsymbol{S}))\right| \\ & \le L\cdot\left(|x_n - \hat{x}_n| + |s_{n-1} - \hat{s}_{n-1}|\right). \end{aligned} \tag{3.18}$$

Note that from (3.5),

$$
\begin{aligned}
&\left|J_n(x_n, s_{n-1}, i_n^1, (\boldsymbol{F}, \boldsymbol{S})) - J_n(\hat{x}_n, \hat{s}_{n-1}, i_n^1, (\boldsymbol{F}, \boldsymbol{S}))\right| \\
&\quad \leq c_H \cdot (|x_n - \hat{x}_n| + |s_{n-1} - \hat{s}_{n-1}|) \\
&\qquad + \sum_{\ell=n}^{N} c_H \cdot (|x_n - \hat{x}_n| + |s_{n-1} - \hat{s}_{n-1}|)\,,
\end{aligned}
\tag{3.19}
$$

which implies (3.18) in view of $N < \infty$. □

REMARK 3.1 In classical inventory models with convex cost, the value function is convex in the initial inventory level (see Bensoussan, Crouhy, and Proth [2]). Theorem 3.1 states that this classical result remains valid for the problem with dual delivery modes and forecast updates.

In view of (3.14), we can write the dynamic programming equation corresponding to the problem as follows:

$$
\begin{aligned}
&U_\ell\left(x_\ell, s_{\ell-1}, i_\ell^1\right) \\
&\quad = H_\ell(x_\ell) + \inf_{\substack{\phi \geq 0 \\ \sigma \geq 0}} \left\{ C_\ell^f(\phi) + C_\ell^s(\sigma) \right. \\
&\qquad \left. + \mathsf{E}\left[U_{\ell+1}\left(x_\ell + \phi + s_{\ell-1} - g_\ell(i_\ell^1, I_\ell^2, v_\ell), \sigma, I_{\ell+1}^1\right) \right] \right\}, \\
&\quad \ell = 1, \ldots, N-1,
\end{aligned}
\tag{3.20}
$$

$$
\begin{aligned}
&U_N\left(x_N, s_{N-1}, i_N^1\right) \\
&\quad = H_N(x_N) + \inf_{\substack{\phi \geq 0 \\ \sigma \geq 0}} \left\{ C_N^f(\phi) + C_N^s(\sigma) \right. \\
&\qquad \left. + \mathsf{E}\left[H_{N+1}\left(x_N + s_{N-1} + \phi - g_N(i_N^1, I_N^2, v_N)\right) \right] \right\}.
\end{aligned}
\tag{3.21}
$$

REMARK 3.2 In the dynamic programming equations (3.20)–(3.21), the inventory cost is also charged for the initial inventory level. In some inventory literature, this cost for the initial inventory level is not charged. This means that $H_\ell(x_\ell)$ and $H_N(x_N)$ would be absent from (3.20)–(3.21), respectively. But this charge is of no consequence.

Next, we state the following theorem, which gives the relationship between the value function and the dynamic programming equation.

THEOREM 3.2 *Assume that* (3.1) *and* (3.3)–(3.7) *hold. Then the value functions* $V_k(x_k, s_{k-1}, i_k^1), 1 \leq k \leq N$, *defined in* (3.13) *and* (3.15), *satisfy the dynamic programming equations* (3.20)–(3.21).

Proof It follows from the definition of $V_N(x_N, s_{N-1}, i_N^1)$ that it satisfies the last equation in (3.20)–(3.21). Suppose that $V_\ell(x_\ell, s_{\ell-1}, i_\ell^1)$ ($\ell = k+1, ..., N$) satisfies (3.20)–(3.21). Now we show that $V_k(x_k, s_{k-1}, i_k^1)$ and $V_{k+1}(x_{k+1}, s_k, i_{k+1}^1)$ also satisfy the first equation in (3.20)–(3.21). That is,

$$\begin{aligned} & V_k\left(x_k, s_{k-1}, i_k^1\right) \\ & = H_k(x_k) + \inf_{\substack{\phi \geq 0 \\ \sigma \geq 0}} \left\{ C_k^f(\phi) + C_k^s(\sigma) + \mathsf{E}\Big[V_{k+1}\left(X_{k+1}, \sigma, I_{k+1}^1\right)\Big]\right\}, \end{aligned} \tag{3.22}$$

where

$$X_{k+1} = x_k + s_{k-1} + \phi - g_k(i_k^1, I_k^2, v_k).$$

By the definition of $V_k(x_k, s_{k-1}, i_k^1)$ and the history dependence of $(\boldsymbol{F}, \boldsymbol{S})$, we have

$$\begin{aligned} & V_k\left(x_k, s_{k-1}, i_k^1\right) \\ & = H_k(x_k) + \inf_{(\boldsymbol{F},\boldsymbol{S}) \in \mathcal{A}_k} \left\{ \mathsf{E}\Big[\sum_{\ell=k}^{N} \Big(C_\ell^f(F_\ell) + C_\ell^s(S_\ell) \right. \\ & \qquad \left. + H_{\ell+1}(X_{\ell+1}) \Big) \Big] \right\} \\ & = H_k(x_k) + \inf_{(\boldsymbol{F},\boldsymbol{S}) \in \mathcal{A}_k} \left\{ C_k^f(F_k) + C_k^s(S_k) \right. \\ & \qquad \left. + \mathsf{E}\Big[H_{k+1}(X_{k+1}) + \sum_{\ell=k+1}^{N} \Big(C_\ell^f(F_\ell) + C_\ell^s(S_\ell) + H_{\ell+1}(X_{\ell+1}) \Big) \Big] \right\}. \end{aligned} \tag{3.23}$$

It follows from (3.1) that

$$\begin{aligned} & \mathsf{E}\Big[H_{k+1}(X_{k+1}) + \sum_{\ell=k+1}^{N} \Big(C_\ell^f(F_\ell) + C_\ell^s(S_\ell) + H_{\ell+1}(X_{\ell+1}) \Big) \Big] \\ & = \mathsf{E}\Big\{ \mathsf{E}\Big[H_{k+1}(X_{k+1}) \\ & \quad + \sum_{\ell=k+1}^{N} \Big(C_\ell^f(F_\ell) + C_\ell^s(S_\ell) + H_{\ell+1}(X_{\ell+1}) \Big) \Big| (I_k^2, I_{k+1}^1) \Big] \Big\} \end{aligned}$$

$$= \mathsf{E}\Big\{ H_{k+1}(X_{k+1}) + C^f_{k+1}(F_{k+1}) + C^s_{k+1}(S_{k+1}) + H_{k+2}(X_{k+2})$$
$$+\mathsf{E}\Big[\sum_{\ell=k+2}^{N} \Big(C^f_\ell(F_\ell) + C^s_\ell(S_\ell) + H_{\ell+1}(X_{\ell+1}) \Big) \Big| (I^2_k, I^1_{k+1}) \Big] \Big\}.$$

By the induction on the index $(k+1)$,

$$\inf_{(\boldsymbol{F},\boldsymbol{S}) \in \mathcal{A}_k} \Big\{ \mathsf{E}\Big[H_{k+1}(X_{k+1})$$
$$+ \sum_{\ell=k+1}^{N} \Big(C^f_\ell(F_\ell) + C^s_\ell(S_\ell) + H_{\ell+1}(X_{\ell+1}) \Big) \Big] \Big\}$$
$$= \mathsf{E}\left[V_{k+1}\left(X_{k+1}, S_k, I^1_{k+1} \right) \right]. \tag{3.24}$$

Therefore, (3.23) and (3.24) complete the proof. □

REMARK 3.3 Compared with the dynamic programming equation of the rolling-horizon problem studied by Sethi and Sorger [14], the dynamic programming equations (3.20)–(3.21) have one more decision variable $s_{\ell-1}$—that is, ordering from the slow source. On the other hand, compared with the dynamic programming equations of the dual-source production-inventory problem studied by Scheller-Wolf and Tayur [12], the dynamic programming equations (3.20)–(3.21) have one more state variable i^1_ℓ—that is, the updated demand information.

Next, we discuss how an optimal solution of our periodic-review inventory model with fast and slow orders and demand-information updates could be found.

Assumptions (3.6)–(3.7) imply that there exists an upper-bound order quantity $Q > 0$ such that

$$\inf_{\substack{\phi \geq 0 \\ \sigma \geq 0}} \Big\{ C^f_\ell(\phi) + C^s_\ell(\sigma)$$
$$+\mathsf{E}\left[V_{\ell+1}\left(x_\ell + \phi + s_{\ell-1} - g_\ell(i^1_\ell, I^2_\ell, v_\ell), \sigma, I^1_{\ell+1} \right) \right] \Big\}$$
$$= \inf_{0 \leq \phi, \sigma \leq Q} \Big\{ C^f_\ell(\phi) + C^s_\ell(\sigma)$$
$$+\mathsf{E}\left[V_{\ell+1}\left(x_\ell + \phi + s_{\ell-1} - g_\ell(i^1_\ell, I^2_\ell, v_\ell), \sigma, I^1_{\ell+1} \right) \right] \Big\}$$

for $\ell = 1, ..., N-1$, and

$$\inf_{\substack{\phi \geq 0 \\ \sigma \geq 0}} \Big\{ C_N^f(\phi) + C_N^s(\sigma) + \mathsf{E}\left[H_{N+1}(x_N + s_{N-1} + \phi - g_N(i_N^1, I_N^2, v_N))\right] \Big\}$$

$$= \inf_{0 \leq \phi, \sigma \leq Q} \Big\{ C_N^f(\phi) + C_N^s(\sigma) + \mathsf{E}\left[H_{N+1}(x_N + s_{N-1} + \phi - g_N(i_N^1, I_N^2, v_N))\right] \Big\}.$$

By Theorem 3.9 in the appendix to this chapter and Theorem 3.1, in view of the discussion leading to (3.10), there exist Borel-measurable functions $\bar{\phi}_N(x_N, s_{N-1}, i_N^1)$ and $\bar{\sigma}_N(x_N, s_{N-1}, i_N^1)(= 0)$ such that

$$\begin{aligned} & C_N^f(\bar{\phi}_N(x_N, s_{N-1}, i_N^1)) + C_N^s(0) \\ & + \mathsf{E}\left[H_{N+1}\left(x_N + s_{N-1} + \bar{\phi}_N(x_N, s_{N-1}, i_N^1) - g_N(i_N^1, I_N^2, v_N)\right)\right] \\ & \quad = \inf_{0 \leq \phi, \sigma \leq Q} \Big\{ C_N^f(\phi) + C_N^s(\sigma) \\ & \qquad + \mathsf{E}\left[H_{N+1}\left(x_N + s_{N-1} + \phi - g_N(i_N^1, I_N^2, v_N)\right)\right] \Big\}, \end{aligned} \tag{3.25}$$

and there exist Borel-measurable functions

$$\left(\bar{\phi}_\ell(x_\ell, s_{\ell-1}, i_\ell^1), \bar{\sigma}_\ell(x_\ell, s_{\ell-1}, i_\ell^1)\right), \quad 1 \leq \ell \leq N-1, \tag{3.26}$$

such that

$$\begin{aligned} & C_\ell^f(\bar{\phi}_\ell(x_\ell, s_{\ell-1}, i_\ell^1)) + C_\ell^s(\bar{\sigma}_\ell(x_\ell, s_{\ell-1}, i_\ell^1)) \\ & \quad + \mathsf{E}\big[V_{\ell+1}\left(x_\ell + \bar{\phi}_\ell(x_\ell, s_{\ell-1}, i_\ell^1) + s_{\ell-1} - g_\ell(i_\ell^1, I_\ell^2, v_\ell),\right. \\ & \qquad\qquad \left.\bar{\sigma}_\ell(x_\ell, s_{\ell-1}, i_\ell^1), I_{\ell+1}^1\right)\big] \\ & \quad = \inf_{0 \leq \phi, \sigma \leq Q} \Big\{ C_\ell^f(\phi) + C_\ell^s(\sigma) \\ & \qquad + \mathsf{E}\left[V_{\ell+1}\left(x_\ell + \phi + s_{\ell-1} - g_\ell(i_\ell^1, I_\ell^2, v_\ell), \sigma, I_{\ell+1}^1\right)\right] \Big\}. \end{aligned} \tag{3.27}$$

Define

$$\begin{aligned}
\bar{X}_1 &= x_1, && (3.28)\\
\bar{F}_1 &= \bar{\phi}_1(x_1, s_0, i_1^1), && (3.29)\\
\bar{S}_1 &= \bar{\sigma}_1(x_1, s_0, i_1^1), && (3.30)
\end{aligned}$$

and

$$\begin{aligned}
\bar{X}_\ell &= \bar{X}_{\ell-1} + \bar{F}_{\ell-1} + \bar{S}_{\ell-2} - g_{\ell-1}(I_{\ell-1}^1, I_{\ell-1}^2, v_{\ell-1}), && (3.31)\\
\bar{F}_\ell &= \bar{\phi}_\ell(\bar{X}_\ell, \bar{S}_{\ell-1}, I_\ell^1), && (3.32)\\
\bar{S}_\ell &= \bar{\sigma}_\ell(\bar{X}_\ell, \bar{S}_{\ell-1}, I_\ell^1), && (3.33)
\end{aligned}$$

for $2 \leq \ell \leq N-1$, where $\bar{S}_0 = s_0$. Finally, define

$$\begin{aligned}
\bar{X}_N &= \bar{X}_{N-1} + \bar{F}_{N-1} + \bar{S}_{N-2} - g_{N-1}(I_{N-1}^1, I_{N-1}^2, v_{N-1}), && (3.34)\\
\bar{F}_N &= \bar{\phi}_N(\bar{X}_N, \bar{S}_{N-1}, I_N^1), && (3.35)\\
\bar{S}_N &= 0. && (3.36)
\end{aligned}$$

Using the dynamic programming equations (3.20)–(3.21), we can prove the following result.

THEOREM 3.3 (VERIFICATION THEOREM) *Assume that* (3.1) *and* (3.3)–(3.7) *hold. Then*

$$((\bar{F}_1, ..., \bar{F}_N), (\bar{S}_1, ..., \bar{S}_N))$$

given in (3.28)–(3.36) *is an optimal solution to the problem. That is,*

$$H_1(x_1) + \mathsf{E}\left[\sum_{\ell=1}^{N} \left(C_\ell^f(\bar{F}_\ell) + C_\ell^s(\bar{S}_\ell) + H_{\ell+1}(\bar{X}_{\ell+1})\right)\right]$$
$$= V_1(x_1, s_0, i_1^1). \qquad (3.37)$$

REMARK 3.4 Theorems 3.2 and 3.3 establish the existence of an optimal nonanticipative policy—that is, there exists a policy in the class of all history-dependent policies whose objective function value equals the value function defined in (3.13), and there exists a nonanticipative policy defined by (3.28)–(3.36) that provides the same value for the objective function.

Proof of Theorem 3.3 By (3.27) we know that

$$((\bar{F}_1, ..., \bar{F}_N), (\bar{S}_1, ..., \bar{S}_N)) \in \mathcal{A}_1$$

—that is, it is a history-dependent policy. Next, we show that equation (3.37) holds. It suffices to show that for any

$$((F_1, ..., F_N), (S_1, ..., S_N)) \in \mathcal{A}_1,$$

we have

$$H_1(x_1) + \mathsf{E}\left[\sum_{\ell=1}^{N} \left(C_\ell^f(F_\ell) + C_\ell^s(S_\ell) + H_{\ell+1}(X_{\ell+1})\right)\right]$$
$$\geq H_1(x_1) + \mathsf{E}\left[\sum_{\ell=1}^{N} \left(C_\ell^f(\bar{F}_\ell) + C_\ell^s(\bar{S}_\ell) + H_{\ell+1}(\bar{X}_{\ell+1})\right)\right], \tag{3.38}$$

where X_ℓ $(1 \leq \ell \leq N)$ is defined to be the same as $\bar{X}_\ell$ in (3.31)–(3.33) and (3.34)–(3.35) with the exception that $\bar{F}_\ell$ and $\bar{S}_\ell$ are replaced by F_ℓ and S_ℓ, respectively. By the definition of $(\bar{F}_1, \bar{S}_1)$ and (3.27), it is possible to obtain

$$C_1^f(\bar{F}_1) + C_1^s(\bar{S}_1) + \mathsf{E}\left[V_2(\bar{X}_2, \bar{S}_1, I_2^1)\right]$$
$$\leq C_1^f(F_1) + C_1^s(S_1) + \mathsf{E}\left[V_2(X_2, S_1, I_2^1)\right]. \tag{3.39}$$

Furthermore, from the history-dependent property of the decisions, we know that $\bar{F}_1$, $\bar{S}_1$, F_1, and S_1 are constants and that $(\bar{F}_2, \bar{S}_2)$ and (F_2, S_2) are dependent on $\{I_1^2, I_2^1\}$. Thus by (3.27),

$$V_2\left(X_2, S_1, I_2^1\right)$$
$$= H_2(X_2) + \inf_{0 \leq \phi, \sigma \leq Q} \left\{C_2^f(\phi) + C_2^s(\sigma)\right.$$
$$\left. + \mathsf{E}\left[V_3\left(X_2 + \phi + S_1 - g_2(I_2^1, I_2^2, v_2), \sigma, I_3^1\right) \middle| (I_1^2, I_2^1)\right]\right\}$$
$$\leq H_2(X_2) + C_2^f(F_2) + C_2^s(S_2) + \mathsf{E}\left[V_3\left(X_3, S_2, I_3^1\right) \middle| (I_1^2, I_2^1)\right], \tag{3.40}$$

and

$$V_2\left(\bar{X}_2, \bar{S}_1, I_2^1\right)$$
$$= H_2(\bar{X}_2) + C_2^f(\bar{F}_2) + C_2^s(\bar{S}_2) + \mathsf{E}\left[V_3\left(\bar{X}_3, \bar{S}_2, I_3^1\right) \middle| (I_1^2, I_2^1)\right]. \tag{3.41}$$

Therefore, it follows from (3.41) that

$$\mathsf{E}[V_2(\bar{X}_2, \bar{S}_1, I_2^1)]$$
$$= \mathsf{E}\left\{H_2(\bar{X}_2) + C_2^f(\bar{F}_2) + C_2^s(\bar{S}_2) + \mathsf{E}\left[V_3(\bar{X}_3, \bar{S}_2, I_3^1) \middle| (I_1^2, I_2^1)\right]\right\}, \tag{3.42}$$

and from (3.40) that

$$\begin{aligned}&\mathsf{E}[V_2(X_2, S_1, I_2^1)]\\&\quad\leq \mathsf{E}\left\{H_2(X_2) + C_2^f(F_2) + C_2^s(S_2) + \mathsf{E}\left[V_3(X_3, S_2, I_3^1)\Big|(I_1^2, I_2^1)\right]\right\}.\end{aligned} \tag{3.43}$$

Combining (3.39) and (3.42)–(3.43) yields

$$\begin{aligned}&H_1(x_1) + \mathsf{E}\left[\sum_{\ell=1}^{2}\left(C_\ell^f(\bar{F}_\ell) + C_\ell^s(\bar{S}_\ell)\right) + H_2(\bar{X}_2)\right]\\&\quad+\mathsf{E}\left[V_3(\bar{X}_3, \bar{S}_2, I_3^1)\right]\\&\quad\leq H_1(x_1) + \mathsf{E}\left[\sum_{\ell=1}^{2}\left(C_\ell^f(F_\ell) + C_\ell^s(S_\ell)\right) + H_2(X_2)\right]\\&\qquad+\mathsf{E}\left[V_3(X_3, S_2, I_3^1)\right].\end{aligned} \tag{3.44}$$

Repeating (3.42) and (3.43), we finally prove that (3.38) holds. □

3.4. Optimality of Base-Stock Policies

For a further analysis of the problem, it is convenient to recast the dynamic programming equations (3.20)–(3.21) involving order quantities ϕ and σ as decision variables to those involving order-up-to levels y and z as decision variables. Such a transformation is standard in the inventory literature (see Fukuda [6], and Whittemore and Saunders [17], for example). Moreover, since the ordering decisions ϕ and σ in any period ℓ will turn out to depend on x_ℓ and $s_{\ell-1}$ through their sum $x_\ell + s_{\ell-1}$, known as the inventory position in period ℓ (denoted by q_ℓ in the following), the dynamic programming equations (3.20)–(3.21) can be rewritten in terms of the state variables q_ℓ and i_ℓ^1, replacing the state variables $x_\ell, s_{\ell-1}$ and i_ℓ^1. Finally, since $H_\ell(x)$ in (3.20)–(3.21) is outside the infimum operation, it is possible to modify the dynamic programming equations (3.20)–(3.21) as follows:

$$\begin{aligned}&\hat{U}_k(q_k, i_k^1)\\&\quad= \inf_{\substack{y\geq q_k\\ z\geq y}}\left\{C_k^f(y - q_k) + C_k^s(z - y) + \mathsf{E}\left[H_{k+1}(y - g_k(i_k^1, I_k^2, v_k))\right]\right.\\&\qquad\left.+\mathsf{E}\left[\hat{U}_{k+1}(z - g_k(i_k^1, I_k^2, v_k), I_{k+1}^1)\right]\right\},\\&\qquad k = 1, \dots, N-1,\end{aligned} \tag{3.45}$$

$$\hat{U}_N(q_N, i_N^1) = \inf_{\substack{y \geq q_N \\ z \geq y}} \Big\{ C_N^f(y - q_N) + C_N^s(z - y) + \mathsf{E}\left[H_{N+1}(y - g_N(i_N^1, I_N^2, v_N))\right] \Big\}. \quad (3.46)$$

Let $\hat{V}_k(q_k, i_k^1)$ $(1 \leq k \leq N)$ be a solution of (3.45)–(3.46). In a way similar to Theorem 3.1, it is possible to show that $\hat{V}_k(q_k, i_k^1)$ $(1 \leq k \leq N)$ is convex in q_k. Furthermore, by the fact that the set $\{(y, z) | y \geq q_k \text{ and } z \geq y\}$ is a convex set, it follows from (3.45)–(3.46) that there exist minimizing functions

$$\hat{\phi}_k(q_k, i_k^1) \quad (1 \leq k \leq N), \quad (3.47)$$

and

$$\hat{\sigma}_\ell(q_\ell, i_\ell^1) \quad (1 \leq \ell \leq N) \quad (3.48)$$

such that for $k = 1, ..., N-1$,

$$\begin{aligned} \hat{V}_k(q_k, i_k^1) &= C_k^f(\hat{\phi}_k(q_k, i_k^1) - q_k) + C_k^s(\hat{\sigma}_k(q_k, i_k^1) - \hat{\phi}_k(q_k, i_k^1)) \\ &\quad + \mathsf{E}\left[H_{k+1}\left(\hat{\phi}_k(q_k, i_k^1) - g_k(i_k^1, I_k^2, v_k)\right)\right] \\ &\quad + \mathsf{E}\left[\hat{V}_{k+1}(\hat{\sigma}_k(q_k, i_k^1) - g_k(i_k^1, I_k^2, v_k), I_{k+1}^1)\right], \end{aligned} \quad (3.49)$$

and

$$\begin{aligned} \hat{V}_N(q_N, i_N^1) &= C_N^f(\hat{\phi}_N(q_N, i_N^1) - q_N) + C_N^s(\hat{\sigma}_N(q_N, i_N^1) - \hat{\phi}_N(q_N, i_N^1)) \\ &\quad + \mathsf{E}\left[H_{N+1}(\hat{\phi}_N(q_N, i_N^1) - g_N(i_N^1, I_N^2, v_N))\right]. \end{aligned} \quad (3.50)$$

In view of (3.46), we know that

$$\hat{\phi}_N(q_N, i_N^1) = \hat{\sigma}_N(q_N, i_N^1).$$

Let

$$\begin{aligned} q_1 &= x_1 + s_0 \\ q_k &= \hat{\sigma}_{k-1}(q_{k-1}, I_{k-1}^1) - g_{k-1}(I_{k-1}^1, I_{k-1}^2, v_{k-1}), \ k = 2, ..., N. \end{aligned}$$

Define

$$\hat{F}_1 = \hat{\phi}_1(q_1, I_1^1) - q_1, \quad (3.51)$$
$$\hat{S}_1 = \hat{\sigma}_1(q_1, I_1^1) - \hat{\phi}_1(q_1, I_1^1), \quad (3.52)$$

and

$$\hat{F}_k = \hat{\phi}_k(q_k, I_k^1) - q_k, \; k = 2, ..., N, \quad (3.53)$$

$$\hat{S}_\ell = \hat{\sigma}_\ell(q_\ell, I_\ell^1) - \hat{\phi}_k(q_k, I_k^1), \; \ell = 2, ..., N-1, \quad (3.54)$$

$$\hat{S}_N = 0. \quad (3.55)$$

From Theorem 3.3 and the revised dynamic programming equations (3.45)–(3.46), we have the following theorem.

THEOREM 3.4 *Assume that* (3.1) *and* (3.3)–(3.7) *hold. Then*

$$V_n(x_n, s_{n-1}, i_n^1) = \hat{V}_n(x_n + s_{n-1}, i_n^1) + H_n(x_n), \quad n = 1, ..., N,$$

and

$$\left(\left(\hat{F}_1, ..., \hat{F}_N\right), \left(\hat{S}_1, ..., \hat{S}_N\right)\right) \quad (3.56)$$

is also an optimal policy for the problem over $\langle 1, N \rangle$.

REMARK 3.5 Theorem 3.4 states that the dynamic programming equations given by (3.45)–(3.46) are equivalent to those given by (3.20)–(3.21). Hence, to derive the optimal nonanticipative policy, it is sufficient to solve the dynamic programming equations (3.45)–(3.46).

Proof of Theorem 3.4 First we show that

$$V_N(x_N, s_{N-1}, i_N^1) = H_N(x_N) + \hat{V}_N(x_N + s_{N-1}, i_N^1). \quad (3.57)$$

From the definition of $\hat{V}_N(q_N, i_N^1)$, we have

$$\begin{aligned}
&\hat{V}_N\left(x_N + s_{N-1}, i_N^1\right) \\
&= \inf_{\substack{y \geq x_N + s_{N-1} \\ z \geq y}} \Big\{ C_N^f(y - (x_N + s_{N-1})) + C_N^s(z - y) \\
&\quad + \mathsf{E}\left[H_{N+1}\left(y - g_N(i_N^1, I_N^2, v_N)\right)\right] \Big\} \\
&= \inf_{\substack{\phi \geq 0 \\ \sigma \geq 0}} \Big\{ C_N^f(\phi) + C_N^s(\sigma) \\
&\quad + \mathsf{E}\left[H_{N+1}\left(x_N + s_{N-1} + \phi - g_N(i_N^1, I_N^2, v_N)\right)\right] \Big\} \\
&= H_N(x_N) + \inf_{\substack{\phi \geq 0 \\ \sigma \geq 0}} \Big\{ C_N^f(\phi) + C_N^s(\sigma) \\
&\quad + \mathsf{E}\left[H_{N+1}\left(x_N + s_{N-1} + \phi - g_N(i_N^1, I_N^2, v_N)\right)\right] \Big\} - H_N(x_N) \\
&= V_N\left(x_N, s_{N-1}, i_N^1\right) - H_N(x_N), \quad (3.58)
\end{aligned}$$

which is equivalent to (3.57). Furthermore, from the second equality of (3.58) and the definitions of $\bar{\phi}_N(x_N, s_{N-1}, i_N^1)$ and $\hat{\phi}_N(x_N + s_{N-1}, i_N^1)$ (see (3.25) and (3.50)), we have

$$\bar{\phi}_N(x_N, s_{N-1}, i_N^1) = \hat{\phi}_N(x_N + s_{N-1}, i_N^1) - (x_N + s_{N-1}), \tag{3.59}$$

$$\bar{\sigma}_N(x_N, s_{N-1}, i_N^1) = \hat{\sigma}_N(x_N + s_{N-1}, i_N^1) - \hat{\phi}_N(x_N + s_{N-1}, i_N^1). \tag{3.60}$$

Now suppose that for $j = N, ..., k+1$,

$$V_j(x_j, s_{j-1}, i_j^1) = \hat{V}_j(x_j + s_{j-1}, i_j^1) + H_j(x_j), \tag{3.61}$$

$$\bar{\phi}_j(x_j, s_{j-1}, i_j^1) = \hat{\phi}_j(x_j + s_{j-1}, i_j^1) - (x_j + s_{j-1}), \tag{3.62}$$

$$\bar{\sigma}_j(x_j, s_{j-1}, i_j^1) = \hat{\sigma}_j(x_j + s_{j-1}, i_j^1) - \hat{\phi}_j(x_j + s_{j-1}, i_j^1), \tag{3.63}$$

where $\bar{\phi}_j(x_j, s_{j-1}, i_j^1)$ and $\bar{\sigma}_j(x_j, s_{j-1}, i_j^1)$ are given by (3.25)and (3.27), respectively. Then we show that (3.61)–(3.63) hold for $j = k$. By the definition of $\hat{V}_k(q_k, i_k^1)$,

$$\begin{aligned}
&\hat{V}_k\left(x_k + s_{k-1}, i_k^1\right) \\
&= \inf_{\substack{y \geq x + s_{k-1} \\ z \geq y}} \Big\{ C_k^f(y - x_k - s_{k-1}) + C_k^s(z - y) \\
&\qquad + \mathsf{E}\left[H_{k+1}\left(y - g_k(i_k^1, I_k^2, v_k)\right)\right] \\
&\qquad + \mathsf{E}\left[\hat{V}_{k+1}\left(z - g_k(i_k^1, I_k^2, v_k), I_{k+1}^1\right)\right] \Big\} \\
&= \inf_{\substack{\phi \geq 0 \\ \sigma \geq 0}} \Big\{ C_k^f(\phi) + C_k^s(\sigma) \\
&\qquad + \mathsf{E}\left[H_{k+1}\left(x_k + s_{k-1} + \phi - g_k(i_k^1, I_k^2, v_k)\right)\right] \\
&\qquad + \mathsf{E}\left[\hat{V}_{k+1}\left(x_k + s_{k-1} + \phi - g_k(i_k^1, I_k^2, v_k) + \sigma, I_{k+1}^1\right)\right] \Big\} \\
&= \inf_{\substack{\phi \geq 0 \\ \sigma \geq 0}} \Big\{ C_k^f(\phi) + C_k^s(\sigma) \\
&\qquad + \mathsf{E}\left[V_{k+1}\left(x_k + s_{k-1} + \phi - g_k(i_k^1, I_k^2, v_k), \sigma, I_{k+1}^1\right)\right] \Big\}
\end{aligned}$$

$$= H_k(x_k) + \inf_{\substack{\phi \geq 0 \\ \sigma \geq 0}} \Big\{ C_k^f(\phi) + C_k^s(\sigma)$$

$$+\mathsf{E}\left[V_{k+1}\left(x_k + s_{k-1} + \phi - g_k(i_k^1, I_k^2, v_k), \sigma, I_{k+1}^1\right)\right] \Big\} - H_k(x_k)$$

$$= V_k\left(x_k, s_{k-1}, i_k^1\right) - H_k(x_k), \tag{3.64}$$

where in establishing the third equality of (3.64), we have applied the first equation of (3.61)–(3.63) for $j = k+1$. Thus, we obtain the first equation of (3.61)–(3.63) for $j = k$. At the same time, from the second equality in (3.64), we have the second and the third equations of (3.61)–(3.63) for $j = k$. Therefore, by induction, the theorem is established. □

Let $(\bar{y}_k, \bar{z}_k)$ be a minimum point of the function

$$C_k^f(y - x_k - s_{k-1}) + C_k^s(z - y) + \mathsf{E}\left[H_{k+1}(y - g_k(i_k^1, I_k^2, v_k))\right]$$

$$+\mathsf{E}\left[\hat{V}_{k+1}(z - g_k(i_k^1, I_k^2, v_k), I_{k+1}^1)\right] \tag{3.65}$$

on the region $\{(y, z) : y \geq x_k + s_{k-1} \text{ and } z \geq y\}$. Based on Theorem 3.4, we have the following corollary.

COROLLARY 3.1 *Assume that* (3.1) *and* (3.3)–(3.7) *hold. If the initial inventory level at the beginning of period k is x_k, the slow-order quantity in period $(k-1)$ is denoted by s_{k-1}, and the observed value of I_k^1 is i_k^1 in period $(k-1)$, then the optimal fast-order quantity f_k and the optimal slow-order quantity s_k in period k can be expressed as follows:*

$$(f_k, s_k) = (\bar{y}_k - x_k - s_{k-1}, \bar{z}_k - \bar{y}_k).$$

Proof Note that from (3.47)–(3.48),

$$\bar{y}_k = \hat{\phi}_k(x_k + s_{k-1}, i_k^1), \quad \bar{z}_k = \hat{\sigma}_k(x_k + s_{k-1}, i_k^1). \tag{3.66}$$

Hence, using the definitions of $\hat{F}_k$ and $\hat{S}_k$ given by (3.53)–(3.55) and the assumptions of the corollary, we have

$$\bar{y}_k - x_k - s_{k-1} = \hat{F}_k, \quad \bar{z}_k - \bar{y}_k = \hat{S}_k.$$

Consequently, the corollary follows from Theorem 3.4. □

The corollary says that there are two inventory levels—fast and slow—with the fast-level $\bar{y}_k$ being smaller than the slow-level $\bar{z}_k$. The optimal policy is to order up to $\bar{y}_k$ via the fast mode and to order an additional amount from $\bar{y}_k$ up to $\bar{z}_k$ via the slow mode. In particular, when the inventory position is strictly smaller than $\bar{y}_k$ and $\bar{y}_k < \bar{z}_k$, both slow and fast orders will be given. When the inventory position is equal to $\bar{y}_k$ but strictly smaller than $\bar{z}_k$, then only the slow order will be issued. When the inventory position is strictly smaller than $\bar{y}_k$ and $\bar{y}_k = \bar{z}_k$, then only the fast order will be issued. Finally, when the inventory position is equal to $\bar{z}_k$, no orders—fast or slow—will be given.

It is worth noting that solving for $(\bar{y}_k, \bar{z}_k)$ is a two-variable optimization problem, which, in general, is more complicated than a single-variable optimization problem. In a particularly important case, it is possible to reduce the problem to solving a pair of single-variable optimization problems and obtain a modified base-stock policy. In this case, we make the assumption that the ordering costs are linear—that is,

$$C_k^f(t) = c_k^f \cdot t, \ c_k^f > 0, \ 1 \le k \le N, \tag{3.67}$$

$$C_k^s(t) = c_k^s \cdot t, \ 0 < c_k^s < c_{k+1}^f, \ 1 \le k \le N-1. \tag{3.68}$$

In view of (3.67)–(3.68), (3.45)–(3.46) can be written as

$$\begin{aligned} \tilde{U}_k(q_k, i_k^1) = \inf_{\substack{y \ge q_k \\ z \ge y}} \Big\{ & c_k^f \cdot (y - q_k) + c_k^s z - c_k^s y \\ & + \mathsf{E}\left[H_{k+1}(y - g_k(i_k^1, I_k^2, v_k))\right] \\ & + \mathsf{E}\left[\tilde{U}_{k+1}(z - g_k(i_k^1, I_k^2, v_k), I_{k+1}^1)\right] \Big\}, \end{aligned} \tag{3.69}$$

$$k = 1, ..., N-1,$$

$$\begin{aligned} \tilde{U}_N(q_N, i_N^1) = \inf_{\substack{y \ge q_N \\ z \ge y}} \Big\{ & c_N^f \cdot (y - q_N) + c_N^s \cdot (z - y) \\ & + \mathsf{E}\left[H_{N+1}(y - g_N(i_N^1, I_N^2, v_N))\right] \Big\}. \end{aligned} \tag{3.70}$$

Let $\tilde{V}_k(q_k, i_k^1)$ $(1 \le k \le N)$ be a solution of (3.69)–(3.70), and let y_N^* be the value of y that minimizes

$$c_N^f y + \mathsf{E}[H_{N+1}(y - g_N(i_N^1, I_N^2, v_N))],$$

and let y_k^* be the value that minimizes the function

$$c_k^f y - c_k^s y + \mathsf{E}[H_{k+1}(y - g_k(i_k^1, I_k^2, v_k))]$$

in y. Define

$$L_k(t) = \Bigg\{ c_k^f \cdot (t - x_k - s_{k-1}) - c_k^s t + \mathsf{E}\left[H_{k+1}(t - g_k(i_k^1, I_k^2, v_k))\right] - \Bigg(c_k^f \cdot (y_k^* - x_k - s_{k-1}) - c_k^s y_k^* + \mathsf{E}\left[H_{k+1}(y_k^* - g_k(i_k^1, I_k^2, v_k))\right] \Bigg) \Bigg\} \cdot \delta(y_k^* - t) + c_k^s t + \mathsf{E}\left[\tilde{V}_{k+1}(t - g_k(i_k^1, I_k^2, v_k), I_{k+1}^1)\right].$$

Let z_k^* be a minimum of $L_k(t)$. Then the optimal policy can be written as follows. Of course, the base-stock levels y_k^* and z_k^* depend on the current forecast information i_k^1 but independent of $(x_k + s_{k-1})$.

THEOREM 3.5 *Assume that* (3.1), (3.3)–(3.5), *and* (3.67)–(3.68) *hold. If the initial inventory level at the beginning of period k is x_k, the slow-order quantity in period $(k-1)$ is denoted by s_{k-1}, and the observed value of I_k^1 is i_k^1 in period $(k-1)$, then the optimal fast-order quantity f_k^* and the optimal slow-order quantity s_k^* in period k are given by the following expressions:*

(i) *when* $y_k^* < z_k^*$,

$$(f_k^*, s_k^*) = \begin{cases} (y_k^* - x_k - s_{k-1}, z_k^* - y_k^*), & \text{if } x_k + s_{k-1} \le y_k^*, \\ (0, z_k^* - x_k - s_{k-1}), & \text{if } y_k^* < x_k + s_{k-1} \le z_k^*, \\ (0, 0), & \text{if } z_k^* < x_k + s_{k-1}; \end{cases}$$

(ii) *when* $y_k^* \ge z_k^*$,

$$(f_k^*, s_k^*) = \begin{cases} (z_k^* - x_k - s_{k-1}, 0), & \text{if } x_k + s_{k-1} \le z_k^*, \\ (0, 0), & \text{if } z_k^* < x_k + s_{k-1} \le y_k^*, \\ (0, 0), & \text{if } y_k^* < x_k + s_{k-1}. \end{cases}$$

Proof First we show (i). Here we consider the case $x_k + s_{k-1} \le y_k^* < z_k^*$. The other cases in (i) can be treated in a similar way. It suffices to show that for all $y \ge x_k + s_{k-1}$ and $z \ge y$,

$$\begin{aligned} & c_k^f \cdot (y - x_k - s_{k-1}) + c_k^s z - c_k^s y + \mathsf{E}\left[H_{k+1}(y - g_k(i_k^1, I_k^2, v_k))\right] \\ & + \mathsf{E}\left[\tilde{V}_{k+1}(z - g_k(i_k^1, I_k^2, v_k), I_{k+1}^1)\right] \\ & \ge c_k^f \cdot (y_k^* - x_k - s_{k-1}) + c_k^s z_k^* - c_k^s y_k^* \\ & + \mathsf{E}\left[H_{k+1}(y_k^* - g_k(i_k^1, I_k^2, v_k))\right] \\ & \quad + \mathsf{E}\left[\tilde{V}_{k+1}(z_k^* - g_k(i_k^1, I_k^2, v_k), I_{k+1}^1)\right]. \end{aligned} \tag{3.71}$$

By the definition of z_k^*, we have for $t \in (y_k^*, \infty)$,

$$c_k^s t + \mathsf{E}\left[\tilde{V}_{k+1}(t - g_k(i_k^1, I_k^2, v_k), I_{k+1}^1)\right]$$
$$\geq c_k^s z_k^* + \mathsf{E}\left[\tilde{V}_{k+1}(z_k^* - g_k(i_k^1, I_k^2, v_k), I_{k+1}^1)\right]. \tag{3.72}$$

From the convexity of $c_k^s t + \mathsf{E}[\tilde{V}_{k+1}(t - g_k(i_k^1, I_k^2, v_k), I_{k+1}^1)]$, we know that the point z_k^* also minimizes the function

$$c_k^s t + \mathsf{E}\left[\tilde{V}_{k+1}(t - g_k(i_k^1, I_k^2, v_k), I_{k+1}^1)\right]$$

on the interval $[0, +\infty)$ at the same time. Thus from the definition of y_k^*, we have (3.71).

Now we take up case (ii) when $y_k^* \geq z_k^*$. We give the proof only for the case $x_k + s_{k-1} \leq z_k^*$. The proof for the other cases is similar. To this end, it suffices to show that for all $y \geq x_k + s_{k-1}$ and $z \geq y$,

$$c_k^f \cdot (y - x_k - s_{k-1}) + c_k^s z - c_k^s y + \mathsf{E}\left[H_{k+1}(y - g_k(i_k^1, I_k^2, v_k))\right]$$
$$+\mathsf{E}\left[\tilde{V}_{k+1}(z - g_k(i_k^1, I_k^2, v_k), I_{k+1}^1)\right]$$
$$\geq c_k^f \cdot (z_k^* - x_k - s_{k-1}) + \mathsf{E}\left[H_{k+1}(z_k^* - g_k(i_k^1, I_k^2, v_k))\right]$$
$$+\mathsf{E}\left[\tilde{V}_{k+1}(z_k^* - g_k(i_k^1, I_k^2, v_k), I_{k+1}^1)\right]. \tag{3.73}$$

First, we have that for $t \leq y_k^*$,

$$c_k^f \cdot (z_k^* - x_k - s_{k-1}) + \mathsf{E}\left[H_{k+1}(z_k^* - g_k(i_k^1, I_k^2, v_k))\right]$$
$$+\mathsf{E}\left[\tilde{V}_{k+1}(z_k^* - g_k(i_k^1, I_k^2, v_k), I_{k+1}^1)\right]$$
$$= \Big\{c_k^f \cdot (z_k^* - x_k - s_{k-1}) - c_k^s z_k^* + \mathsf{E}\left[H_{k+1}(z_k^* - g_k(i_k^1, I_k^2, v_k))\right]$$
$$-\Big(c_k^f \cdot (y_k^* - x_k - s_{k-1}) - c_k^s y_k^*$$
$$+\mathsf{E}\left[H_{k+1}(y_k^* - g_k(i_k^1, I_k^2, v_k))\right]\Big)\Big\}$$
$$+c_k^s z_k^* + \mathsf{E}\left[\tilde{V}_{k+1}(z_k^* - g_k(i_k^1, I_k^2, v_k), I_{k+1}^1)\right]$$
$$+\Big(c_k^f \cdot (y_k^* - x_k - s_{k-1}) - c_k^s y_k^* + \mathsf{E}\left[H_{k+1}(y_k^* - g_k(i_k^1, I_k^2, v_k))\right]\Big)$$

$$\leq \left\{ c_k^f \cdot (t - x_k - s_{k-1}) - c_k^s t + \mathsf{E}\left[H_{k+1}(t - g_k(i_k^1, I_k^2, v_k))\right] \right.$$
$$- \left(c_k^f \cdot (y_k^* - x_k - s_{k-1}) - c_k^s y_k^* \right.$$
$$\left. \left. + \mathsf{E}\left[H_{k+1}(y_k^* - g_k(i_k^1, I_k^2, v_k))\right] \right) \right\}$$
$$+ c_k^s t + \mathsf{E}\left[\tilde{V}_{k+1}(t - g_k(i_k^1, I_k^2, v_k), I_{k+1}^1)\right]$$
$$+ \left(c_k^f \cdot (y_k^* - x_k - s_{k-1}) - c_k^s y_k^* + \mathsf{E}\left[H_{k+1}(y_k^* - g_k(i_k^1, I_k^2, v_k))\right] \right)$$
$$= c_k^f \cdot (t - x_k - s_{k-1}) + \mathsf{E}\left[H_{k+1}(t - g_k(i_k^1, I_k^2, v_k))\right]$$
$$+ \mathsf{E}\left[\tilde{V}_{k+1}(t - g_k(i_k^1, I_k^2, v_k), I_{k+1}^1)\right], \qquad (3.74)$$

where the above inequality makes use of the definition of z_k^*. Let $(\bar{y}_k, \bar{z}_k)$ be defined by (3.65) with

$$C_k^f(y - x_k - s_{k-1}) = c_k^f \cdot (y - x_k - s_{k-1}),$$
$$C_k^s(z - y) = c_k^s \cdot (z - y).$$

There are two cases to consider: $\bar{z}_k \leq y_k^*$ and $\bar{z}_k > y_k^*$. Later, we prove that the second case $\bar{z}_k > y_k^*$ does not arise. Thus to complete the proof, it suffices to consider the case $\bar{z}_k \leq y_k^*$. Using (3.74) with $t = \bar{z}_k$,

$$c_k^f \cdot (z_k^* - x_k - s_{k-1}) + \mathsf{E}\left[H_{k+1}(z_k^* - g_k(i_k^1, I_k^2, v_k))\right]$$
$$+ \mathsf{E}\left[\tilde{V}_{k+1}(z_k^* - g_k(i_k^1, I_k^2, v_k), I_{k+1}^1)\right]$$
$$\leq c_k^f \cdot (\bar{z}_k - x_k - s_{k-1}) + \mathsf{E}\left[H_{k+1}(\bar{z}_k - g_k(i_k^1, I_k^2, v_k))\right]$$
$$+ \mathsf{E}\left[\tilde{V}_{k+1}(\bar{z}_k - g_k(i_k^1, I_k^2, v_k), I_{k+1}^1)\right]. \qquad (3.75)$$

Since

$$c_k^f \cdot (y - x_k - s_{k-1}) - c_k^s y + \mathsf{E}\left[H_{k+1}\left(y - g_k(i_k^1, I_k^2, v_k)\right)\right]$$

is a convex function of y, we know that $(\bar{z}_k, \bar{z}_k)$ also minimizes

$$c_k^f \cdot (y - x_k - s_{k-1}) + c_k^s z - c_k^s y + \mathsf{E}\left[H_{k+1}(y - g_k(i_k^1, I_k^2, v_k))\right]$$
$$+ \mathsf{E}\left[\tilde{V}_{k+1}(z - g_k(i_k^1, I_k^2, v_k), I_{k+1}^1)\right].$$

This indicates that for all $y \geq x_k + s_{k-1}$ and $z \geq y$,

$$c_k^f \cdot (y - x_k - s_{k-1}) + c_k^s z - c_k^s y + \mathsf{E}\left[H_{k+1}(y - g_k(i_k^1, I_k^2, v_k))\right]$$
$$+\mathsf{E}\left[\tilde{V}_{k+1}(z - g_k(i_k^1, I_k^2, v_k), I_{k+1}^1)\right]$$
$$\geq c_k^f \cdot (\bar{z}_k - x_k - s_{k-1}) + \mathsf{E}\left[H_{k+1}(\bar{z}_k - g_k(i_k^1, I_k^2, v_k))\right]$$
$$+\mathsf{E}\left[\tilde{V}_{k+1}(\bar{z}_k - g_k(i_k^1, I_k^2, v_k), I_{k+1}^1)\right]. \tag{3.76}$$

Combining (3.75)–(3.76), (3.73) is established.

Finally, we show that $\bar{z}_k > y_k^*$ does not arise. If $\bar{z}_k > y_k^*$, by the convexity of

$$c_k^f y - c_k^s y + \mathsf{E}\left[H_{k+1}(y - g_k(i_k^1, I_k^2, v_k))\right],$$

we know that $(y_k^*, \bar{z}_k)$ also minimizes

$$c_k^f y + c_k^s \cdot (z - y) + \mathsf{E}\left[H_{k+1}(y - g_k(i_k^1, I_k^2, v_k))\right]$$
$$+\mathsf{E}\left[\tilde{V}_{k+1}(z - g_k(i_k^1, I_k^2, v_k), I_{k+1}^1)\right]$$

on the region $\{(y, z) : y \geq x_k + s_{k-1} \text{ and } z \geq y\}$. By the convexity of $c_k^s t + \mathsf{E}[\tilde{V}_{k+1}(t - g_k(i_k^1, I_k^2, v_k), I_{k+1}^1)]$ and the nonnegativity of

$$c_k^f \cdot (t - x_k - s_{k-1}) - c_k^s t + \mathsf{E}\left[H_{k+1}(t - g_k(i_k^1, I_k^2, v_k))\right]$$
$$-\left(c_k^f \cdot (y_k^* - x_k - s_{k-1}) - c_k^s y_k^* + \mathsf{E}\left[H_{k+1}(y_k^* - g_k(i_k^1, I_k^2, v_k))\right]\right),$$

$\bar{z}_k$ would minimize $c_k^s t + \mathsf{E}[\tilde{V}_{k+1}(t - g_k(i_k^1, I_k^2, v_k), I_{k+1}^1)]$ over $[0, \infty)$. This would mean that for all $z \in [0, \infty)$,

$$c_k^s z + \mathsf{E}\left[\tilde{V}_{k+1}(z - g_k(i_k^1, I_k^2, v_k), I_{k+1}^1)\right]$$
$$\geq c_k^s \bar{z}_k + \mathsf{E}\left[\tilde{V}_{k+1}(\bar{z}_k - g_k(i_k^1, I_k^2, v_k), I_{k+1}^1)\right]. \tag{3.77}$$

On the other hand, by the definition of z_k^*, z_k^* minimizes

$$c_k^s t + \mathsf{E}[\tilde{V}_{k+1}(t - g_k(i_k^1, I_k^2, v_k), I_{k+1}^1)]$$

over $[y_k^*, \infty)$. That is, for all $\tilde{z}_k \in [y_k^*, \infty)$,

$$c_k^s z_k^* + \mathsf{E}\left[\tilde{V}_{k+1}(z_k^* - g_k(i_k^1, I_k^2, v_k), I_{k+1}^1)\right]$$
$$< c_k^s \tilde{z}_k + \mathsf{E}\left[\tilde{V}_{k+1}(\tilde{z}_k - g_k(i_k^1, I_k^2, v_k), I_{k+1}^1)\right],$$

which contradicts (3.77). As a result, the case $\bar{z}_k > y_k^*$ does not arise. □

The theorem says that in case (i), there are two base-stock levels—fast and slow—with the fast base-stock level y_k^* being smaller than the slow base-stock level z_k^*. Moreover, when the inventory position is too low (that is, smaller than y_k^*), then we order up to y_k^* via the fast mode and order an additional amount from y_k^* up to z_k^* via the slow mode. On the other hand, when the inventory position is too high (that is, larger than z_k^*), then we order nothing. Finally, if the inventory position is neither too low nor too high (that is, when it is between the levels y_k^* and z_k^*), then we simply order up to z_k^* via the slow mode. In case (ii), there is only one base-stock level z_k^*, and if the inventory position is too low (that is, smaller than z_k^*), then we order up to z_k^* via the fast mode and order nothing via the slow mode. On the other hand, if the inventory position is too high (that is, larger than z_k^*), we order altogether nothing.

3.5. The Nonstationary Infinite-Horizon Problem

We now consider an infinite-horizon version of the problem formulated in Section 3.2. By letting $N = \infty$ and $(\boldsymbol{F}, \boldsymbol{S}) = ((F_n, S_n), (F_{n+1}, S_{n+1}), \ldots)$, the extended real-valued objective function of the problem is

$$J_n^\infty(x_n, s_{n-1}, i_n^1, (\boldsymbol{F}, \boldsymbol{S})) = H_n(x_n) + \sum_{k=n}^{\infty} \alpha^{k-n} \mathsf{E}\left[C_k^f(F_k) + C_k^s(S_k) + \alpha H_{k+1}(X_{k+1})\right], \tag{3.78}$$

where α is a given discount factor, $0 < \alpha < 1$,

$$X_{n+1} = x_n + s_{n-1} + F_n - g_n(i_n^1, I_n^2, v_n),$$

and X_k $(k > n+1)$ are defined by (3.8). Similar to (3.20)–(3.21), the dynamic programming equations for the infinite-horizon problem are

$$\begin{aligned} U_n^\infty(x_n, s_{n-1}, i_n^1) &= H_n(x_n) + \inf_{\substack{\phi \geq 0 \\ \sigma \geq 0}} \Big\{ C_n^f(\phi) + C_n^s(\sigma) \\ &\quad + \alpha \mathsf{E}\left[U_{n+1}^\infty(x_n + s_{n-1} + \phi - g_n(i_n^1, I_n^2, v_n), \sigma, I_{n+1}^1)\right] \Big\}, \\ n &= 1, 2, \ldots. \end{aligned} \tag{3.79}$$

In what follows, we show that there exists a solution of (3.79) that is continuous and convex in x_n. Furthermore, similar to Theorems 3.2 and 3.3, we show that the value function of the infinite-horizon problem is a solution of (3.79) and the decision that attains the infimum in (3.79) is an optimal nonanticipative policy. Our method is that of successive approximation of the infinite-horizon problem by longer and longer finite-horizon problems.

Let us, therefore, examine the finite-horizon approximation $J_{n,k}(x_n, s_{n-1}, i_n^1)$ of (3.78), which is obtained by the first k-period truncation of the infinite-horizon problem. The objective function for this problem is to minimize

$$J_{n,k}(x_n, s_{n-1}, i_n^1, (\boldsymbol{F}, \boldsymbol{S})) = H_n(x_n) + \sum_{\ell=n}^{n+k} \alpha^{\ell-n} \mathsf{E}\left[C_\ell^f(F_\ell) + C_\ell^s(S_\ell) + \alpha H_{\ell+1}(X_{\ell+1})\right]. \tag{3.80}$$

Let $V_{n,k}(x_n, s_{n-1}, i_n^1)$ be the value function of the truncated problem—that is,

$$V_{n,k}(x_n, s_{n-1}, i_n^1) = \inf_{(\boldsymbol{F},\boldsymbol{S})\in\mathcal{A}_{n,k}} \left\{J_{n,k}(x_n, s_{n-1}, i_n^1, (\boldsymbol{F}, \boldsymbol{S}))\right\}. \tag{3.81}$$

Since (3.80) is a finite-horizon problem on the interval $\langle n, n+k\rangle$, we can apply Theorem 3.2 to prove that $V_{n,k}(x_n, s_{n-1}, i_n^1)$ satisfies the dynamic programming equations

$$\begin{aligned} &U_{n+\ell,k-\ell}(x_{n+\ell}, s_{n+\ell-1}, i_{n+\ell}^1) \\ &\quad = H_{n+\ell}(x_{n+\ell}) + \inf_{\substack{\phi\geq 0\\ \sigma\geq 0}} \Big\{C_{n+\ell}^f(\phi) + C_{n+\ell}^s(\sigma) \\ &\qquad + \alpha\mathsf{E}\Big[U_{n+\ell+1,k-\ell-1}(Z_{n+\ell+1}(x_{n+\ell}+\phi), \sigma, I_{n+\ell+1}^1)\Big]\Big\}, \\ &\quad \ell = 0, ..., k-1, \end{aligned} \tag{3.82}$$

$$\begin{aligned} &U_{n+k,0}(x_{n+k}, s_{n+k-1}, i_{n+k}^1) \\ &\quad = H_{n+k}(x_{n+k}) + \inf_{\substack{\phi\geq 0\\ \sigma\geq 0}} \Big\{C_{n+k}^f(\phi) + C_{n+k}^s(\sigma) \\ &\qquad + \alpha\mathsf{E}\Big[H_{n+k+1}(Z_{n+k+1}(x_{n+k}+\phi))\Big]\Big\}, \end{aligned} \tag{3.83}$$

where

$$Z_{n+\ell+1}(t) = t + s_{n+\ell-1} - g_{n+\ell}(i_{n+\ell}^1, I_{n+\ell}^2, v_{n+\ell}).$$

To get the optimal policy for the infinite-horizon problem, we assume that there exist constants $c > 0$ and $M > 0$ such that for all $k \geq 1$,

$$\left|C_k^f(x_1) - C_k^f(x_2)\right| \leq c \cdot |x_1 - x_2|, \tag{3.84}$$

$$|C_k^s(x_1) - C_k^s(x_2)| \leq c \cdot |x_1 - x_2|, \tag{3.85}$$

$$|H_k(x_1) - H_k(x_2)| \leq c \cdot |x_1 - x_2|, \tag{3.86}$$

$$\mathsf{E}\left[g_k(I_k^1, I_k^2, v_k)\right] < M. \tag{3.87}$$

Furthermore, we assume that

$$C_k^f(t) + \mathsf{E}\left[H_{k+1}(t - g_k(I_k^1, I_k^2, v_k))\right] \to \infty \text{ as } t \to \infty, \tag{3.88}$$

$$C_k^s(t) + \mathsf{E}\left[H_{k+1}(t - g_k(I_k^1, I_k^2, v_k))\right] \to \infty \text{ as } t \to \infty, \tag{3.89}$$

uniformly hold with respect to k.

It follows from (3.84)–(3.89) that for any (x_n, s_{n-1}) and $(\hat{x}_n, \hat{s}_{n-1})$,

$$\begin{aligned}
&\left|J_{n,k}(x_n, s_{n-1}, i_n^1, (\boldsymbol{F}, \boldsymbol{S})) - J_{n,k}(\hat{x}_n, \hat{s}_{n-1}, i_n^1, (\boldsymbol{F}, \boldsymbol{S}))\right| \\
&\leq c \cdot |x_n - \hat{x}_n| + \sum_{\ell=n}^{n+k} \alpha^{\ell-n+1}\left(c \cdot |x_n - \hat{x}_n| + c \cdot |s_{n-1} - \hat{s}_{n-1}|\right) \\
&\leq \frac{c}{1-\alpha}\left(|x_n - \hat{x}_n| + |s_{n-1} - \hat{s}_{n-1}|\right).
\end{aligned} \tag{3.90}$$

Therefore, we have

$$\begin{aligned}
&\left|V_{n,k}(x_n, s_{n-1}, i_n^1) - V_{n,k}(\hat{x}_n, \hat{s}_{n-1}, i_n^1)\right| \\
&\quad \leq \frac{c}{1-\alpha}\left(|x_n - \hat{x}_n| + |s_{n-1} - \hat{s}_{n-1}|\right).
\end{aligned} \tag{3.91}$$

THEOREM 3.6 *Assume that* (3.1), (3.4)–(3.5), *and* (3.84)–(3.89) *hold. Then the limit of* $V_{n,k}(x_n, s_{n-1}, i_n^1)$ *exists as* $k \to \infty$. *Letting the limit be denoted by* $V_n^\infty(x_n, s_{n-1}, i_n^1)$, *we have*

(i) $V_n^\infty(x_n, s_{n-1}, i_n^1)$ *is convex and Lipschitz continuous in* (x_n, s_{n-1}) *on* $(-\infty, +\infty) \times [0, +\infty)$;

(ii) $V_n^\infty(x_n, s_{n-1}, i_n^1)$ *is a solution of* (3.79);

(iii) *there exist functions* $\bar{F}_n(x_n, s_{n-1}, i_n^1)$ *and* $\bar{S}_n(x_n, s_{n-1}, i_n^1)$ *which provide the infima in* (3.79) *with* $U_n^\infty(x_n, s_{n-1}, i_n^1) = V_n^\infty(x_n, s_{n-1}, i_n^1)$, *and*

$$(\bar{\boldsymbol{F}}, \bar{\boldsymbol{S}}) = \left\{(\bar{F}_n(x_n, s_{n-1}, i_n^1), \bar{S}_n(x_n, s_{n-1}, i_n^1)),\ n \geq 1\right\}$$

is an optimal nonanticipative policy—that is,

$$\begin{aligned}
V_1^\infty(x_1, s_0, i_1^1) &= J_1^\infty(x_1, s_0, i_1^1, (\bar{\boldsymbol{F}}, \bar{\boldsymbol{S}})) \\
&= \inf_{(\boldsymbol{F},\boldsymbol{S}) \in \mathcal{A}}\left\{J_1^\infty(x_1, s_0, i_1^1, (\boldsymbol{F}, \boldsymbol{S}))\right\}.
\end{aligned}$$

Proof First we show that there exists a function $V_n^\infty(x_n, s_{n-1}, i_n^1)$ such that

$$\lim_{k\to\infty} V_{n,k}(x_n, s_{n-1}, i_n^1) = V_n^\infty(x_n, s_{n-1}, i_n^1). \tag{3.92}$$

Let

$$\begin{aligned}
&(\bar{\boldsymbol{F}}_{n,k}, \bar{\boldsymbol{S}}_{n,k}) \\
&= \Big(\Big(\bar{F}_n^k(x_n, s_{n-1}, i_n^1), \ldots, \bar{F}_{n+k}^k(x_{n+k}, s_{n+k-1}, i_{n+k}^1)\Big), \\
&\quad \Big(\bar{S}_n^k(x_n, s_{n-1}, i_n^1), \ldots, \bar{S}_{n+k}^k(x_{n+k}, s_{n+k-1}, i_{n+k}^1)\Big)\Big)
\end{aligned} \tag{3.93}$$

attain the infimum on the right-hand side of (3.82)–(3.83). Note that

$$\bar{S}_{n+k}^k(x_{n+k}, s_{n+k-1}, i_{n+k}^1) = 0.$$

Thus,

$$\begin{aligned}
V_{n,k}(x_n, s_{n-1}, i_n^1) &= J_{n,k}(x_n, s_{n-1}, i_n^1, (\bar{\boldsymbol{F}}_{n,k}, \bar{\boldsymbol{S}}_{n,k})) \\
&\geq J_{n,k-1}(x_n, s_{n-1}, i_n^1, (\bar{\boldsymbol{F}}_{n,k-1}, \bar{\boldsymbol{S}}_{n,k-1})) \\
&\geq \inf_{(\boldsymbol{F},\boldsymbol{S})\in\mathcal{A}} \Big\{J_{n,k-1}(x_n, s_{n-1}, i_n^1, (\boldsymbol{F}, \boldsymbol{S}))\Big\} \\
&= V_{n,k-1}(x_n, s_{n-1}, i_n^1),
\end{aligned} \tag{3.94}$$

which implies that for fixed x_n, s_{n-1}, and i_n^1, $V_{n,k}(x_n, s_{n-1}, i_n^1)$ is an increasing sequence in k. On the other hand, for any k,

$$V_{n,k}(x_n, s_{n-1}, i_n^1) \leq J_n^\infty(x_n, s_{n-1}, i_n^1, \boldsymbol{O}), \tag{3.95}$$

where $\boldsymbol{O}$ is a policy of ordering nothing at each period by both fast and slow modes. From (3.86) and (3.87), $J_n^\infty(x_n, s_{n-1}, i_n^1, \boldsymbol{O}) < \infty$. Consequently, (3.92) follows from (3.94).

By Theorem 3.1, we know that for each k, $V_{n,k}(x_n, s_{n-1}, i_n^1)$ is convex and Lipschitz continuous in (x_n, s_{n-1}) on $(-\infty, +\infty) \times [0, +\infty)$. Hence (i) follows from (3.92).

Next, we show that $V_n^\infty(x_n, s_{n-1}, i_n^1)$ is a solution of (3.79). Using Theorem 3.2,

$$\begin{aligned}
&V_{n,k}(x_n, s_{n-1}, i_n^1) \\
&\quad = H_n(x_n) + \inf_{\substack{\phi\geq 0\\ \sigma\geq 0}} \Big\{C_n^f(\phi) + C_n^s(\sigma) \\
&\qquad + \alpha \mathsf{E}\left[V_{n+1,k-1}(x_n + s_{n-1} + \phi - g_n(i_n^1, I_n^2, v_n), \sigma, I_{n+1}^1)\right]\Big\}
\end{aligned}$$

$$\leq H_n(x_n) + C_n^f(\phi) + C_n^s(\sigma)$$
$$+\alpha \mathsf{E}\left[V_{n+1,k-1}\left(x_n + s_{n-1} + \phi - g_n(i_n^1, I_n^2, v_n), \sigma, I_{n+1}^1\right)\right]. \tag{3.96}$$

Taking limits on both sides of (3.96) with respect to k, we get

$$V_n^\infty(x_n, s_{n-1}, i_n^1)$$
$$\leq H_n(x_n) + \inf_{\substack{\phi \geq 0 \\ \sigma \geq 0}} \Big\{ C_n^f(\phi) + C_n^s(\sigma)$$
$$+\alpha \mathsf{E}\left[V_{n+1}^\infty(x_n + s_{n-1} + \phi - g_n(i_n^1, I_n^2, v_n), \sigma, I_{n+1}^1)\right] \Big\}. \tag{3.97}$$

Using (3.88)–(3.89), there exists a $Q > 0$ such that for all n and k

$$\bar{F}_{n+\ell}^k(x_{n+\ell}, s_{n+\ell-1}, i_{n+\ell}^1) \leq Q, \quad 0 \leq \ell \leq k, \tag{3.98}$$
$$\bar{S}_{n+\ell}^k(x_{n+\ell}, s_{n+\ell-1}, i_{n+\ell}^1) \leq Q, \quad 0 \leq \ell \leq k-1, \tag{3.99}$$

where $\bar{F}_{n+\ell}^k(x_{n+\ell}, s_{n+\ell-1}, i_{n+\ell}^1)$ and $\bar{S}_{n+\ell}^k(x_{n+\ell}, s_{n+\ell-1}, i_{n+\ell}^1)$ are defined by (3.93). Furthermore, by (3.94), for any $\ell < k$,

$$V_{n,k}(x_n, s_{n-1}, i_n^1)$$
$$= H_n(x_n) + C_n^f(\bar{F}_n^k(x_n, s_{n-1}, i_n^1)) + C_n^s(\bar{S}_n^k(x_n, s_{n-1}, i_n^1))$$
$$+\alpha \mathsf{E}\Big[V_{n+1,k-1}\Big(x_n + s_{n-1} + \bar{F}_n^k(x_n, s_{n-1}, i_n^1) - g_n(i_n^1, I_n^2, v_n),$$
$$\bar{S}_n^k(x_n, s_{n-1}, i_n^1), I_{n+1}^1\Big)\Big]$$
$$\geq H_n(x_n) + C_n^f(\bar{F}_n^k(x_n, s_{n-1}, i_n^1)) + C_n^s(\bar{S}_n^k(x_n, s_{n-1}, i_n^1))$$
$$+\alpha \mathsf{E}\Big[V_{n+1,\ell}\Big(x_n + s_{n-1} + \bar{F}_n^k(x_n, s_{n-1}, i_n^1) - g_n(i_n^1, I_n^2, v_n),$$
$$\bar{S}_n^k(x_n, s_{n-1}, i_n^1), I_{n+1}^1\Big)\Big]. \tag{3.100}$$

Fixed ℓ and let $k \to \infty$. In view of (3.98) and (3.99), we can, for any given n, x_n, s_{n-1}, and i_n^1, extract a converging subsequence

$$\left(\bar{F}_n^k(x_n, s_{n-1}, i_n^1), \bar{S}_n^k(x_n, s_{n-1}, i_n^1)\right).$$

Let

$$\lim_{k \to \infty} \left(\bar{F}_n^k(x_n, s_{n-1}, i_n^1), \bar{S}_n^k(x_n, s_{n-1}, i_n^1)\right)$$
$$= \left(\bar{F}_n^\infty(x_n, s_{n-1}, i_n^1), \bar{S}_n^\infty(x_n, s_{n-1}, i_n^1)\right). \tag{3.101}$$

From the uniform integrability of

$$V_{n+1,\ell}(x_n + s_{n-1} + \phi - g_n(i_n^1, I_n^2, v_n), \sigma, I_{n+1}^1) \text{ (see (3.91))},$$

we can pass to the limit on the right-hand side of (3.100). We obtain (noting that the left-hand side converges as well)

$$\begin{aligned} & V_{n,k}(x_n, s_{n-1}, i_n^1) \\ & \geq H_n(x_n) + C_n^f(\bar{F}_n^\infty(x_n, s_{n-1}, i_n^1)) + C_n^s(\bar{S}_n^\infty(x_n, s_{n-1}, i_n^1)) \\ & \quad + \alpha \mathsf{E}\left[V_{n+1,\ell}\left(x_n + s_{n-1} + \bar{F}_n^\infty(x_n, s_{n-1}, i_n^1) - g_n(i_n^1, I_n^2, v_n),\right.\right. \\ & \qquad \left.\left.\bar{S}_n^\infty(x_n, s_{n-1}, i_n^1), I_{n+1}^1\right)\right]. \end{aligned} \tag{3.102}$$

Furthermore,

$$\begin{aligned} & H_n(x_n) + C_n^f(\bar{F}_n^\infty(x_n, s_{n-1}, i_n^1)) + C_{n+i}^s(\bar{S}_n^\infty(x_n, s_{n-1}, i_n^1)) \\ & \quad + \alpha \mathsf{E}\left[V_{n+1,\ell}\left(x_n + s_{n-1} + \bar{F}_n^\infty(x_n, s_{n-1}, i_n^1) - g_n(i_n^1, I_n^2, v_n),\right.\right. \\ & \qquad \left.\left.\bar{S}_n^\infty(x_n, s_{n-1}, i_n^1), I_{n+1}^1\right)\right] \\ & = H_n(x_n) + C_n^f(\bar{F}_n^\infty(x_n, s_{n-1}, i_n^1)) + C_n^s(\bar{S}_n^\infty(x_n, s_{n-1}, i_n^1)) \\ & \quad + \alpha \mathsf{E}\left[V_{n+1}^\infty\left(x_n + s_{n-1} + \bar{F}_n^\infty(x_n, s_{n-1}, i_n^1) - g_n(i_n^1, I_n^2, v_n),\right.\right. \\ & \qquad \left.\left.\bar{S}_n^\infty(x_n, s_{n-1}, i_n^1), I_{n+1}^1\right)\right] \\ & \geq H_n(x_n) + \inf_{\substack{\phi \geq 0 \\ \sigma \geq 0}} \left\{ C_n^f(\phi) + C_n^s(\sigma) \right. \\ & \quad \left. + \alpha \mathsf{E}\left[V_{n+1}^\infty\left(x_n + s_{n-1} + \phi - g_n(i_n^1, I_n^2, v_n), \sigma, I_{n+1}^1\right)\right] \right\}. \end{aligned} \tag{3.103}$$

Therefore, by (3.102) and (3.103),

$$\begin{aligned} & V_n^\infty(x_n, s_{n-1}, i_n^1) \\ & \geq H_n(x_n) + \inf_{\substack{\phi \geq 0 \\ \sigma \geq 0}} \left\{ C_n^f(\phi) + C_n^s(\sigma) \right. \\ & \quad \left. + \alpha \mathsf{E}\left[V_{n+1}^\infty\left(x_n + s_{n-1} + \phi - g_n(i_n^1, I_n^2, v_n), \sigma, I_{n+1}^1\right)\right] \right\}, \end{aligned} \tag{3.104}$$

which and (3.97) imply (ii) of the theorem. (iii) can be proved along the line of the proof of Theorem 3.3. The detail is omitted here. □

REMARK 3.6 Theorem 3.6 does not imply that there is a unique solution of the dynamic programming equations (3.79). In addition, it is possible to show that the value function is the minimal positive solution of (3.79). Furthermore, it is also possible to obtain a uniqueness proof, provided that the cost functions $C_n^f(\cdot)$, $C_n^s(\cdot)$ and $H_n(\cdot)$ are subject to additional conditions.

To derive the optimality of a base-stock policy in the same way as in Section 3.4, we still make assumptions (3.67)–(3.68). Let

$$G_n(t) = c_n^f \cdot (t - x_n - s_{n-1}) - c_n^s t + \mathsf{E}[H_{n+1}(t - g_n(i_n^1, I_n^2, v_n))].$$

Let y_n^* be a minimum of the function $G_n(t)$. Furthermore, let

$$L_n(t) = c_n^s t + \mathsf{E}[V_{n+1}^\infty(t - g_n(i_n^1, I_n^2, v_n))] + [G_n(t) - G_n(y_n^*)] \cdot \delta(y_n^* - t),$$

and let z_n^* be a minimum of the function $L_n(t)$. Similar to Theorem 3.5, we have the following result.

THEOREM 3.7 *Assume that* (3.1), (3.4)–(3.5), (3.67)–(3.68), *and* (3.86)–(3.87) *hold. Then the policy* (f_n^*, s_n^*) *given by the following is an optimal nonanticipative policy*:

(i) *when* $y_n^* < z_n^*$,

$$(f_n^*, s_n^*) = \begin{cases} (y_n^* - x_n - s_{n-1}, z_n^* - y_n^*), & \textit{if } x_n + s_{n-1} \le y_n^*, \\ (0, z_n^* - x_n - s_{n-1}), & \textit{if } y_n^* < x_n + s_{n-1} \le z_n^*, \\ (0, 0), & \textit{if } z_n^* < x_n + s_{n-1}; \end{cases}$$

(ii) *when* $y_n^* \ge z_n^*$,

$$(f_n^*, s_n^*) = \begin{cases} (z_n^* - x_n - s_{n-1}, 0), & \textit{if } x_n + s_{n-1} \le z_n^*, \\ (0, 0), & \textit{if } z_n^* < x_n + s_{n-1} \le y_n^*, \\ (0, 0), & \textit{if } y_n^* < x_n + s_{n-1}. \end{cases}$$

REMARK 3.7 When $C_n^f(u) = c_k^f \cdot u$ and $C_n^s(u) = c_k^s \cdot u$ with $c_n^f > 0$ and $c_n^s > 0$, (3.84)–(3.85), and (3.88)–(3.89) hold. Thus we do not need to specify that (3.84)–(3.85), and (3.88)–(3.89) hold in the theorem.

Proof of Theorem 3.7 The proof is similar to the proof of Theorem 3.5, and is omitted. □

3.6. An Example

In this section, we use an example to illustrate the structure properties of the cost function and solutions in detail. For simplicity, we consider $N = 2$,

$$C_1^f(u) = \begin{cases} +\infty, & \text{if } u > 0, \\ 0, & \text{if } u = 0, \end{cases} \qquad C_1^s(u) = c_1^s \cdot u, \tag{3.105}$$

$$C_2^f(u) = c_2^f \cdot u, \quad g_1(i_1^1, I_1^2, v_1) = 0, \tag{3.106}$$

and

$$\begin{aligned} H_1(x) = H_2(x) &= H_3(x) \\ &= \begin{cases} hx, & \text{if } x > 0, \\ -px, & \text{if } x \leq 0. \end{cases} \end{aligned} \tag{3.107}$$

Suppose that $x_1 = s_0 = 0$.

Let information I_2^1 be uniformly distributed within the interval of width a centered at v_2 with $v_2 > a$. Formally, the density function denoted by $\lambda_2(i_2^1)$ is given by

$$\lambda_2(i_2^1) = \frac{1}{a}, \quad i_2^1 \in \left[v_2 - \frac{a}{2}, \; v_2 + \frac{a}{2}\right]. \tag{3.108}$$

For an observed $i_2^1 \in [v_2 - a/2, \; v_2 + a/2]$, the conditional density function of $g_2(i_2^1, I_2^2, v_2)$ denoted by $\psi_2(i_2^2 | i_2^1)$ is given by

$$\psi_2(i_2^2 | i_2^1) = \begin{cases} \frac{1}{\varepsilon a}, & \text{if } i_2^2 \in \left[i_2^1 - \frac{\varepsilon a}{2}, \; i_2^1 + \frac{\varepsilon a}{2}\right], \\ 0, & \text{otherwise,} \end{cases} \tag{3.109}$$

where $0 < \varepsilon < 1$.

With these given parameters, (3.11) can be written as

$$\begin{aligned} & J_1(0, 0, i_1^1, ((F_1, F_2), S_1)) \\ &= \mathsf{E}\left[C_1^f(F_1) + C_1^s(S_1) + H_2(X_2) + C_2^f(F_2) + H_3(X_3)\right], \end{aligned} \tag{3.110}$$

where

$$\begin{aligned} X_2 &= x_1 + s_0 + F_1 - g_1(i_1^1, I_1^2, v_1) \\ &= F_1, \end{aligned}$$

and

$$X_3 = X_2 + S_1 + F_2 - g_2(I_2^1, I_2^2, v_2).$$

Similarly, (3.45)–(3.46) can be written as

$$\hat{U}_1(q_1, i_1^1)$$
$$= \inf_{\substack{y \geq q_1 \\ z \geq y}} \left\{ C_1^f(y - q_1) + C_1^s(z - y) + \mathsf{E}\left[H_2(y - g_1(i_1^1, I_1^2, v_1))\right] \right.$$
$$\left. + \mathsf{E}\left[\hat{U}_2(z - g_1(i_1^1, I_1^2, v_1), I_2^1)\right] \right\}, \tag{3.111}$$

$$\hat{U}_2(q_2, i_2^1)$$
$$= \inf_{y \geq q_2} \left\{ C_2^f(y - q_2) + \mathsf{E}\left[H_3(y - g_2(i_2^1, I_2^2, v_2))\right] \right\}. \tag{3.112}$$

Let

$$\beta = \frac{p - c_2^f}{p + h}.$$

Now we find the optimal order quantity for the fast order at period 2 when I_2^1 is observed with $I_2^1 = i_2^1$. To this end, solving equation

$$c_2^f + \frac{\partial \mathsf{E}[H_3(y - g_2(i_2^1, I_2^2, v_2))]}{\partial y} = 0, \tag{3.113}$$

we get the solution

$$y^* = \varepsilon a \cdot (\beta - \frac{1}{2}) + i_2^1.$$

Hence, the optimal order quantity for the fast order when I_2^1 is observed with $I_2^1 = i_2^1$, is as follows:

$$f_2^* = \begin{cases} \varepsilon a \cdot (\beta - \frac{1}{2}) + i_2^1 - q_2, & \text{if } \varepsilon a \cdot (\beta - \frac{1}{2}) + i_2^1 \geq q_2, \\ 0, & \text{otherwise.} \end{cases} \tag{3.114}$$

In (3.114), the optimal order quantity for the fast order at period 2, f_2^*, is a piecewise function of the observed information i_2^1 and the inventory position q_2. Therefore, the value function $\hat{V}_2(q_2, i_2^1)$ is also a piecewise function of q_2. By (3.114), the manufacturer needs to make a fast order at period 2 only when the inventory position is lower than the ordering point—that is, $f_2^* \geq 0$ if $q_2 \leq \varepsilon a(\beta - 1/2) + i_2^1$; otherwise, $f_2^* = 0$. Further, in view of (3.108) and (3.109), $g_2(i_2^1, I_2^2, v_2) \leq i_2^1 + \varepsilon a/2$ *w.p.*1 when I_2^1 is observed; therefore, no penalty cost arises if the quantity of the inventory position is sufficiently large—that is, $q_2 \geq i_2^1 + \varepsilon a/2$. Thus, we present the value function in the following cases:

$$\hat{V}_2(q_2, i_2^1) = \begin{cases} -h \cdot i_2^1 + h \cdot q_2, & \text{if } i_2^1 \le q_2 - \frac{\varepsilon a}{2}, \\ \frac{h+p}{2\varepsilon a}(i_2^1)^2 + Y_1(q_2) \cdot i_2^1 + Y_2(q_2), & \\ & \text{if } q_2 - \frac{\varepsilon a}{2} \le i_2^1 \le q_2 - \varepsilon a(\beta - \frac{1}{2}), \\ c_2^f \cdot i_2^1 - c_2^f \cdot q_2 + Y, & \text{if } q_2 - \varepsilon a(\beta - \frac{1}{2}) \le i_2^1, \end{cases} \tag{3.115}$$

where $Y_1(q_2)$, $Y_2(q_2)$ and Y are defined as

$$\begin{aligned} Y_1(q_2) &= \frac{p-h}{2} - \frac{h+p}{\varepsilon a} q_2, \\ Y_2(q_2) &= \frac{h+p}{2\varepsilon a}(q_2)^2 + \frac{h-p}{2} q_2 + \frac{h+p}{8}\varepsilon a, \\ Y &= \varepsilon a \left[c_2^f \cdot (\beta - \frac{1}{2}) + \frac{h\beta^2 + p(1-\beta)^2}{2} \right]. \end{aligned}$$

Based on $\hat{V}_2(q_2, i_2^1)$ given by (3.115), we find the optimal fast-order quantity f_1^* and the optimal slow-order quantity s_1^* at period 1. First, it follows from $C_1^f(u) = \infty$ that

$$f_1^* = 0.$$

Let

$$G_1(z) = c_1^s \cdot z + \mathsf{E}[\hat{V}_2(z, I_2^1)],$$

and

$$\alpha_1 = p + h, \quad \alpha_2 = p - c_2^f.$$

It follows from (3.115) that if $0 \le z \le v_2 - \dfrac{a}{2} + \varepsilon a(\beta - \dfrac{1}{2})$, taking the derivative of $G_1(z)$, then $q_2 \ge 0$ and

$$\frac{\mathrm{d}G_1(z)}{\mathrm{d}z} = c_1^s - c_2^f; \tag{3.116}$$

if $v_2 - \frac{a}{2} + \varepsilon a(\beta - \frac{1}{2}) \leq z < v_2 - \frac{a}{2}(1-\varepsilon)$, then

$$\begin{aligned}\frac{\mathrm{d}G_1(z)}{\mathrm{d}z} &= \frac{\alpha_1}{2\varepsilon a^2}z^2 + \frac{1}{2\varepsilon a^2}[a\alpha_1 + \varepsilon a(\alpha_1 - 2\alpha_2) - 2\alpha_1 v_2]z \\ &+ \frac{\alpha_1}{2\varepsilon a^2}v_2^2 + \frac{1}{2\varepsilon a^2}[-a\alpha_1 + \varepsilon a(p-h) - 2\varepsilon a c_2^f]v_2 \\ &+ c_1^s - \frac{c_2^f + p}{2} + \frac{\varepsilon\beta\alpha_2}{2} - \frac{\varepsilon\alpha_2}{2} + \frac{\alpha_1}{4} + \frac{\varepsilon\alpha_1}{8} + \frac{\alpha_1}{8\varepsilon};\end{aligned} \tag{3.117}$$

if $v_2 - \frac{a}{2}(1-\varepsilon) \leq z < v_2 + \frac{a}{2} + \varepsilon a(\beta - \frac{1}{2})$, then

$$\frac{\mathrm{d}G_1(z)}{\mathrm{d}z} = \frac{h + c_2^f}{a}z - \frac{h + c_2^f}{a}v_2 - \frac{\varepsilon\alpha_2(h + c_2^f)}{2\alpha_1} + \frac{2c_1^s + h - c_2^f}{2}; \tag{3.118}$$

if $v_2 + \frac{a}{2} + \varepsilon a(\beta - \frac{1}{2}) \leq z < v_2 + \frac{a}{2} + \frac{\varepsilon a}{2}$, then

$$\begin{aligned}\frac{\mathrm{d}G_1(z)}{\mathrm{d}z} &= -\frac{\alpha_1}{2\varepsilon a^2}z^2 + \frac{\alpha_1}{2\varepsilon a^2}[2v_2 + a + \varepsilon a]z - \frac{\alpha_1}{2\varepsilon a^2}v_2^2 \\ &- \frac{\alpha_1}{2\varepsilon a}[1+\varepsilon]v_2 - \frac{\varepsilon\alpha_1}{8} + \frac{4c_1^s + 3h - p}{4} - \frac{\alpha_1}{8\varepsilon};\end{aligned} \tag{3.119}$$

and if $z \geq v_2 + \frac{a}{2} + \frac{\varepsilon a}{2}$, then

$$G_1(z) = c_1^s z + h\int_{v_2 - \frac{a}{2}}^{v_2 + \frac{a}{2}} \frac{z - t}{a}\mathrm{d}t. \tag{3.120}$$

This implies that

$$\frac{\mathrm{d}G_1(z)}{\mathrm{d}z} = c_1^s + h.$$

To demonstrate the cost function graphically, we depict the cost function with three sets of parameters, where the parameters are for the base case, $h = 0.1, p = 5.3, c_1^s = 1.0, c_2^f = 2, a = 15, \varepsilon = 0.5, v_2 = 50$; for the higher holding cost, $h = 0.5$; and for the higher penalty cost, $p = 5.5$. The graphs appear in Figure 3.2.

Now we discuss some insights into the relationship of the optimal order quantity and other parameters. By Theorem 3.1, we know that the cost function $G_1(z)$ is convex in z. Therefore, it is sufficient to find the optimal order quantity from the first-order condition.

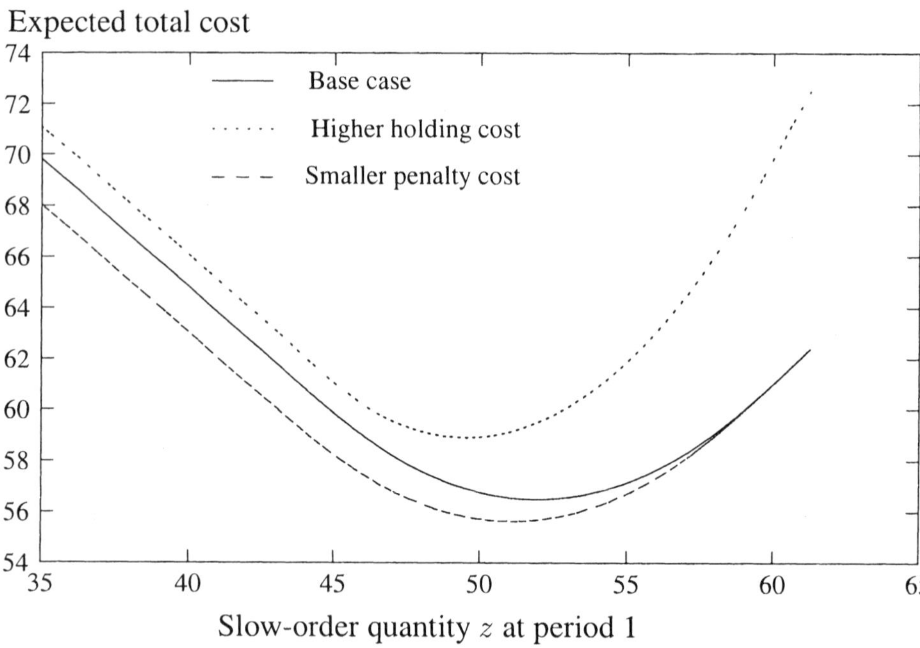

Figure 3.2. Sample cost curves with different cost parameters

LEMMA 3.1 *Assume that* (3.105)–(3.109) *hold. Then we have*
(i) *when* $c_1^s \geq c_2^f$, $s_1^* = 0$ *is optimal*;
(ii) *when* $c_1^s < c_2^f$, *the optimal* $s_1^* \geq v_2 - \frac{a}{2} + \varepsilon a(\beta - \frac{1}{2})$.

Proof First we prove (i). By (3.116), we get $\mathrm{d}G_1(z)/\mathrm{d}z \geq 0$. Therefore, the optimal (minimum) value is achieved at the extreme point zero.

Next we establish (ii). The cost function is convex. In addition, $\mathrm{d}G_1(z)/\mathrm{d}z \leq 0$; therefore, in the interval $[0, \; v_2 - a/2 + \varepsilon a\,(\beta - 1/2)]$, the minimum lies on the right boundary—that is, $s_1^* \geq v_2 - a/2 + \varepsilon a(\beta - 1/2)$. □

Lemma 3.1 indicates that actions must be taken at period 1 when the unit slow-order cost is less than the fast-order cost at period 2. From the point of view of just-in-time production, no material should be ordered at period 1. Similarly, the quick-response program rejects ordering materials at this time. In other words, just-in-time production and a quick-response program apply only for the cases where there is no per unit order cost difference between different supply sources. This observation is corroborated by other researchers. For example, Fisher, Hammond, Obermeyer, and Raman [5] observed that "to address the problem of inaccurate forecasts, many manufacturers have turned to one or another popular production-scheduling system. But quick-response programs, just-in-time (JIT) inventory systems, manufacturing resource planning, and the like are simply not up to the task." Lemma 3.1 provides an analytical example of why manufacturers need a better system than a pure just-in-time approach.

Further, it is interesting to consider scenarios with capacity constraints where only one source of raw material is available and the lead time and price are constant. The just-in-time production strategy suggests that no decision on raw-material order quantity or production commitment be made until the latest stage. In many industrial settings, especially in the quick-response systems, the available capacity limits the production lead time. If the production lead time can be reduced, the demand forecast is more accurate at a later time. Lemma 3.1 suggests that decisions on raw-material ordering and production commitments can be made earlier. With some earlier productions, precious capacities could be reserved at a later stage when the forecasts become more accurate. This coincides with the findings of Cohen and Mallik [3]. They argue that by holding excess capacity, a firm has an option to respond to uncertain events and may be able to take advantage of arbitrage opportunities.

We summarize results in this section into the following theorem.

THEOREM 3.8 *Assume that* (3.105)–(3.109) *hold. For any setting of parameters, we have*

(i) *if* $c_1^s > c_2^f$, *then* $s_1^* = 0$;

(ii) *if* $c_1^s = c_2^f$, *then* $s_1^* = x$, *for any* x *that satisfies* $0 \leq x \leq v_2 - a/2 + \varepsilon a(\beta - 1/2)$;

(iii) *let*

$$x = v_2 - \frac{a}{2} + \varepsilon a(\beta - \frac{1}{2}) + \frac{a}{\sqrt{\alpha_1}}\sqrt{2\varepsilon(c_2^f - c_1^s)},$$

if $v_2 - a/2 + \varepsilon a(\beta - 1/2) \leq x \leq v_2 - a/2 + (\varepsilon a)/2$, *then* $s_1^* = x$;

(iv) *let*

$$x = v_2 - \frac{ac_1^s}{c_2^f + h} + \frac{a}{2(c_2^f + h)}\left[c_2^f - h\right] + \frac{\varepsilon a}{2(c_2^f + h)\alpha_1}\left[hp - c_2^f(h - \alpha_2)\right],$$

if $v_2 - a/2 + \varepsilon a/2 \leq x \leq v_2 + a/2 + \varepsilon a(\beta - 1/2)$, *then* $s_1^* = x$;

(v) *let*

$$x = v_2 + \frac{a}{2} + \frac{\varepsilon a}{2} - \frac{a}{\sqrt{\alpha_1}}\sqrt{2\varepsilon(c_1^s + h)},$$

if $v_2 + a/2 + \varepsilon a(\beta - 1/2) \leq x \leq v_2 + a/2 + \varepsilon a/2$, *then* $s_1^* = x$;

(vi) *else,* $s_1^* = v_2 + a/2 + \varepsilon a/2$.

It can be proved that the optimal order quantity is a linear function with respect to either a or ε. Figure 3.3 provides examples of the changes of cost

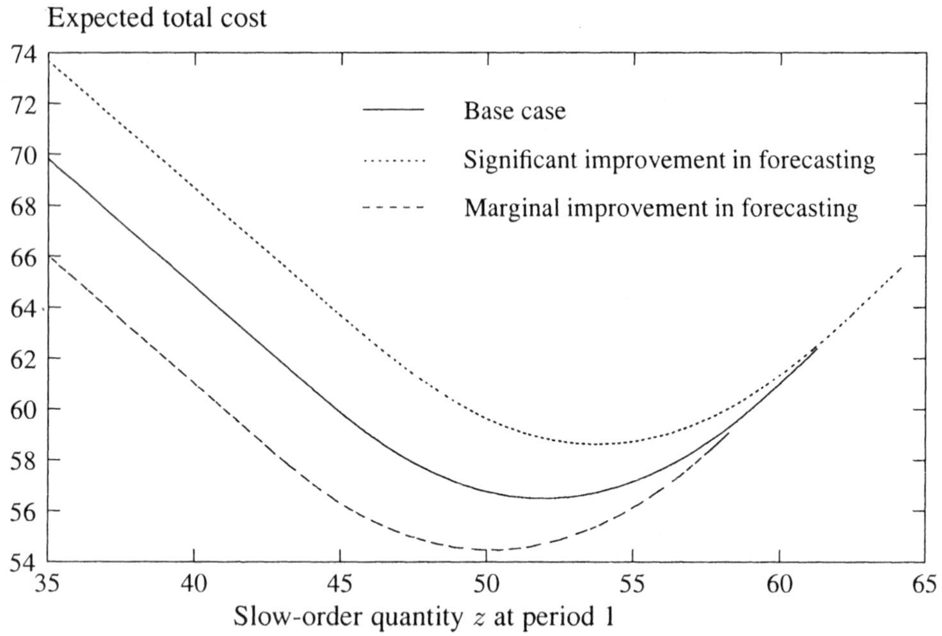

Figure 3.3. Sample cost curves with different forecasting-improvement factors

function with respect to the changes in ε. The smaller ε is, the less that the stage 1 order quantity will be. We find that, similar to the ε, the larger variance requires a larger order quantity. We demonstrate this feature in Figure 3.4, where the larger the variance is, the higher the cost and the larger the order quantity are. On the other hand, it seems that the order quantity is more sensitive to the degree of improvement than the variance itself.

REMARK 3.8 If a is sufficiently small, for the case of $c_1^s < c_2^f$, $s_1^* = v_2$ is nearly optimal.

The analysis given above reveals the existence of optimal purchase policies with respect to cost parameters and demand information. These findings answer questions such as how well the manufacturer can forecast and what the optimal expenditure is. However, the manufacturer would be interested to know the value of information updates and, further, the opportunities for continuous improvement. In this section, we explore these managerial implications by marginal cost/benefit analysis.

It is reasonable to assume that the manufacturer does not have control over cost parameters in the short run. After knowing the optimal material-purchase policies, the manufacturer would look for other directions of further improve-

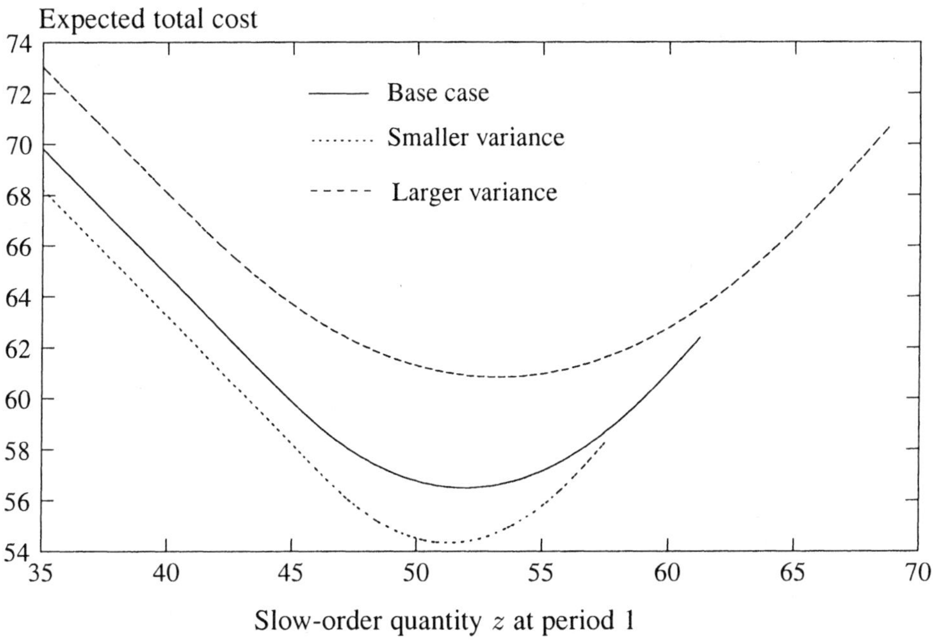

Figure 3.4. Sample cost curves with different forecasting errors

ment, such as improving its demand forecast. Intuitively, improving either stage 1 and stage 2 forecasts results in the cost reduction. The marginal costs of information updates with respect to ε and a provide indications on the value of information updates. Specifically, the marginal benefit of information updates can be expressed as follows. If $v_2 - 1/2 + \varepsilon a(\beta - 1/2) \leq s_1^* \leq v_2 - a/2 + \varepsilon a/2$,

$$\left\{\begin{array}{lll} \frac{\partial G_1^*(s_1^*)}{\partial \varepsilon} & = \frac{(h+c_1^s)(h-c_1^s)+(c_2^f-c_1^s)^2}{2(p+h)} & - \frac{\sqrt{2}}{3}\frac{(c_2^f-c_1^s)^{3/2}}{\sqrt{p+h}}\frac{\sqrt{a}}{\sqrt{\varepsilon}}, \\ \frac{\partial G_1^*(s_1^*)}{\partial a} & = \frac{c_2^f-c_1^s}{2} - \frac{\sqrt{2}}{3}\frac{(c_2^f-c_1^s)^{3/2}}{\sqrt{p+h}}\frac{\sqrt{\varepsilon}}{\sqrt{a}}; & \end{array}\right. \tag{3.121}$$

if $v_2 - a/2 + \varepsilon a/2 \leq s_1^* \leq v_2 + a/2 + \varepsilon a(\beta - 1/2)$,

$$\left\{\begin{array}{ll} \frac{\partial G_1^*(s_1^*)}{\partial \varepsilon} & = \frac{(h+c_2^f)^3\varepsilon}{12(p+h)^2 a} + \frac{(p-c_2^f)(h+c_1^s)}{2(p+h)}, \\ \frac{\partial G_1^*(s_1^*)}{\partial a} & = -\frac{(h+c_2^f)^3\varepsilon^2}{24(p+h)^2 a^2} + \frac{(c_2^f-c_1^s)(h+c_1^s)}{2(h+c_2^f)}; \end{array}\right. \tag{3.122}$$

and if $v_2 + a/2 + a\varepsilon(\beta - 1/2) \leq s_1^* \leq v_2 + a/2 + (a\varepsilon)/2$,

$$\begin{cases} \frac{\partial G_1^*(s_1^*)}{\partial \varepsilon} & = \frac{h+c_1^s}{2} - \frac{\sqrt{2}}{3}\frac{(h+c_1^s)^{3/2}}{\sqrt{p+h}}\frac{\sqrt{a}}{\sqrt{\varepsilon}}, \\ \frac{\partial G_1^*(s_1^*)}{\partial a} & = \frac{h+c_1^s}{2} - \frac{\sqrt{2}}{3}\frac{(h+c_1^s)^{3/2}}{\sqrt{p+h}}\frac{\sqrt{\varepsilon}}{\sqrt{a}}. \end{cases} \tag{3.123}$$

Based on equations (3.121), (3.122), and (3.123), the manufacturer is able to determine the impact that one unit of demand forecast improvement in either stage 1 or stage 2 has on its cost structure and further to determine whether its effort in improving demand information is worthwhile. From these equations, when c_2^f is large, improving the first-stage forecasts would yield a larger payoff; similarly, when the difference of c_1^s and c_2^f is small, improving the second-stage forecasts would be much beneficial. More important, the notion of equal principle (Samuelson and Nordhaus [11]) suggests that the company should put its last dollar to the place where a higher return is expected. Equations (3.121), (3.122), and (3.123) provide formulas for calculating marginal return with respect to demand-forecast improvement. Comparing $\partial G_1^*(s_1^*)/\partial a$ and $\partial G_1^*(s_1^*)/\partial \varepsilon$ indicates the potential return. For example, if $\partial G_1^*(s_1^*)/\partial a \geq \partial G_1^*(s_1^*)/\partial \varepsilon$, the manufacturer should concentrate its effort to improve the demand forecasting at stage 1.

3.7. Concluding Remarks

In this chapter, we consider a discrete-time, periodic-review inventory system with dual supply modes and demand-information updates. We demonstrate that the optimal inventory-replenishment policy is a base-stock policy for both finite and discounted infinite-horizon problems. Recently, Gallego, Sethi, Wang, Yan, and Zhang [8] have developed an algorithm to compute the optimal base-stock level with dual-supply modes but without demand-information updates. Extension of their work to the demand-information updates is still in progress. Our model generalizes several special cases in the literature. The extension of our model to include fixed order cost is discussed in the next chapter.

3.8. Notes

The main material in the chapter is based on Sethi, Yan, and Zhang [15]. The example in Section 3.6 is based on Yan, Liu, and Hsu [18].

Bensoussan, Crouhy, and Proth [2] consider an inventory model with two supply modes—one instantaneous and the other with a one-period lead time. They allow for fixed as well as variable costs associated with ordering decisions. They obtain an optimal policy, which represents a generalization of the well-known (s, S) policy. Hausmann, Lee, and Zhang [10] study an inventory system with two supply modes—fast and slow—under the assumption of stationary demand. Explicit formulas for optimal ordering decisions are developed. Scheller-Wolf and Tayur [12] study a Markovian dual-source production inventory model but without the consideration of advance-demand information. They prove the optimality of a state-dependent base-stock policy. For other instances of state-dependent policies, see Scheller-Wolf and Tayur [12], Song and Zipkin [16], and Sethi and Cheng [13].

The distinctive feature of our model is the treatment of a multiperiod inventory model allowing for both information updates and multiple sourcing partners. It differs from Fisher, Hammond, Obermeyer, and Raman [5], Hausmann, Lee, and Zhang [10], Scheller-Wolf and Tayur [12] in the sense that we make use of demand-forecast updates in making decisions. In contrast to Barnes-Schuster, Bassok, and Anupindi [1], Yan, Liu, and Hsu [18], Donohue [4], and Gurnani and Tang [9], we consider an N-period inventory model, $N \leq \infty$. Furthermore, our model of the demand-updating process covers, as a special case, the additive demand-updating process employed in Gallego and Özer [7].

3.9. Appendix

In this appendix, we introduce the selection theorem that is used to establish the existence of the optimal nonanticipative policy. First, we introduce the definition of *lower semicontinuous functions.*

DEFINITION 3.1 *Let $J(\cdot)$ be a function defined on R^n. We say $J(\cdot)$ is* lower semicontinuous *if for any $x \in R^n$,*

$$\liminf_{x_n \to x} J(x_n) \geq J(x).$$

THEOREM 3.9 (SELECTION THEOREM) *Let $J(x, y)$ be a function on $R^n \times R^m$, lower semicontinuous, and bounded from below, and let $\mathcal{K}$ be a compact set of R^m. Then there exists a Borel-measurable function $B(x)$ defined on R^n such that*

$$J(x, B(x)) = \inf_{y \in \mathcal{K}} \{J(x, y)\}.$$

REMARK 3.9 We can weaken the condition of lower semicontinuity of both variables by imposing Lebesgue measurability. In this book, however, the lower semicontinuous condition is enough for us to carry out the existence of the optimal nonanticipative policy. For a proof, see Bensoussan, Crouhy, and Proth [2].

References

[1] D. Barnes-Schuster, Y. Bassok, and R. Anupindi. Coordination and flexibility in supply contracts with options. *Manufacturing and Service Operations Management*, 4:171–207, 2002.

[2] A. Bensoussan, M. Crouhy, and J.M. Proth. *Mathematical Theory of Production Planning*. North-Holland, Amsterdam, 1983.

[3] M.A. Cohen and S. Mallik. Global supply chains: Research and applications. *Production and Operations Management*, 6:193–210, 1997.

[4] K.L. Donohue. Efficient supply contracts for fashion goods with forecast updating and two production modes. *Management Science*, 46:1397–1411, 2000.

[5] M. Fisher, J.H. Hammond, W.R. Obermeyer, and A. Raman. Making supply meet demand in an uncertain world. *Harvard Business Review*, May-June, 83–93, 1994.

[6] Y. Fukuda. Optimal policies for the inventory problem with negotiable lead time. *Management Science*, 10:690–708, 1964.

[7] G. Gallego and Ö. Özer. Integrating replenishment decisions with advance demand information. *Management Science*, 47:1344–1360, 2001.

[8] G. Gallego, S. Sethi, Z. Wang, H. Yan, and H. Zhang. Stationary policies for multiple procurement modes. Working Paper, Columbia University, New York, 2003.

[9] H. Gurnani and C.S. Tang. Note: optimal ordering decisions with uncertain cost and demand forecast updating. *Management Science*, 45:1456–1462, 1999.

[10] W.H. Hausman, H.L. Lee, and V.L. Zhang. Optimal ordering for an inventory system with dual lead times. Working Paper, Stanford University, Stanford, CA, 1993.

[11] P.A. Samuelson and W.D. Nordhaus. *Economics*, 16th edition, Irwin, Chicago, 1998.

[12] A. Scheller-Wolf and S. Tayur. A markovian dual-source production-inventory model with order bands. Working Paper, Carnegie Mellon University, Pittsburgh, PA, 1998.

[13] S.P. Sethi and F. Cheng. Optimality of (s,S) policies in inventory models with Markovian demand. *Operations Research*, 45:931–939, 1997.

[14] S.P. Sethi and G. Sorger. A theory of rolling-horizon decision making. *Annals of Operations Research*, 29:387–416, 1991.

[15] S.P. Sethi, H. Yan, and H. Zhang. Peeling layers of an onion: Inventory model with multiple delivery modes and forecast updates. *Journal of Optimization Theory and Applications*, 108:253–281, 2001.

[16] J. Song and P. Zipkin. Inventory control in a fluctuating demand environment. *Operations Research*, 41:351–370, 1993.

[17] A.S. Whittemore and S.C. Saunders. Optimal inventory under stochastic demand with two supply options. *SIAM Journal of Applied Mathematics*, 32:293–305, 1977.

[18] H. Yan, K. Liu, and A. Hsu. Optimal ordering in a dual-supplier system with demand forecast updates. *Production and Operations Management*, 12:30–45, 2003.

Chapter 4

INVENTORY MODELS WITH TWO CONSECUTIVE DELIVERY MODES AND FIXED COST

4.1. Introduction

In this chapter, we introduce fixed order costs in the model studied in Chapter 3. Thus, we consider forecast updates, two delivery modes, and a fixed cost associated with each delivery mode in a periodic-review inventory system. The demand in any one period (hidden in the core of an onion) has a number of sources of randomness (layers of the onion), and these uncertainties are resolved successively in periods leading to the period in which the demand materializes (peeling the layers to get to the core). Two delivery modes are fast and slow. A fast order that is issued at the beginning of a period is delivered at the end of the period, whereas a slow order that is issued at the beginning of a period is delivered at the end of the next period. In addition to fixed order costs, there are variable costs. Fast orders are assumed to be more expensive than slow orders. In words, at the beginning of each period, the inventory or backlog level is reviewed, and the forecast of the demand to be realized at the end of the period is updated. Also known at the time is the slow order that was issued in the previous period—an order that will be delivered at the end of the current period. With these data in hand, decisions are made about the amounts to be ordered this period by fast and slow modes. At the end of the period, the slow order issued in the previous period and the fast order issued at the beginning of the current period are delivered. Then the demand for the current period materializes, and the inventory or backlog level, or simply the inventory level, at the beginning of the next period gets determined. The total cost in each period consists of procurement costs and inventory or backlog costs. The objective is to make ordering decisions that minimize total costs over the problem horizon.

The remainder of this chapter is organized as follows. In Section 4.2, we provide the required notation and the model formulation. Dynamic programming equations for the problem are developed in Section 4.3. In Section 4.4, we obtain the optimality of an (s, S) type policy for the finite-horizon problem. Section 4.5 looks into some monotonicity properties of the optimal policy parameters. Section 4.6 is devoted to extending the optimality results to the infinite-horizon case. The chapter is concluded in Sections 4.7 and 4.8.

4.2. Notation and Model Formulation

As described in Chapter 3, the dynamics of the system contains two parts: the material flows and the information flows. The inbound material flows come from two supply sources (fast and slow), and the outbound material flows go to customers. Orders are made at the beginning of a period. An order from the fast source arrives at the end of the current period, whereas an order from the slow and possibly cheaper source arrives at the end of the next period. The information flows include the initial demand forecast, periodical demand-forecast updates, and the realized customer demand. The decision variables are the quantities ordered from the fast and slow sources at the beginning of each period. The decisions are based on the past history, although we show that it is sufficient to decide on the basis of the current inventory position and the current (updated) demand information. A time line of the system dynamics and the ordering decisions is illustrated in Figure 3.1.

In addition to the notation introduced in Chapter 3, we also use the following notation in this chapter:

$$
\begin{aligned}
K_k^f &= \text{the fixed fast-order cost in period } k;\\
K_k^s &= \text{the fixed slow-order cost in period } k;\\
\Lambda_k(\cdot) &= \text{the distribution function of } I_k^1;\\
\Theta_k(\cdot) &= \text{the distribution function of } g_k(I_k^1, I_k^2, v_k);\\
\Psi_k(\cdot|i_k^1) &= \text{the conditional distribution function of } g_k(I_k^1, I_k^2, v_k)\\
&\quad\ \text{given } I_k^1 = i_k^1.
\end{aligned}
$$

For notational convenience, let

$$I_1^1 \equiv i_1^1$$

be a deterministic constant. We make the following assumptions in this chapter:

$$\{(I_k^1, I_k^2), 1 \le k \le N\} \ \text{ is a sequence of independent random vectors,} \tag{4.1}$$

$$\mathsf{E}[D_k] = \mathsf{E}\left[g_k(I_k^1, I_k^2, v_k)\right] < \infty, \ \ 1 \le k \le N, \tag{4.2}$$

$$C_k^f(u) \text{ and } C_k^s(u) \text{ are increasing, nonnegative and convex,} \tag{4.3}$$

$$\begin{cases} H_k(x) \text{ is convex and} \\ |H_k(x) - H_k(\hat{x})| \le c_H \cdot |x - \hat{x}|, \;\; 1 \le k \le N+1, \end{cases} \tag{4.4}$$

for some $c_H > 0$, and

$$\min\{K_k^f, K_k^s\} \ge \max\{K_{k+1}^f, K_{k+1}^s\}, \qquad k = 1, ..., N-1. \tag{4.5}$$

REMARK 4.1 Assumption (4.1) does not preclude the case in which I_k^2 depends on I_k^1. Assumption (4.5) is similar in spirit to those used in Sethi and Cheng [15] and Beyer, Sethi, and Taksar [3].

Similar to (3.6) and (3.7), we assume that

$$C_k^f(t) + \mathsf{E}[H_{k+1}(t - g_k(I_k^1, I_k^2, v_k))] \to \infty \text{ as } t \to \infty, \tag{4.6}$$
$$C_k^s(t) + \mathsf{E}[H_{k+1}(t - g_k(I_k^1, I_k^2, v_k))] \to \infty \text{ as } t \to \infty. \tag{4.7}$$

The inventory-balance equations are defined as

$$X_{k+1} = X_k + F_k + S_{k-1} - g_k(I_k^1, I_k^2, v_k), \; 1 \le k \le N, \tag{4.8}$$

where $S_0 = s_0$, a possible existing slow order to be delivered in period 1, and

$$X_1 = x_1, \text{ the initial inventory level.} \tag{4.9}$$

The objective is to choose a sequence of orders from the fast and slow sources over time to minimize the total expected value of the costs incurred during the interval $\langle 1, N \rangle$. Thus the objective function is

$$\begin{aligned} &J_1(x_1, s_0, i_1^1, (\boldsymbol{F}, \boldsymbol{S})) \\ &= H_1(x_1) + \mathsf{E}\Bigg[\sum_{\ell=1}^{N} \Bigg(K_\ell^f \cdot \delta(F_\ell) + C_\ell^f(F_\ell) \\ &\qquad\qquad + K_\ell^s \cdot \delta(S_\ell) + C_\ell^s(S_\ell) + H_{\ell+1}(X_{\ell+1})\Bigg)\Bigg], \end{aligned} \tag{4.10}$$

where x_1 is the initial on-hand inventory at the beginning of period 1, s_0 is an outstanding slow order to be delivered at the end of period 1, and $(\boldsymbol{F}, \boldsymbol{S}) = ((F_1, ..., F_N), (S_1, ..., S_N))$ with F_k and S_k being fast-order and slow-order quantities in period k. The decisions $(\boldsymbol{F}, \boldsymbol{S})$ are *history-dependent* or *nonanticipative* to be admissible—that is, (F_k, S_k) are nonnegative real-valued functions of the history of the demand information up to period $(k-1)$ given by

$\{(I_\ell^1, I_\ell^2), 0 \le \ell \le k-1\}$ and I_k^1, and F_N and S_N are a nonnegative real-valued functions of the history of the demand information up to period $(N-1)$ given by $\{(I_\ell^1, I_\ell^2), 0 \le \ell \le N-1\}$ and I_N^1. In view of (3.10), we still have $S_N = 0$ here.

Let $\mathcal{A}_1$ denote the class of all admissible decisions for the problem over $\langle 1, N \rangle$. Then the value function for the problem over $\langle 1, N \rangle$ with the initial inventory level x_1 can be defined by

$$V_1(x_1, s_0, i_1^1) = \inf_{(\boldsymbol{F},\boldsymbol{S})\in\mathcal{A}_1} \left\{ J_1\left(x_1, s_0, i_1^1, (\boldsymbol{F}, \boldsymbol{S})\right) \right\}. \tag{4.11}$$

Note that the existence of an optimal policy is not required to define the value function. Of course, once the existence is established, the "inf" in (4.11) can be replaced by "min".

4.3. Dynamic Programming and Optimal Nonanticipative Policy

In this section, we first give the dynamic programming equation satisfied by the value function. We then provide a verification theorem that states the cost associated with the nonanticipative policy obtained from the solution of the dynamic programming equations equals the value function of the problem on $\langle 1, N \rangle$.

As in Section 3.3, we define the problem over $\langle n, N \rangle$. Let

$$\begin{aligned} &J_n(x_n, s_{n-1}, i_n^1, (\boldsymbol{F}, \boldsymbol{S})) \\ &= H_n(x_n) + \mathsf{E}\Bigg[\sum_{\ell=n}^{N} \Big(K_\ell^f \cdot \delta(F_\ell) + C_\ell^f(F_\ell) \\ &\qquad + K_\ell^s \cdot \delta(S_\ell) + C_\ell^s(S_\ell) + H_{\ell+1}(X_{\ell+1})\Big)\Bigg], \end{aligned} \tag{4.12}$$

where

$$(\boldsymbol{F}, \boldsymbol{S}) = ((F_n, ..., F_N), (S_n, ..., S_N))$$

is a *history-dependent* or *nonanticipative admissible decision* for the problem defined over periods $\langle n, N \rangle$ (see Section 3.3). Here s_{n-1} has the same meaning as s_0 and is an outstanding slow order to be delivered at the end of period n. Define the value function associated with the problem over periods $\langle n, N \rangle$ as follows:

$$V_n(x_n, s_{n-1}, i_n^1) = \inf_{(\boldsymbol{F},\boldsymbol{S})\in\mathcal{A}_n} \left\{ J_n\left(x_n, s_{n-1}, i_n^1, (\boldsymbol{F}, \boldsymbol{S})\right) \right\}, \tag{4.13}$$

where $\mathcal{A}_n$ denotes the class of all history-dependent admissible decisions for the problem over $\langle n, N\rangle$. We have the following theorem on the property of the value function $V_n(x_n, s_{n-1}, i_n^1)$.

Using the principle of optimality, we can, accordingly, write the following dynamic programming equations for our problem:

$$\begin{cases} U_\ell(x_\ell, s_{\ell-1}, i_\ell^1) \\ \quad = H_\ell(x_\ell) + \inf_{\substack{\phi\geq 0\\ \sigma\geq 0}} \Big\{ K_\ell^f \cdot \delta(\phi) + C_\ell^f(\phi) + K_\ell^s \cdot \delta(\sigma) + C_\ell^s(\sigma) \\ \quad +\mathsf{E}\left[U_{\ell+1}\left(x_\ell + \phi + s_{\ell-1} - g_\ell(i_\ell^1, I_\ell^2, v_\ell), \sigma, I_{\ell+1}^1\right)\right] \Big\}, \\ \qquad\qquad \ell = 1, ..., N-1, \\ U_N(x_N, s_{N-1}, i_N^1) \\ \quad = H_N(x_N) + \inf_{\substack{\phi\geq 0\\ \sigma\geq 0}} \Big\{ K_N^f \cdot \delta(\phi) + C_N^f(\phi) + K_N^s \cdot \delta(\sigma) \\ \quad +C_N^s(\sigma) + \mathsf{E}\left[H_{N+1}\left(x_N + s_{N-1} + \phi - g_N(i_N^1, I_N^2, v_N)\right)\right] \Big\}. \end{cases} \tag{4.14}$$

Note that if I_ℓ^2 depends on I_ℓ^1, then the expectation on the right-hand side will be understood to be conditional expectation given $I_\ell^1 = i_\ell^1$. The dynamic programming equations (4.14) involve three variables, namely, x_ℓ, $s_{\ell-1}$, and i_ℓ^1, (inventory level, slow order made in the previous period, and the demand-information update).

We now state the relevant existence results in the following two theorems. The methodology and the technique of proving these theorems are quite similar to those in the previous chapter, which deals with the case of multiple supply modes and demand-information updates without a fixed order cost. Hence, we omit their proofs and direct the interested readers to Theorems 3.1 and 3.3.

THEOREM 4.1 *Assume that* (4.1)–(4.7) *hold. Then the value function* $V_1(x_1, s_0, i_1^1)$ *is convex and Lipschitz continuous in* x_1 *on* $(-\infty, +\infty)$*, and the value functions* $V_\ell(x_\ell, s_{\ell-1}, i_\ell^1), 2 \leq \ell \leq N$*, are Lipschitz continuous in* $(x_\ell, s_{\ell-1})$ *on* $(-\infty, +\infty) \times [0, +\infty)$*. At the same time, they are also the solutions of the dynamic programming equations* (4.14)*. Moreover, there exist functions* $(\bar\phi_N(x_N, s_{N-1}, i_N^1), 0)$ *that provide the infimum in the last equation of* (4.14)*, and functions*

$$\left(\bar\phi_\ell(x_\ell, s_{\ell-1}, i_\ell^1), \bar\sigma_\ell(x_\ell, s_{\ell-1}, i_\ell^1)\right), \quad 1 \leq \ell \leq N-1,$$

that provide the infima in the first $(N-1)$ *equations of* (4.14) *with* $U_\ell(x_\ell, s_{\ell-1}, i_\ell^1) = V_\ell(x_\ell, s_{\ell-1}, i_\ell^1)$.

To solve the problem of minimizing $J_1(x_1, s_0, i_1^1, (\boldsymbol{F}, \boldsymbol{S}))$, we use $\bar{\phi}_\ell(\cdot)$ and $\bar{\sigma}_\ell(\cdot)$ of Theorem 4.1 to define

$$\begin{cases} \bar{X}_1 = x_1, \\ \bar{F}_1 = \bar{\phi}_1(x_1, s_0, i_1^1), \\ \bar{S}_1 = \bar{\sigma}_1(x_1, s_0, i_1^1), \end{cases} \tag{4.15}$$

and

$$\begin{cases} \bar{X}_\ell = \bar{X}_{\ell-1} + \bar{F}_{\ell-1} + \bar{S}_{\ell-2} - g_{\ell-1}(I_{\ell-1}^1, I_{\ell-1}^2, v_{\ell-1}), \\ \bar{F}_\ell = \bar{\phi}_\ell(\bar{X}_\ell, \bar{S}_{\ell-1}, I_\ell^1), \\ \bar{S}_\ell = \bar{\sigma}_\ell(\bar{X}_\ell, \bar{S}_{\ell-1}, I_\ell^1), \end{cases} \tag{4.16}$$

for $2 \leq \ell \leq N-1$, where $\bar{S}_0 = s_0$ and

$$\begin{cases} \bar{X}_N = \bar{X}_{N-1} + \bar{F}_{N-1} + \bar{S}_{N-2} - g_{N-1}(I_{N-1}^1, I_{N-1}^2, v_{N-1}), \\ \bar{F}_N = \bar{\phi}_N(\bar{X}_N, \bar{S}_{N-1}, I_N^1), \\ \bar{S}_N = 0. \end{cases} \tag{4.17}$$

THEOREM 4.2 (VERIFICATION THEOREM) *Assume that* (4.1)–(4.7) *hold. Then*

$$((\bar{F}_1, ..., \bar{F}_N), (\bar{S}_1, ..., \bar{S}_N))$$

described in (4.15)–(4.17) *is an optimal solution to the problem. Hence, the minimum cost* $V_1(x_1, s_0, i_1^1)$ *is*

$$H_1(x_1) + \mathsf{E}\Bigg[\sum_{\ell=1}^{N} \Big(K_\ell^f \cdot \delta(\bar{F}_\ell) + C_\ell^f(\bar{F}_\ell) + K_\ell^s \cdot \delta(\bar{S}_\ell)$$
$$+ C_\ell^s(\bar{S}_\ell) + H_{\ell+1}(\bar{X}_{\ell+1})\Big)\Bigg]. \tag{4.18}$$

Theorems 4.1 and 4.2 establish the existence of an optimal nonanticipative policy. Specifically, there exists a nonanticipative policy defined by equations (4.15)–(4.17), which provides a value of the objective function equal to the value function.

4.4. Optimality of (s, S) Ordering Policies

For a further analysis of the problem, it is convenient to recast the dynamic programming equations (4.14) involving order quantities ϕ and σ as decision variables to those involving order-up-to levels y $(= x_\ell + s_{\ell-1} + \phi)$ and z $(=$

$x_\ell + s_{\ell-1} + \phi + \sigma$) as decision variables. Moreover, since the ordering decisions ϕ and σ in any period ℓ depend on x_ℓ and $s_{\ell-1}$ through their sum $x_\ell + s_{\ell-1}$, known as the inventory position y_ℓ in period ℓ, the dynamic programming equation (4.14) can be, similar to Section 3.4 in Chapter 3, rewritten as follows in terms of the state variables y_ℓ and i_ℓ^1, instead of the state variables $x_\ell, s_{\ell-1}$ and i_ℓ^1:

$$\left\{\begin{array}{l} \hat{U}_\ell(y_\ell, i_\ell^1) \\ = \inf_{\substack{y \geq y_\ell \\ z \geq y}} \Big\{ K_\ell^f \cdot \delta(y - y_\ell) + C_\ell^f(y - y_\ell) + K_\ell^s \cdot \delta(z - y) \\ \qquad + C_\ell^s(z - y) + \mathsf{E}\left[H_{\ell+1}\left(y - g_\ell(i_\ell^1, I_\ell^2, v_\ell)\right)\right] \\ \qquad + \mathsf{E}\left[\hat{U}_{\ell+1}\left(z - g_\ell(i_\ell^1, I_\ell^2, v_\ell), I_{\ell+1}^1\right)\right] \Big\}, \\ \qquad \ell = 1, ..., N-1, \\ \hat{U}_N(y_N, i_N^1) \\ = \inf_{\substack{y \geq y_N \\ z \geq y}} \Big\{ K_N^f \cdot \delta(y - y_N) + C_N^f(y - y_N) + K_N^s \cdot \delta(z - y) \\ \qquad + C_N^s(z - y) + \mathsf{E}\left[H_{N+1}\left(y - g_N(i_N^1, I_N^2, v_N)\right)\right] \Big\}. \end{array}\right. \tag{4.19}$$

Let $\hat{V}_\ell(y_\ell, i_\ell^1)$ be the solution of (4.19). By the fact that the set $\{(y, z) | y \geq y_\ell \text{ and } z \geq y\}$ is convex, using the selection theorem (see Theorem 3.9), it is straightforward to get the following Theorem 4.3.

THEOREM 4.3 *Let the assumptions* (4.1)–(4.7) *hold. Then there exist functions* $(\hat{\phi}_N(y_N, i_N^1), \hat{\sigma}(y_N, i_N^1))$ *that provide the infimum in the last equation of* (4.19) *and functions*

$$\left(\hat{\phi}_\ell(y_\ell, i_\ell^1), \hat{\sigma}_\ell(y_\ell, i_\ell^1)\right), \quad 1 \leq \ell \leq N-1,$$

that provide the infima in the first $(N-1)$ *equations of* (4.19) *with* $\hat{U}_\ell(y_\ell, i_\ell^1) = \hat{V}_\ell(y_\ell, i_\ell^1)$.

It follows from Theorem 4.3 that

$$\begin{aligned} \hat{V}_\ell(y_\ell, i_\ell^1) &= K_\ell^f \cdot \delta\left(\hat{\phi}_\ell(y_\ell, i_\ell^1) - y_\ell\right) + C_\ell^f\left(\hat{\phi}_\ell(y_\ell, i_\ell^1) - y_\ell\right) \\ &\quad + K_\ell^s \cdot \delta\left(\hat{\sigma}_\ell(y_\ell, i_\ell^1) - \hat{\phi}_\ell(y_\ell, i_\ell^1)\right) \end{aligned}$$

$$
\begin{aligned}
&+C_\ell^s\left(\hat{\sigma}_\ell(y_\ell, i_\ell^1) - \hat{\phi}_\ell(y_\ell, i_\ell^1)\right) \\
&+\mathsf{E}\left[H_{\ell+1}\left(\hat{\phi}_\ell(y_\ell, i_\ell^1) - g_\ell(i_\ell^1, I_\ell^2, v_\ell)\right)\right] \\
&+\mathsf{E}\left[\hat{V}_{\ell+1}\left(\hat{\sigma}_\ell(y_\ell, i_\ell^1) - g_\ell(i_\ell^1, I_\ell^2, v_\ell), I_{\ell+1}^1\right)\right] \\
&\qquad \ell = 1, ..., N-1,
\end{aligned} \tag{4.20}
$$

$$
\begin{aligned}
\hat{V}_N(y_N, i_N^1) \;=\;& K_N^f \cdot \delta\left(\hat{f}_N(y_N, i_N^1) - y_N\right) + C_N^f\left(\hat{\phi}_N(y_N, i_N^1) - y_N\right) \\
&+K_N^s \cdot \delta\left(\hat{\sigma}_N(y_N, i_N^1) - \hat{\phi}_N(y_N, i_N^1)\right) \\
&+C_N^s\left(\hat{\sigma}_N(y_N, i_N^1) - \hat{\phi}_N(y_N, i_N^1)\right) \\
&+\mathsf{E}\left[H_{N+1}\left(\hat{\phi}_N(y_N, i_N^1) - g_N(i_N^1, I_N^2, v_N)\right)\right].
\end{aligned} \tag{4.21}
$$

Let

$$
\begin{cases}
Y_1 = y_1 = x_1 + s_0, \\
Y_k = \hat{s}_{k-1}(Y_{k-1}, I_{k-1}^1) - g_{k-1}(I_{k-1}^1, I_{k-1}^2, v_{k-1}), \; k = 2, ..., N.
\end{cases}
$$

Define

$$
\begin{cases}
\hat{F}_1 = \hat{f}_1(y_1, I_1^1) - Y_1, \\
\hat{S}_1 = \hat{\sigma}_1(y_1, I_1^1) - \hat{\phi}_1(y_1, I_1^1),
\end{cases} \tag{4.22}
$$

and

$$
\begin{cases}
\hat{F}_k = \max\{\hat{\phi}_k(Y_k, I_k^1) - Y_k, 0\}, \; k = 2, ..., N, \\
\hat{S}_\ell = \hat{\sigma}_\ell(Y_\ell, I_\ell^1) - \hat{\phi}_\ell(Y_\ell, I_\ell^1), \; \ell = 2, ..., N-1, \\
\hat{S}_N = 0.
\end{cases} \tag{4.23}
$$

Formally, (4.19) is derived from (4.14) by taking $U_\ell(x_\ell, s_{\ell-1}, i_\ell^1) - H_\ell(x_\ell) = \hat{U}_\ell(y_\ell, i_\ell^1)$ in (4.14). Note that $U_\ell(x_\ell, s_{\ell-1}, i_\ell^1) - H_\ell(x_\ell)$ involves three state variables x_ℓ, i_ℓ^1 and $s_{\ell-1}$, while $\hat{U}_\ell(y_\ell, i_\ell^1)$ involves only two state variables y_ℓ and i_ℓ^1. Hence in this sense, (4.19) is a simplified version of (4.14). In the next theorem, we show that the policies (4.22) and (4.23) derived from the simplified dynamic programming equation (4.19) also provide an optimal policy.

THEOREM 4.4 *Assume that* (4.1)-(4.7) *hold. Then*

$$
V_1(x_1, s_0, i_1^1) = \hat{V}_1(x_1 + s_0, i_1^1) + H_1(x_1),
$$

and

$$\left(\left(\hat{F}_1, ..., \hat{F}_N\right), \left(\hat{S}_1, ..., \hat{S}_N\right)\right) \tag{4.24}$$

given in (4.22)–(4.23) *is an optimal policy for the problem over* $\langle 1, N \rangle$.

Proof Let $\{V_n(x_n, s_{n-1}, i_n^1); 1 \le n \le N\}$ and $\{\hat{V}_n(y_n, i_n^1); 1 \le n \le N\}$ be functions defined by (4.13) and (4.20)–(4.21), respectively. First, we show that

$$V_N(x_N, s_{N-1}, i_N^1) = H_N(x_N) + \hat{V}_N(x_N + s_{N-1}, i_N^1). \tag{4.25}$$

From the definition of $\hat{V}_N(y_N, i_N^1)$, we have

$$\begin{aligned}
&\hat{V}_N\left(x_N + s_{N-1}, i_N^1\right)\\
&= \inf_{\substack{y \ge x_N + s_{N-1}\\ z \ge y}} \Big\{ K_N^f \cdot \delta(y - [x_N + s_{N-1}]) + C_N^f(y - (x_N + s_{N-1}))\\
&\quad + K_N^s \cdot \delta(z - y) + C_N^s(z - y)\\
&\quad + \mathsf{E}\left[H_{N+1}\left(y - g(i_N^1, I_N^2, v_N)\right)\right] \Big\}\\
&= \inf_{\substack{\phi \ge 0\\ \sigma \ge 0}} \Big\{ K_N^f \cdot \delta(\phi) + C_N^f(\phi) + K_N^s \cdot \delta(\sigma) + C_N^s(\sigma)\\
&\quad + \mathsf{E}\left[H_{N+1}\left(x_N + s_{N-1} + \phi - g_N(i_N^1, I_N^2, v_N)\right)\right] \Big\}\\
&= H_N(x_N) + \inf_{\substack{\phi \ge 0\\ \sigma \ge 0}} \Big\{ K_N^f \cdot \delta(\phi) + C_N^f(\phi) + K_N^s \cdot \delta(\sigma) + C_N^s(\sigma)\\
&\quad + \mathsf{E}\left[H_{N+1}\left(x_N + s_{N-1} + \phi - g_N(i_N^1, I_N^2, v_N)\right)\right] \Big\}\\
&\quad - H_N(x_N)\\
&= V_N\left(x_N, s_{N-1}, i_N^1\right) - H_N(x_N),
\end{aligned} \tag{4.26}$$

which is equivalent to (4.25). Furthermore, from the second equality of (4.26), we have

$$\bar{\phi}_N(x_N, s_{N-1}, i_N^1) = \hat{\phi}_N(x_N + s_{N-1}, i_N^1) - (x_N + s_{N-1}), \tag{4.27}$$

where $\bar{\phi}_N(x_N, s_{N-1}, i_N^1)$ is given by Theorem 4.1. Now suppose that for $j = N, ..., k+1$,

$$\begin{cases}
V_j\left(x_j, s_{j-1}, i_j^1\right) = \hat{V}_j\left(x_j + s_{j-1}, i_j^1\right) + H_j(x_j),\\
\bar{\phi}_j\left(x_j, s_{j-1}, i_j^1\right) = \hat{\phi}_j\left(x_j + s_{j-1}, i_j^1\right) - \left(x_j + s_{j-1}\right),\\
\bar{\sigma}_j\left(x_j, s_{j-1}, i_j^1\right) = \hat{\sigma}_j\left(x_j + s_{j-1}, i_j^1\right) - \hat{\phi}_j\left(x_j + s_{j-1}, i_j^1\right),
\end{cases} \tag{4.28}$$

where $\bar{\phi}_j(x_j, s_{j-1}, i_j^1)$ and $\bar{\sigma}_j(x_j, s_{j-1}, i_j^1)$ are given by Theorem 4.1. Next we show that (4.28) holds for $j = k$. By the definition of $\hat{V}_k(y_k, i_k^1)$,

$$
\begin{aligned}
&\hat{V}_k\left(x_k + s_{k-1}, i_k^1\right) \\
&= \inf_{\substack{y \geq x_k + s_{k-1} \\ z \geq y}} \Big\{ K_k^f \cdot \delta\left(y - x_k - s_{k-1}\right) \\
&\qquad + C_k^f\left(y - x_k - s_{k-1}\right) + K_k^s \cdot \delta(z - y) + C_k^s(z - y) \\
&\qquad + \mathsf{E}\left[H_{k+1}\left(y - g_k(i_k^1, I_k^2, v_k)\right)\right] \\
&\qquad + \mathsf{E}\left[\hat{V}_{k+1}\left(z - g_k(i_k^1, I_k^2, v_k), I_{k+1}^1\right)\right] \Big\} \\
&= \inf_{\substack{\phi \geq 0 \\ \sigma \geq 0}} \Big\{ K_k^f \cdot \delta(\phi) + C_k^f(\phi) + K_k^s \cdot \delta(\sigma) + C_k^s(\sigma) \\
&\qquad + \mathsf{E}\left[H_{k+1}\left(x_k + s_{k-1} + \phi - g_k(i_k^1, I_k^2, v_k)\right)\right] \\
&\qquad + \mathsf{E}\left[\hat{V}_{k+1}\left(x_k + s_{k-1} + \phi - g_k(i_k^1, I_k^2, v_k) + \sigma, I_{k+1}^1\right)\right] \Big\} \\
&= \inf_{\substack{\phi \geq 0 \\ \sigma \geq 0}} \Big\{ K_k^f \cdot \delta(\phi) + C_k^f(\phi) + K_k^s \cdot \delta(\sigma) + C_k^s(\sigma) \\
&\qquad + \mathsf{E}\left[V_{k+1}\left(x_k + s_{k-1} + \phi - g_k(i_k^1, I_k^2, v_k), \sigma, I_{k+1}^1\right)\right] \Big\} \\
&= H_k(x_k) + \inf_{\substack{\phi \geq 0 \\ \sigma \geq 0}} \Big\{ K_k^f \cdot \delta(\phi) + C_k^f(\phi) + K_k^s \cdot \delta(\sigma) + C_k^s(\sigma) \\
&\qquad + \mathsf{E}\left[V_{k+1}\left(x_k + s_{k-1} + \phi - g_k(i_k^1, I_k^2, v_k), \sigma, I_{k+1}^1\right)\right] \Big\} \\
&\quad - H_k(x_k) \\
&= V_k\left(x_k, s_{k-1}, i_k^1\right) - H_k(x_k),
\end{aligned}
\tag{4.29}
$$

where in establishing the third equality of (4.29), we apply the first equation of (4.28) for $j = k + 1$. Thus, we obtain the first equation of (4.28) for $j = k$. At the same time, from the second equality in (4.29), we have the second and the third equations of (4.28) for $j = k$. Therefore, by induction, the theorem is established. □

For an (s, S) type policy to be optimal, we assume that the slow- and fast-order cost functions $C_k^s(\cdot)$ and $C_k^f(\cdot)$ are linear—that is,

$$\begin{cases} C_k^f(t) = c_k^f \cdot t, \;\; C_k^s(t) = c_k^s \cdot t, \;\; k = 1, ..., N, \\ c_k^f \geq c_{k-1}^s, \;\; 2 \leq k \leq N. \end{cases} \tag{4.30}$$

Furthermore, we assume that for $k = 1, ..., N$,

$$\lim_{y \to +\infty} \left\{ (c_k^f \wedge c_{k-1}^s) y + \mathsf{E}\left[H_{k+1}\left(y - g_k(i_k^1, I_k^1, v_k) \right) \right] \right\} = +\infty \tag{4.31}$$

with $c_0^s = 1$.

REMARK 4.2 Condition (4.31) rules out some trivial or unrealistic cases. It requires that both holding and ordering costs are not negligible. Thus with (4.31), placing an infinitely large order from either the slow or the fast mode cannot be optimal. Condition (4.31) extends a similar condition required in Sethi and Cheng [15], which, in turn, generalizes the classical assumption of a strictly positive unit holding cost made by Scarf [13].

Next we define K-convex functions required for further analysis of the problem. Some well-known results on K-convex functions are collected in Lemma 4.1.

DEFINITION 4.1 *A function $h(x) : R \to R$ is said to be K-*convex *if it satisfies*

$$K + h(z + y) \geq h(y) + z \frac{h(y) - h(y - x)}{x}$$

for any $y \in R$, $z \geq 0$ and $x > 0$.

LEMMA 4.1 (i) *If $h(x) : R \to R$ is K-convex, it is M-convex for any $M \geq K$. In particular, if $h(x)$ is convex—that is, 0-convex—it is also K-convex for any $K \geq 0$.*

(ii) *If $h_1(x)$ is K-convex and $h_2(x)$ is M-convex, then for $a \geq 0$ and $b \geq 0$, $ah_1(x) + bh_2(x)$ is $(aK + bM)$-convex.*

(iii) *If $h(x)$ is K-convex and ξ is a random variable such that $\mathsf{E}|h(x - \xi)| < \infty$, then $\mathsf{E}[h(x - \xi)]$ is also K-convex.*

Proof Their proofs can be found in Bensoussan, Crouhy, and Proth [2]. Here for the completeness, we present the proofs. (i) is trivial from the definition. For (ii), note that for any $y \in R$, $z \geq 0$ and $x > 0$,

$$K + h_1(z + y) \geq h_1(y) + z \frac{h_1(y) - h_1(y - x)}{x}, \tag{4.32}$$

and

$$M + h_2(z+y) \geq h_2(y) + z\frac{h_2(y) - h_2(y-x)}{x}. \tag{4.33}$$

Consequently,

$$\begin{aligned}
&(aK + bM) + [a \cdot h_1(z+y) + b \cdot h_2(z+y)] \\
&\geq [a \cdot h_1(y) + b \cdot h_2(y)] \\
&\quad + z\frac{[a \cdot h_1(y) + b \cdot h_2(y)] - [a \cdot h_1(y-x) + b \cdot h_2(y-x)]}{x},
\end{aligned}$$

which implies that $a \cdot h_1(x) + b \cdot h_2(x)$ is $(aK + bM)$-convex.

(iii) follows directly from

$$K + h(z + y - \xi) \geq h(y - \xi) + z\frac{h(y-\xi) - h(y - x - \xi)}{x}.$$

□

Let

$$h^* = \inf_{-\infty < x < +\infty} h(x).$$

Define s and S with $s \leq S$ as follows:

$$S = \begin{cases} \inf\{x : h(x) = h^*\}, & \text{if } \{x : h(x) = h^*\} \neq \emptyset, \\ +\infty, & \text{if } \{x : h(x) = h^*\} = \emptyset \text{ and } h(-\infty) \neq h^*, \\ -\infty, & \text{if } \{x : h(x) = h^*\} = \emptyset \text{ and } h(-\infty) = h^*, \end{cases}$$

and

$$s = \begin{cases} \inf\{x : h(x) = K + h(S) \text{ and } x < S\}, & \\ \qquad\qquad \text{if } \{x : h(x) = K + h(S) \text{ and } x < S\} \neq \emptyset, & \\ -\infty, \quad \text{otherwise}, & \end{cases}$$

where

$$h(+\infty) = \liminf_{x \to +\infty} h(x) \text{ and } h(-\infty) = \liminf_{x \to -\infty} h(x). \tag{4.34}$$

Then we have the following lemma, which represents an extension of Proposition 4.2 of Sethi and Cheng [15] requiring a stronger condition $\lim_{y \to +\infty} h(y) = +\infty$.

LEMMA 4.2 *If $h(x)$ is a continuous K-convex function, then*

(i) $h(S) = \inf\{h(x)\}$ *with the convention given in* (4.34) *when* $S = +\infty$ *or* $S = -\infty$;

(ii) $h(x) \le K + h(y)$ *for any x and y with $s \le x \le y$;*
(iii) *the function*

$$\begin{aligned} m(x) &= \inf_{y \ge x} \left\{ K \cdot \delta(y-x) + h(y) \right\} \\ &= \begin{cases} K + h(S), & \textit{for} \;\; x \le s, \\ h(x), & \textit{for} \;\; x > s, \end{cases} \end{aligned}$$

is also K-convex.
Furthermore, if $s > -\infty$, then we have
(iv) $h(s) = K + h(S)$;
(v) $h(\cdot)$ *is strictly decreasing on* $(-\infty,\ s]$.

Proof From the definitions of S and s and the continuity of $h(\cdot)$, it is easy to see that (i) and (iv) hold. Now we prove (ii). Let us consider all possible cases as follows:

Case ii.1: $[S = -\infty]$

In this case, clearly we have $s = -\infty$. If $-\infty = s = S < x \le y$, then from the definition of S, there exists a z with $z < x$, such that

$$h(z) \le h(x). \tag{4.35}$$

By the K-convexity of $h(\cdot)$,

$$K + h(y) \ge h(x) + \frac{y-x}{x-z} \left[h(x) - h(z) \right].$$

Therefore, using (4.35) we get

$$h(x) \le h(y) + K, \tag{4.36}$$

which is (ii).

Case ii.2: $[S > -\infty$ and $s = -\infty]$

The proof for this case is divided into two subcases according to $x \le S$ and $x > S$.

Subcase ii.2.1: $[x \le S]$

We first show that for any $x \le S$,

$$h(x) \le K + h(S). \tag{4.37}$$

Noting that $s = -\infty$, we know that if

$$\{x : h(x) = K + h(S) \text{ and } x < S\} = \emptyset,$$

then for all $x \le S$,

$$h(x) < K + h(S), \tag{4.38}$$

and if

$$\{x : h(x) = K + h(S) \text{ and } x < S\} \neq \emptyset, \tag{4.39}$$

then

$$\liminf_{x \to -\infty} h(x) \leq K + h(S). \tag{4.40}$$

Thus to prove (4.37) from (4.38), it suffices to show that (4.37) holds under (4.39). Suppose to the contrary that under (4.39) there is an x_0 with $x_0 < S$, such that

$$h(x_0) > K + h(S). \tag{4.41}$$

It follows from the continuity of $h(\cdot)$ and (4.40) that there are $\bar{x}_0$ $(< x_0)$ and $\hat{x}_0$ $(> x_0)$ such that

$$h(x_0) > h(\bar{x}_0) \geq K + h(\hat{x}_0). \tag{4.42}$$

Using (4.42) and the K-convexity of $h(\cdot)$, we have

$$h(\bar{x}_0) \geq K + h(\hat{x}_0) \geq h(x_0) + \frac{\hat{x}_0 - x_0}{x_0 - \bar{x}_0}\left[h(x_0) - h(\bar{x}_0)\right]. \tag{4.43}$$

Consequently,

$$(\hat{x}_0 - \bar{x}_0) \cdot h(\bar{x}_0) \geq (\hat{x}_0 - \bar{x}_0) \cdot h(x_0), \tag{4.44}$$

implying that

$$h(\bar{x}_0) \geq h(x_0),$$

which contradicts (4.42). Therefore, we have (4.37) under (4.39). Hence if $x \leq S$, by the definition of S and (4.37),

$$h(x) \leq K + h(S) \leq K + h(y), \tag{4.45}$$

which is (ii).

Subcase ii.2.2: $[S < x \leq y]$

By the K-convexity of $h(\cdot)$,

$$K + h(y) \geq h(x) + \frac{y - x}{x - S}\left[h(x) - h(S)\right].$$

Again by the definition of S, we have

$$h(x) \leq K + h(y), \tag{4.46}$$

which is (ii).

Combining Subcases ii.2.1 and ii.2.2, we know that (ii) holds in Case ii.2.

Case ii.3: $[s > -\infty]$

The proof of (ii) in this case is divided into two subcases according to $S = +\infty$ and $S < +\infty$.

Subcase ii.3.1: $[S = +\infty]$

For any given x $(x > s)$, from the definition of S, there exists a sequence $\{y_n : n \geq 1\}$ such that

$$\begin{cases} y_n > x, \quad \text{for all } n \geq 1, \\ \lim_{n\to\infty} y_n = +\infty, \\ h(y_n) = h(S) + \frac{1}{2^n}. \end{cases}$$

It follows from the convexity of $h(\cdot)$ and the definition of s that for all $n \geq 1$,

$$h(s) + \frac{1}{2^n} = K + h(y_n) \geq h(x) + \frac{y_n - x}{x - s}\left[h(x) - h(s)\right]. \tag{4.47}$$

Consequently, for all $n \geq 1$,

$$h(s) + \frac{x - s}{2^n(y_n - s)} \geq h(x), \tag{4.48}$$

which implies that

$$h(x) \leq h(s). \tag{4.49}$$

Therefore,

$$h(x) \leq K + h(S) \leq K + h(y).$$

This is (ii).

Subcase ii.3.2: $[S < +\infty]$

The proof for this subcase is the same as the proof of (2.12) in Bensoussan, Crouhy, and Proth [2], p.318. For the completeness, here we present it. It holds when $x = y$, or $x = s$, or $x = S$. So it suffices to consider

$$S < x < y \text{ or } s < x < S.$$

If $S < x < y$, by K-convexity,

$$\begin{aligned} K + h(y) &\geq h(x) + \frac{y - x}{x - S}[h(x) - h(S)] \\ &\geq h(x). \end{aligned}$$

Hence (ii) holds.

If $s < x < S$, by again K-convexity,

$$\begin{aligned} h(s) &= K + h(S) \\ &\geq h(x) + \frac{S-x}{x-s}[h(x) - h(s)], \end{aligned}$$

which implies

$$h(s) \cdot \left(1 + \frac{S-x}{x-s}\right) \geq h(x) \cdot \left(1 + \frac{S-x}{x-s}\right).$$

Consequently, $h(s) \geq h(x)$. Therefore,

$$h(x) \leq K + h(S) \leq h(y) + K,$$

this is equivalent to (ii).

With (ii) in hand, the proof of (iii) is similar to the proof of Proposition 2.2 in Bensoussan, Crouhy, and Proth [2], p. 319. The following is a proof basically borrowed from Bensoussan, Crouhy, and Proth [2].

If $s = -\infty$, for any $y \geq x$,

$$K + h(y) \geq K + h(S) > h(x).$$

Thus, $m(x) = h(x)$, this gives the K-convexity of $m(x)$. So it suffices to consider $s > -\infty$. If $x < s$, we have

$$g(x) > K + h(S).$$

But $S \geq x$; therefore,

$$\begin{aligned} &\inf_{y \geq x} \left\{ K \cdot \delta(y-x) + h(y) \right\} \\ &= \inf_{y > x} \left\{ K \cdot \delta(y-x) + h(y) \right\}. \end{aligned} \tag{4.50}$$

It follows from the definition of S that

$$\inf_{y > x} \left\{ K \cdot \delta(y-x) + h(y) \right\} = K + h(S). \tag{4.51}$$

If $x \geq s$, it follows from (ii) that

$$\inf_{y \geq x} \left\{ K \cdot \delta(y-x) + h(y) \right\} = h(x). \tag{4.52}$$

Combining (4.51)–(4.52) we get the first part of (iii).

Now we show the K-convexity of $m(\cdot)$. By the definition of K-convexity, it suffices to show that for any $y \in R$, $z \geq 0$ and $x > 0$,

$$K + m(z+y) \geq m(y) + z\frac{m(y) - m(y-x)}{x}. \tag{4.53}$$

The proof will be divided into three cases.

Case iii.1: $[\, y \geq s]$

If $y - x \geq s$, then

$$m(z+y) = h(z+y), \quad m(y) = h(y), \quad m(y-x) = h(y-x).$$

This clearly implies (4.53) by the K-convexity of $h(\cdot)$.

If $y - x < s$, then (4.53) is reduced into

$$K + h(z+y) \geq h(y) + z\frac{h(y) - h(s)}{x}. \tag{4.54}$$

On the other hand, it follows from the K-convexity of $h(\cdot)$ and (ii) that

$$K + h(z+y) \geq h(y) + z\frac{h(y) - h(s)}{y - s}, \tag{4.55}$$

$$K + h(y+z) \geq h(y). \tag{4.56}$$

Thus if $h(y) \geq h(s)$, then (4.54) follows from $x \geq y - s$ and (4.55). If $h(y) < h(s)$, then (4.54) directly follows from (4.56).

Case iii.2: $[y + z \geq s \geq y]$

Note that

$$K + h(y+z) \geq K + h(S) = h(s),$$

and

$$m(z+y) = h(z+y), \quad m(y) = h(s), \quad m(y-x) = h(s).$$

These obviously imply (4.53).

Case iii.3: $[y + z \leq s]$

Note that

$$m(z+y) = h(s), \quad m(y) = h(s), \quad m(y-x) = h(s).$$

Then (4.53) amounts to

$$K + h(s) \geq h(s).$$

This is trivially true.

Finally, we prove (v). It suffices to show that for any x and y with $-\infty < x < y \leq s$,

$$h(x) > h(y). \tag{4.57}$$

It follows from the K-convexity of $h(\cdot)$ that for any $x < y \leq s$,

$$K + h(S) \geq h(y) + \frac{S-y}{y-x}\left[h(y) - h(x)\right]. \tag{4.58}$$

On the other hand, by the definition of s we have that if $y < s$,

$$h(y) > K + h(S). \tag{4.59}$$

Consequently, by (4.58) and (4.59), we know that if $x < y < s$, then (4.57) holds. Hence to prove (4.57), it suffices to show that if $x < y = s$, then

$$h(x) > h(s). \tag{4.60}$$

Similar to (4.58), we have

$$K + h(S) \geq h(s) + \frac{S-s}{s-x}\left[h(s) - h(x)\right]. \tag{4.61}$$

Thus from the definition of s, we obtain that if $s \neq S$, then (4.60) holds. However, if $s = S$, then $h(\cdot)$ is convex. Consequently, we also have (4.60). Therefore, we get (v). □

In view of (4.30), (4.19) can be written as

$$\left\{\begin{array}{l} \tilde{U}_k(y_k, i_k^1) \\ = \inf_{\substack{y \geq y_k \\ z \geq y}} \Big\{ K_k^f \cdot \delta(y - y_k) + c_k^f \cdot [y - y_k] + K_k^s \cdot \delta(z - y) \\ \qquad + c_k^s \cdot [z - y] + \mathsf{E}\left[H_{k+1}\left(y - g_k(i_k^1, I_k^2, v_k)\right)\right] \\ \qquad + \mathsf{E}\left[\tilde{U}_{k+1}\left(z - g_k(i_k^1, I_k^2, v_k), I_{k+1}^1\right)\right] \Big\}, \\ \qquad k = 1, ..., N-1, \\ \\ \tilde{U}_N(y_N, i_N^1) \\ = \inf_{\substack{y \geq y_N \\ z \geq y}} \Big\{ K_N^f \cdot \delta(y - y_N) + c_N^f \cdot [y - y_N] \\ \qquad + K_N^s \cdot \delta(z - y) + c_N^s \cdot [z - y] \\ \qquad + \mathsf{E}\left[H_{N+1}\left(y - g_N(i_N^1, I_N^2, v_N)\right)\right] \Big\}. \end{array}\right. \tag{4.62}$$

Let $\tilde{V}_k(y_k, i_k^1)$ be the solution of (4.62). Then we have the following result.

THEOREM 4.5 *Assume that* (4.1)–(4.5), *and* (4.30)–(4.31) *hold. Assume that the initial inventory level at the beginning of period* k $(1 \leq k \leq N-1)$ *is* x_k, *the slow-order quantity in period* $(k-1)$ *is* s_{k-1}, *and the observed value of* I_k^1 *is* i_k^1 *in period* $(k-1)$. *Then there exist numbers* ϕ_k, Φ_k, σ_k, *and* Σ_k *with* $\phi_k \leq \Phi_k$ *and* $\sigma_k \leq \Sigma_k$, *which do not depend on the inventory position* $x_k + s_{k-1}$, *such that the optimal fast-order quantity* $\tilde{F}_k$ *and the optimal slow-order quantity* $\tilde{S}_k$ *in period* k *can be determined by the following expressions*:

$$\tilde{F}_k = \begin{cases} \Phi_k - (x_k + s_{k-1}), & \text{if } x_k + s_{k-1} \leq \phi_k, \\ 0, & \text{if } x_k + s_{k-1} > \phi_k, \end{cases}$$

$$\tilde{S}_k = \begin{cases} \Sigma_k - (x_k + s_{k-1} + \tilde{F}_k), & \text{if } x_k + s_{k-1} + \tilde{F}_k \leq \sigma_k, \\ 0, & \text{if } x_k + s_{k-1} + \tilde{F}_k > \sigma_k. \end{cases}$$

Finally, if the initial inventory level at the beginning of the last period is x_N, *the slow-order quantity in period* $(N-1)$ *is* s_{N-1}, *and the observed value of* I_N^1 *is* i_N^1 *in period* $(N-1)$, *then there exist numbers* ϕ_N *and* Φ_N *with* $\phi_N \leq \Phi_N$, *which do not depend on the inventory position* $x_N + s_{N-1}$, *such that the optimal fast-order quantity in the last period is given by*

$$\tilde{F}_N = \begin{cases} \Phi_N - (x_N + s_{N-1}), & \text{if } x_N + s_{N-1} \leq \phi_N, \\ 0, & \text{if } x_N + s_{N-1} > \phi_N. \end{cases}$$

Proof First, we consider period N. By (4.4) and Lemma 4.1, we know that the function

$$c_N^f y + \mathsf{E}\left[H_{N+1}(y - g_N(i_N^1, I_N^2, v_N))\right]$$

is convex in y. Furthermore, from (4.31),

$$\lim_{y \to +\infty} \left(c_N^f \cdot y + \mathsf{E}\left[H_{N+1}(y - g_N(i_N^1, I_N^2, v_N))\right] \right) = +\infty. \tag{4.63}$$

By the last equation of (4.62),

$$\tilde{V}_N(y_N, i_N^1) = -c_N^f y_N + \inf_{y \geq y_N} \Big\{ K_N^f \cdot \delta(y - y_N) + c_N^f y + \mathsf{E}\left[H_{N+1}\left(y - g_N(i_N^1, I_N^2, v_N)\right)\right] \Big\}, \tag{4.64}$$

where $y_N = x_N + s_{N-1}$. It follows from (4.63) and Lemma 4.2 (iii) that there are ϕ_N and Φ_N with $\phi_N \le \Phi_N < \infty$, which are independent of y_N, such that

$$\inf_{y \ge y_N} \left\{ K_N^f \cdot \delta(y - y_N) + c_N^f y + \mathsf{E}\left[H_{N+1}(y - g_N(i_N^1, I_N^2, v_N))\right] \right\}$$
$$= \begin{cases} K_N^f + c_N^f \Phi_N + \mathsf{E}\left[H_{N+1}(\Phi_N - g_N(i_N^1, I_N^2, v_N))\right], \\ \qquad \text{if } y_N \le \phi_N, \\ c_N^f y_N + \mathsf{E}\left[H_{N+1}(y_N - g_N(i_N^1, I_N^2, v_N))\right], \\ \qquad \text{if } y_N > \phi_N. \end{cases} \tag{4.65}$$

Therefore, from (4.64) and Theorem 4.4, we have that the optimal fast-order quantity $\tilde{F}_N$ in period N is given by

$$\tilde{F}_N = \begin{cases} \Phi_N - y_N, & \text{if } y_N \le \phi_N, \\ 0, & \text{otherwise,} \end{cases} \tag{4.66}$$

and

$$\tilde{V}_N(y_N, i_N^1)$$
$$= \begin{cases} K_N^f + c_N^f \cdot (\Phi_N - y_N) + \mathsf{E}\left[H_{N+1}(\Phi_N - g_N(i_N^1, I_N^2, v_N))\right], \\ \qquad \text{if } y_N \le \phi_N, \\ \mathsf{E}\left[H_{N+1}(y_N - g_N(i_N^1, I_N^2, v_N))\right], \\ \qquad \text{if } y_N > \phi_N. \end{cases} \tag{4.67}$$

This proves the theorem in period N. At the same time, by Lemma 4.2 we also know that

$$\tilde{V}_N(y, i_N^1) \text{ is a nonnegative, continuous } K_N^f\text{-convex function of } y. \tag{4.68}$$

By (4.65) we know that ϕ_N and Φ_N depend on i_N^1. When we need to stress this dependence, we sometimes write ϕ_N and Φ_N as $\phi_N(i_N^1)$ and $\Phi_N(i_N^1)$, respectively.

Now consider period $(N-1)$. First, by (4.67)–(4.68),

$$\lim_{z \to +\infty} \left(c_{N-1}^s z + \mathsf{E}\left[\tilde{V}_N(z - g_{N-1}(i_{N-1}^1, I_{N-1}^2, v_{N-1}), I_N^1)\right] \right)$$

$$
\begin{aligned}
&= \lim_{z\to+\infty} \Big(c^s_{N-1} z \\
&\qquad + \mathsf{E}\Big\{ \delta\left(\left[z - g_{N-1}(i^1_{N-1}, I^2_{N-1}, v_{N-1}) \right] - \phi_N(I^1_N) \right) \cdot \\
&\qquad \cdot \mathsf{E}\left[H_{N+1}\left(z - g_{N-1}(i^1_{N-1}, I^2_{N-1}, v_{N-1}) \right.\right. \\
&\qquad\qquad \left.\left. - g_N(I^1_N, I^2_N, v_N) \right) \middle| (I^2_{N-1}, I^1_N) \right] \Big\} \\
&\qquad + \mathsf{E}\Big\{ \delta\Big(\phi_N(I^1_N) - \left[z - g_{N-1}(i^1_{N-1}, I^2_{N-1}, v_{N-1}) \right] \Big) \cdot \\
&\qquad \cdot \Big(K^f_N + c^f_N \cdot \Big[\Phi_N(I^1_N) - \left[z - g_{N-1}(i^1_{N-1}, I^2_{N-1}, v_{N-1}) \right] \Big] \\
&\qquad + \mathsf{E}\left[H_{N+1}(\Phi_N(I^1_N) - g_N(I^1_N, I^2_N, v_N)) \middle| (I^2_{N-1}, I^1_N) \right] \Big) \Big\} \Big) \\
&\geq \lim_{z\to+\infty} \Big(c^s_{N-1} z \\
&\qquad + \mathsf{E}\Big\{ \delta\Big(\left[z - g_{N-1}(i^1_{N-1}, I^2_{N-1}, v_{N-1}) \right] - \phi_N(I^1_N) \Big) \cdot \\
&\qquad \cdot \mathsf{E}\left[H_{N+1}\left(z - g_{N-1}(i^1_{N-1}, I^2_{N-1}, v_{N-1}) \right.\right. \\
&\qquad\qquad \left.\left. - g_N(I^1_N, I^2_N, v_N) \right) \middle| (I^2_{N-1}, I^1_N) \right] \Big\} \Big) \\
&\geq \lim_{z\to+\infty} \Big(c^s_{N-1} z + \int_{-\infty}^{+\infty} \mathrm{d}\Lambda_N(i) \cdot \\
&\qquad \cdot \int_0^{z-\phi_N(i)} \mathsf{E}\left[H_{N+1}\left(z - x - g_N(i, I^2_N, v_N) \right) \right] \mathrm{d}\Psi_{N-1}(x | i^1_{N-1}) \Big) \\
&\geq \lim_{z\to+\infty} \int_{-\infty}^{+\infty} \mathrm{d}\Lambda_N(i) \Big\{ \int_0^{z-|\phi_N(i)|} \Big(c^s_{N-1} \cdot [z - x] \\
&\qquad + \mathsf{E}\left[H_{N+1}\left(z - x - g_N(i, I^2_N, v_N) \right) \right] \Big) \mathrm{d}\Psi_{N-1}(x | i^1_{N-1}) \Big\} \\
&= +\infty \quad \text{(by (4.31))}.
\end{aligned}
\tag{4.69}
$$

Consequently, Lemma 4.2 gives that there are σ_{N-1} and Σ_{N-1} with $\sigma_{N-1} \leq \Sigma_{N-1} < +\infty$, such that

$$
\begin{aligned}
\inf_{z\geq y} \Big\{ & K^s_{N-1} \cdot \delta(z - y) + c^s_{N-1} z \\
& + \mathsf{E}\left[\tilde{V}_N\left(z - g_{N-1}(i^1_{N-1}, I^2_{N-1}, v_{N-1}), I^1_N \right) \right] \Big\}
\end{aligned}
$$

$$= \begin{cases} K^s_{N-1} + c^s_{N-1}\Sigma_{N-1} \\ \quad + \mathsf{E}\left[\tilde{V}_N\left(\Sigma_{N-1} - g_{N-1}(i^1_{N-1}, I^2_{N-1}, v_{N-1}), I^1_N\right)\right], \\ \qquad\qquad \text{if } y \le \sigma_{N-1}, \\ c^s_{N-1}y + \mathsf{E}\left[\tilde{V}_N\left(y - g_{N-1}(i^1_{N-1}, I^2_{N-1}, v_{N-1}), I^1_N\right)\right], \\ \qquad\qquad \text{otherwise}. \end{cases} \tag{4.70}$$

We write the left-hand side of (4.70) as $L_{N-1}(y, i^1_{N-1})$. Then by (4.5), (4.68), and Lemma 4.1, we know that

$$L_{N-1}(y, i^1_{N-1}) \text{ is also a nonnegative, continuous } K^s_{N-1}\text{-}\textit{convex function of } y. \tag{4.71}$$

Furthermore, σ_{N-1} and Σ_{N-1} also depend on i^1_{N-1}, written sometimes as $\sigma_{N-1}(i^1_{N-1})$ and $\Sigma_{N-1}(i^1_{N-1})$, respectively. Then

$$\begin{aligned} &\lim_{y\to+\infty}\Big\{(c^f_{N-1} - c^s_{N-1})\cdot y + \mathsf{E}\left[H_N\left(y - g_{N-1}(i^1_{N-1}, I^2_{N-1}, v_{N-1})\right)\right] \\ &\qquad + L_{N-1}(y, i^1_{N-1})\Big\} \\ &\ge \lim_{y\to+\infty}\Big\{(c^f_{N-1} - c^s_{N-1})\cdot y \\ &\qquad + \mathsf{E}\left[H_N\left(y - g_{N-1}(i^1_{N-1}, I^2_{N-1}, v_{N-1})\right)\right] \\ &\qquad + \delta(y - \sigma_{N-1})\cdot L_{N-1}(y, i^1_{N-1})\Big\} \\ &= \lim_{y\to+\infty}\Big\{c^f_{N-1}y + \mathsf{E}\Big[H_N\left(y - g_{N-1}(i^1_{N-1}, I^2_{N-1}, v_{N-1})\right) \\ &\qquad + \tilde{V}_N\left(y - g_{N-1}(i^1_{N-1}, I^2_{N-1}, v_{N-1}), I^1_N\right)\Big]\Big\} \\ &\ge \lim_{y\to+\infty}\Big\{c^f_{N-1}y + \mathsf{E}\left[H_N\left(y - g_{N-1}(i^1_{N-1}, I^2_{N-1}, v_{N-1})\right)\right]\Big\} \\ &= +\infty \ \text{(by (4.31))}. \end{aligned} \tag{4.72}$$

Consequently, Lemma 4.2 implies that there exist ϕ_{N-1} and Φ_{N-1} with $\phi_{N-1} \le \Phi_{N-1} < \infty$, such that

$$\begin{aligned}
&\inf_{y \ge y_{N-1}} \Big\{ K^f_{N-1} \cdot \delta(y - y_{N-1}) + (c^f_{N-1} - c^s_{N-1}) \cdot y + L_{N-1}(y, i^1_{N-1}) \\
&\qquad + \mathsf{E}\left[H_N \left(y - g_{N-1}(i^1_{N-1}, I^2_{N-1}, v_{N-1}) \right) \right] \Big\} \\
&= \begin{cases}
K^f_{N-1} + (c^f_{N-1} - c^s_{N-1}) \cdot \Phi_{N-1} + L_{N-1}(\Phi_{N-1}, i^1_{N-1}) \\
+\mathsf{E}\left[H_N \left(\Phi_{N-1} - g_{N-1}(i^1_{N-1}, I^2_{N-1}, v_{N-1}) \right) \right], \\
\qquad \text{if } y_{N-1} \le \phi_{N-1}, \\
(c^f_{N-1} - c^s_{N-1}) \cdot y_{N-1} + L_{N-1}(y_{N-1}, i^1_{N-1}) \\
+\mathsf{E}\left[H_N \left(y_{N-1} - g_{N-1}(i^1_{N-1}, I^2_{N-1}, v_{N-1}) \right) \right], \\
\qquad \text{otherwise,}
\end{cases}
\end{aligned} \tag{4.73}$$

where $y_{N-1} = x_{N-1} + s_{N-2}$. Clearly, ϕ_{N-1} and Φ_{N-1} also depend on i^1_{N-1}. When we need to stress this dependence, we write ϕ_{N-1} and Φ_{N-1} as $\phi_{N-1}(i^1_{N-1})$ and $\Phi_{N-1}(i^1_{N-1})$, respectively. Note that from the first equation of (4.62) for $k = N - 1$,

$$\begin{aligned}
&\tilde{V}_{N-1}(y_{N-1}, i^1_{N-1}) \\
&\quad = -c^f_{N-1} y_{N-1} + \inf_{y \ge y_{N-1}} \Big\{ K^f_{N-1} \cdot \delta(y - y_{N-1}) \\
&\qquad + (c^f_{N-1} - c^s_{N-1}) \cdot y \\
&\qquad + \mathsf{E}\left[H_N \left(y - g_{N-1}(i^1_{N-1}, I^2_{N-1}, v_{N-1}) \right) \right] \\
&\qquad + \inf_{z \ge y} \Big\{ K^s_{N-1} \cdot \delta(z - y) + c^s_{N-1} z \\
&\qquad + \mathsf{E}\left[\tilde{V}_N \left(z - g_{N-1}(i^1_{N-1}, I^2_{N-1}, v_{N-1}), I^1_N \right) \right] \Big\}\Big\}.
\end{aligned} \tag{4.74}$$

By Theorem 4.4, (4.70), and (4.73), we have that the optimal fast- and slow-order quantities $(\tilde{F}_{N-1}, \tilde{S}_{N-1})$ in period $(N - 1)$ are given, respectively, by

$$\tilde{F}_{N-1} = \begin{cases} \Phi_{N-1} - y_{N-1}, & \text{if } y_{N-1} \le \phi_{N-1}, \\ 0, & \text{otherwise,} \end{cases} \tag{4.75}$$

and

$$\tilde{S}_{N-1} = \begin{cases} \Sigma_{N-1} - (y_{N-1} + \tilde{F}_{N-1}), & \text{if } y_{N-1} + \tilde{F}_{N-1} \leq \sigma_{N-1}, \\ 0, & \text{otherwise}. \end{cases} \tag{4.76}$$

Consequently, we get the required result for period $(N-1)$. Repeating this procedure, we can get the theorem for any period k $(1 \leq k \leq N-2)$. □

The following corollary is an immediate consequence of Theorem 4.5.

COROLLARY 4.1 *Assume that* (4.1)–(4.5), *and* (4.30)–(4.31) *hold. Furthermore, assume that* $K_N^f = 0$ *and* $K_k^f = K_k^s = 0$, $k = 1, ..., N-1$. *Then*

$$\phi_k = \Phi_k \text{ and } \sigma_k = \Sigma_k, \ k = 1, ..., N-1,$$

and

$$\phi_N = \Phi_N.$$

Hence, if the initial inventory level at the beginning of period k $(1 \leq k \leq N-1)$ *is* x_k, *the slow-order quantity in period* $(k-1)$ *is* s_{k-1}, *and the observed value of* I_k^1 *is* i_k^1 *in period* $(k-1)$, *then the optimal fast-order quantity* $\tilde{F}_k$ *and the optimal slow-order quantity* $\tilde{S}_k$ *in period* k *can be determined by the following expressions*:

$$\begin{aligned} \tilde{F}_k &= \max\{\Phi_k - (x_k + s_{k-1}), 0\}, \\ \tilde{S}_k &= \max\{\Sigma_k - (x_k + s_{k-1} + \tilde{F}_k), 0\}. \end{aligned}$$

If the initial inventory level at the beginning of the last period is x_N, *the slow-order quantity in period* $(N-1)$ *is* s_{N-1}, *and the observed value of* I_N^1 *is* i_N^1 *in period* $(N-1)$, *then the optimal fast-order quantity in the last period is given by*

$$\tilde{F}_N = \max\{\Phi_N - (x_N + s_{N-1}), 0\}.$$

REMARK 4.3 The corollary states the optimality of the base-stock policy for fast and slow orders when there are no set-up costs. The base-stock levels Φ_k and Σ_k are independent of the inventory position $x_k + s_{k-1}$. Note that the optimal policy given by the corollary is the same as the one given in Theorem 3.5.

It is illuminating to observe that with respect to the inventory position y_k, the fast-order policy is an (s, S)-type policy with $s = \phi_k(i_k^1)$ and $S = \Phi_k(i_k^1)$.

The policy for the slow order is also an (s, S)-type policy but with respect to the "slow-order inventory position" $y_k + \tilde{F}_k$. Here $s = \sigma_k(i_k^1)$ and $S = \Sigma_k(i_k^1)$. Note that the slow-order quantity $\tilde{S}_k$ is decided after the fast-order quantity $\tilde{F}_k$ has been determined.

REMARK 4.4 It is easy to extend the model to allow (I_k^2, I_{k+1}^1) to depend on (I_{k-1}^2, I_k^1). This extension would require a state variable i_{k-1}^2 representing the value of I_{k-1}^2 in addition to the already existing state variables y_k and i_k^1 in the dynamic programming equation (4.19). Furthermore, extending the model to multiple updates, while straightforward, increases the dimension of the state space and makes the problem computationally more difficult.

From the proof we know that $\Sigma_k(i_k^1)$ is independent of c_k^f, K_k^f and K_k^s but depends on c_k^s; $\sigma_k(i_k^1)$ is independent of c_k^f and K_k^f but depends on c_k^s and K_k^s; $\Phi_k(i_k^1)$ is independent of K_k^f but depends on c_k^f; and $\phi_k(i_k^1)$ depends on c_k^f and K_k^f. Furthermore, we have the following monotonicity result.

THEOREM 4.6 *Assume that* (4.1)–(4.5), *and* (4.30)–(4.31) *hold. Fix k. Let $\Phi_k(i_k^1)$ and $\Sigma_k(i_k^1)$ be the minimum possible order-up-to levels specified in Theorem* 4.5. *For these $\Phi_k(i_k^1)$ and $\Sigma_k(i_k^1)$, let $\phi_k(i_k^1)$ and $\sigma_k(i_k^1)$ be the minimum possible reorder quantities specified in Theorem* 4.5. *Then*

(i) *$\Phi_k(i_k^1)$ and $\Sigma_k(i_k^1)$ are nonincreasing in c_k^f and c_k^s, respectively;*

(ii) *for $K \geq K_k^s$, $\phi_k(i_k^1)$ is nonincreasing in the fast-order fixed cost K, and for $K \geq K_{k+1}^f$, $\sigma_k(i_k^1)$ is nonincreasing in the slow-order fixed cost K.*

Theorem 4.6 (i) presents the intuitive notion that an increase in the fast-order unit cost decreases the fast order-up-to level. The same holds for the slow mode. Theorem 4.6 (ii) says that an increase in K_k^f decreases the fast-reorder point in period k, while a decrease does the opposite provided that the decreased value of K_k^f is not below K_k^s. This qualification is required to preserve the required K_k^f-convexity property of the kth period value function. A similar observation applies for the slow mode, but in this case K_k^s should not fall below K_{k+1}^f for the result to hold.

Proof of Theorem 4.6 Here we give only the proof for $\Sigma_k(i_k^1)$ and $\sigma_k(i_k^1)$. The proofs for $\Phi_k(i_k^1)$ and $\phi_k(i_k^1)$ are similar.

From the proof of Theorem 4.5 and (4.62), we know that $\Sigma_k(i_k^1)$ satisfies

$$c_k^s \cdot \Sigma_k(i_k^1) + \mathsf{E}\left[\tilde{V}_{k+1}\left(\Sigma_k(i_k^1) - g_k(i_k^1, I_k^2, v_k), I_{k+1}^1\right)\right]$$

$$= \inf_{w \geq 0}\left\{c_k^s w + \mathsf{E}\left[\tilde{V}_{k+1}\left(w - g_k(i_k^1, I_k^2, v_k), I_{k+1}^1\right)\right]\right\}, \quad (4.77)$$

and

$$\sigma_k(i_k^1) = \inf\Big\{w < \Sigma_k(i_k^1) : \mathsf{E}\left[\tilde{V}_{k+1}\left(w - g_k(i_k^1, I_k^2, v_k), I_{k+1}^1\right)\right] \\ + c_k^s w = K_k^s + c_k^s \cdot \Sigma_k(i_k^1) \\ + \mathsf{E}\left[\tilde{V}_{k+1}\left(\Sigma_k(i_k^1) - g_k(i_k^1, I_k^2, v_k), I_{k+1}^1\right)\right]\Big\}. \tag{4.78}$$

Furthermore, $c_k^s w + \mathsf{E}[V_{k+1}(w - g_k(i_k^1, I_k^2, v_k), I_{k+1}^1)]$ is decreasing on $(-\infty, \sigma_k(i_k^1))$. For $\varepsilon > 0$, let $\Sigma_k^\varepsilon(i_k^1)$ be the minimum value satisfying

$$(c_k^s + \varepsilon) \cdot \Sigma_k^\varepsilon(i_k^1) + \mathsf{E}\left[\tilde{V}_{k+1}\left(\Sigma_k^\varepsilon(i_k^1) - g_k(i_k^1, I_k^2, v_k), I_{k+1}^1\right)\right] \\ = \inf_{w \geq 0}\left\{(c_k^s + \varepsilon)w + \mathsf{E}\left[\tilde{V}_{k+1}\left(w - g_k(i_k^1, I_k^2, v_k), I_{k+1}^1\right)\right]\right\}, \tag{4.79}$$

and let $\sigma_k^\varepsilon(i_k^1)$ satisfy

$$\sigma_k^\varepsilon(i_k^1) = \inf\Big\{w < \Sigma_k(i_k^1) : \mathsf{E}\left[\tilde{V}_{k+1}\left(w - g_k(i_k^1, I_k^2, v_k), I_{k+1}^1\right)\right] \\ + c_k^s w = (K_k^s + \varepsilon) + c_k^s \cdot \Sigma_k(i_k^1) \\ + \mathsf{E}\left[\tilde{V}_{k+1}\left(\Sigma_k(i_k^1) - g_k(i_k^1, I_k^2, v_k), I_{k+1}^1\right)\right]\Big\}. \tag{4.80}$$

We need to show that

$$\Sigma_k^\varepsilon(i_k^1) \leq \Sigma_k(i_k^1) \text{ and } \sigma_k^\varepsilon(i_k^1) \leq \sigma_k(i_k^1). \tag{4.81}$$

If $\Sigma_k(i_k^1)$ also satisfies (4.79), then the first equation of (4.81) holds. On the other hand, if $\Sigma_k(i_k^1)$ does not satisfy (4.79), then

$$(c_k^s + \varepsilon) \cdot \Sigma_k^\varepsilon(i_k^1) + \mathsf{E}\left[\tilde{V}_{k+1}\left(\Sigma_k^\varepsilon(i_k^1) - g_k(i_k^1, I_k^2, v_k), I_{k+1}^1\right)\right] \\ \leq (c_k^s + \varepsilon) \cdot \Sigma_k(i_k^1) + \mathsf{E}\left[\tilde{V}_{k+1}\left(\Sigma_k(i_k^1) - g_k(i_k^1, I_k^2, v_k), I_{k+1}^1\right)\right]. \tag{4.82}$$

This means that

$$\left\{c_k^s \Sigma_k^\varepsilon(i_k^1) + \mathsf{E}\left[\tilde{V}_{k+1}\left(\Sigma_k^\varepsilon(i_k^1) - g_k(i_k^1, I_k^2, v_k), I_{k+1}^1\right)\right]\right\} \\ - \left\{c_k^s \cdot \Sigma_k(i_k^1) + \mathsf{E}\left[\tilde{V}_{k+1}\left(\Sigma_k(i_k^1) - g_k(i_k^1, I_k^2, v_k), I_{k+1}^1\right)\right]\right\} \\ \leq \varepsilon \cdot \left[\Sigma_k(i_k^1) - \Sigma_k^\varepsilon(i_k^1)\right]. \tag{4.83}$$

But from (4.77), we know that the left-hand side of (4.83) is nonnegative. Therefore, (4.83) implies the first equation of (4.81). The second equation of (4.81) follows from the monotonicity of $c_k^s w + \mathsf{E}[\tilde{V}_{k+1}(w - g_k(i_k^1, I_k^2, v_k), I_{k+1}^1)]$ on $(-\infty, \sigma_k(i_k^1))$. □

4.5. Monotonicity Properties

In this section, we show that the policy parameters $\phi_k(i_k^1)$, $\Phi_k(i_k^1)$, $\sigma_k(i_k^1)$, and $\Sigma_k(i_k^1)$ are monotone with respect to i_k^1, the realization of the first determinant I_k^1 of the demand. In addition to its intuitive appeal, this behavior can be used in numerical analysis of the optimal policy (see Brown and Lee [4]). To proceed, we introduce the notion of a stochastic order (for a detailed discussion, see Ross [12] and Shaked and Shanthikumar [17]).

DEFINITION 4.2 *Let (Z_1, Z_2) be a two-dimensional random vector, and let $\Psi(z_2|z_1)$ be the conditional distribution function of Z_2, given $Z_1 = z_1$—that is,*

$$\Psi(z_2|z_1) = \mathsf{P}(Z_2 \le z_2 | Z_1 = z_1).$$

Then Z_2 is said to be conditionally stochastically decreasing *with respect to Z_1 if for $z_1 \le \hat{z}_1$, we have $\Psi(z_2|z_1) \le \Psi(z_2|\hat{z}_1)$ for any z_2. Likewise, Z_2 is said to be* conditionally stochastically increasing *with respect to Z_1 if for $z_1 \le \hat{z}_1$, we have $\Psi(z_2|z_1) \ge \Psi(z_2|\hat{z}_1)$ for any z_2.*

Thus, if $g_k(I_k^1, I_k^2, v_k)$ is additive—that is, if

$$g_k(I_k^1, I_k^2, v_k) = v_k + I_k^1 + I_k^2$$

—then $g_k(I_k^1, I_k^2, v_k)$ is conditionally stochastically increasing with respect to I_k^1. However, if $g_k(I_k^1, I_k^2, v_k)$ is given as

$$g_k(I_k^1, I_k^2, v_k) = v_k/(1 + I_k^1 + I_k^2),$$

then $g_k(I_k^1, I_k^2, v_k)$ is conditionally stochastically decreasing with respect to I_k^1. We have the following monotonicity result.

THEOREM 4.7 *Assume that* (4.1)–(4.5), *and* (4.30)–(4.31) *hold. Consider a period n in $\{1, ..., N\}$. Let y_n be the initial inventory position at the beginning of period n, let i_n^1 and $\hat{i}_n^1$ be two of the possible realizations of I_n^1 with $i_n^1 < \hat{i}_n^1$.*

If $g_n(I_n^1, I_n^2, v_n)$ is conditionally stochastically increasing with respect to I_n^1, then

(i) *for $n = N$, we have $\phi_N(i_N^1) \le \phi_N(\hat{i}_N^1)$ and $\Phi_N(i_N^1) \le \Phi_N(\hat{i}_N^1)$, and*

(ii) *for $n \ne N$, if $K_N^f = K_k^f = K_k^s = 0$, $k = n, ..., N-1$, we have*

$$\Phi_n(i_n^1) = \phi_n(i_n^1) \le \Phi_n(\hat{i}_n^1) = \phi_n(\hat{i}_n^1)$$

and

$$\Sigma_n(i_n^1) = \sigma_n(i_n^1) \le \Sigma_n(\hat{i}_n^1) = \sigma_n(\hat{i}_n^1).$$

If $g_n(I_n^1, I_n^2, v_n)$ is conditionally stochastically decreasing with respect to I_n^1, then

(iii) *for $n = N$, we have $\phi_N(i_N^1) \ge \phi_N(\hat{i}_N^1)$ and $\Phi_N(i_N^1) \ge \Phi_N(\hat{i}_N^1)$, and*

(iv) *for $n \ne N$, if $K_N^f = K_k^f = K_k^s = 0$, $k = n, ..., N-1$, we have*

$$\Phi_n(i_n^1) = \phi_n(i_n^1) \ge \Phi_n(\hat{i}_n^1) = \phi_n(\hat{i}_n^1)$$

and

$$\Sigma_n(i_n^1) = \sigma_n(i_n^1) \ge \Sigma_n(\hat{i}_n^1) = \sigma_n(\hat{i}_n^1).$$

REMARK 4.5 This result is consistent with intuition. Since a higher value of I_n^1 signals an increased demand, it results in higher order-up-to levels and reorder points. Likewise, the opposite is true when a higher value of I_n^1 signals a decreased demand.

To prove Theorem 4.7, we need the following two lemmas.

LEMMA 4.3 *Let $H(u)$ be a convex function such that $|H(u)| \le C_H \cdot (1+|u|^{k_0})$ for some $C_H > 0$ and $k_0 > 0$. Let $g(I^1, I^2)$ be a nonnegative function of two random variables I^1 and I^2 with $\mathsf{E}[g(i^1, I^2)]^{k_0} < +\infty$ for any observed value i^1 of I^1. If $g(I^1, I^2)$ is conditionally stochastically increasing with respect to I^1, then*

$$\frac{\mathsf{d}\mathsf{E}[H(u - g(i^1, I^2))]}{\mathsf{d}u} \ge \frac{\mathsf{d}\mathsf{E}[H(u - g(\hat{i}^1, I^2))]}{\mathsf{d}u} \tag{4.84}$$

for any $i^1 < \hat{i}^1$, whenever both derivatives exist. Likewise, if $g(I^1, I^2)$ is conditionally stochastically decreasing with respect to I^1, then

$$\frac{\mathsf{d}\mathsf{E}[H(u - g(i^1, I^2))]}{\mathsf{d}u} \le \frac{\mathsf{d}\mathsf{E}[H(u - g(\hat{i}^1, I^2))]}{\mathsf{d}u} \tag{4.85}$$

for any $i^1 < \hat{i}^1$, whenever both derivatives exist.

Proof First, we should note that the integrals appearing below are understood to be in the Lebesgue sense and are therefore well defined on account of the convexity of $H(\cdot)$. With $\Psi(\cdot|i^1)$ as the conditional distribution of $g(I^1, I^2)$

given $I^1 = i^1$,

$$
\begin{aligned}
\mathsf{E}[H(u - g(i^1, I^2))] &= \int_0^\infty H(u-\tau)\mathsf{d}\Psi(\tau|i^1) \\
&= -\int_0^\infty H(u-\tau)\mathsf{d}(1-\Psi(\tau|i^1)) \\
&= -\int_0^\infty \frac{\mathsf{d}[H(u-\tau)(1-\Psi(\tau|i^1))]}{\mathsf{d}\tau}\mathsf{d}\tau \\
&\quad + \int_0^\infty [1-\Psi(\tau|i^1)]\,\frac{\mathsf{d}H(u-\tau)}{\mathsf{d}\tau}\mathsf{d}\tau \\
&= H(u) + \int_0^\infty [1-\Psi(\tau|i^1)]\frac{\mathsf{d}H(u-\tau)}{\mathsf{d}\tau}\mathsf{d}\tau.
\end{aligned}
$$

Taking a derivative, we have

$$
\begin{aligned}
&\frac{\mathsf{d}\mathsf{E}[H(u - g(i^1, I^2))]}{\mathsf{d}u} \\
&= \frac{\mathsf{d}H(u)}{\mathsf{d}u} - \int_0^\infty [1-\Psi(\tau|i^1)]\frac{\mathsf{d}^2H(u-\tau)}{\mathsf{d}\tau^2}\mathsf{d}\tau.
\end{aligned}
\tag{4.86}
$$

From the convexity of $H(\cdot)$, we get that if $g(I^1, I^2)$ is stochastically increasing in I^1, then

$$
[1-\Psi(\tau|i^1)]\frac{\mathsf{d}^2H(u-\tau)}{\mathsf{d}\tau^2} \leq [1-\Psi(\tau|\hat{i}^1)]\frac{\mathsf{d}^2H(u-\tau)}{\mathsf{d}\tau^2},
\tag{4.87}
$$

and if $g(I^1, I^2)$ is stochastically decreasing in I^1, then

$$
[1-\Psi(\tau|i^1)]\frac{\mathsf{d}^2H(u-\tau)}{\mathsf{d}\tau^2} \geq [1-\Psi(\tau|\hat{i}^1)]\frac{\mathsf{d}^2H(u-\tau)}{\mathsf{d}\tau^2}.
\tag{4.88}
$$

Result (4.84) follows from (4.86) and (4.87). Similarly, Result (4.85) follows from (4.86) and (4.88). □

LEMMA 4.4 *Let $W_1(u)$ and $W_2(u)$ be continuous and almost surely differentiable K-convex functions with $\lim_{u\to+\infty} W_i(u) = +\infty$, $i = 1, 2$. Assume that*

$$
\frac{\mathsf{d}W_1(u)}{\mathsf{d}u} \geq \frac{\mathsf{d}W_2(u)}{\mathsf{d}u}
\tag{4.89}
$$

whenever both derivatives exist. Let s_i and S_i be such that for $i = 1, 2$,

$$S_i = \inf\{\tau : W_i(\tau) = \inf\{W_i(u)\}\}, \tag{4.90}$$

$$s_i = \inf\{u : W_i(u) = K + W_i(S_i) \text{ and } u \leq S_i\}. \tag{4.91}$$

Then $s_1 \leq s_2$ and $S_1 \leq S_2$.

Proof By the K-convexity, we know that

$$W_i(u) \text{ is strictly decreasing on } (-\infty, s_i), \text{ and} \tag{4.92}$$
$$W_i(u_1) \leq K + W_i(u_2), \; s_i \leq u_1 \leq u_2. \tag{4.93}$$

The proof of the lemma is trivial if $S_1 = -\infty$. Therefore, in what follows we consider only the case when $S_1 > -\infty$.

First, we prove that $S_1 \leq S_2$. This is trivial if $S_2 = +\infty$ or $W_1(S_2) = \inf\{W_1(u)\}$. Thus we need to prove only that $S_1 \leq S_2$ when $S_2 < +\infty$ and $W_1(S_2) \neq \inf\{W_1(u)\}$. Suppose to the contrary that $S_1 > S_2$. By (4.90), $W_1(S_2) + W_2(S_1) \geq W_1(S_1) + W_2(S_2)$. Since $W_1(S_2) \neq \inf\{W_1(u)\}$, we have

$$W_1(S_2) - W_2(S_2) > W_1(S_1) - W_2(S_1). \tag{4.94}$$

On the other hand, for any $x < S_2$, we have from (4.89)

$$\int_x^{S_1} \mathrm{d}[W_1(u) - W_2(u)] \geq \int_x^{S_2} \mathrm{d}[W_1(u) - W_2(u)].$$

This implies that $W_1(S_1) - W_2(S_1) \geq W_1(S_2) - W_2(S_2)$. Consequently, we get a contradiction with (4.94). This proves $S_1 \leq S_2$.

To prove $s_1 \leq s_2$, we note from (4.89) and (4.92) that

$$0 \geq \frac{\mathrm{d}W_1(u)}{\mathrm{d}u} \geq \frac{\mathrm{d}W_2(u)}{\mathrm{d}u} \quad \text{for almost everywhere } u \in (-\infty, s_1). \tag{4.95}$$

The proof is trivial if $s_1 = -\infty$. Thus, we need to prove $s_1 \leq s_2$ only when $s_1 > -\infty$. This proof is in two parts. In part 1, we consider that

$$\int_\alpha^{s_1} \mathrm{d}W_2(u) = 0 \;\text{ for some } \alpha < s_1. \tag{4.96}$$

This and (4.95) imply that

$$\int_\alpha^{s_1} \mathrm{d}W_1(u) = 0 \;\text{ for some } \alpha < s_1. \tag{4.97}$$

From (4.91) and (4.97), we have

$$W_1(\alpha) = W_1(s_1) = K + W_1(S_1).$$

Since $\alpha < s_1$, this contradicts with the definition of s_1 in (4.91). This proves $s_1 \leq s_2$ in part 1. In part 2, we need only to consider the case when

$$\int_{\alpha}^{s_1} \mathrm{d}W_2(u) < 0 \text{ for all } \alpha < s_1. \tag{4.98}$$

Suppose to the contrary that $s_1 > s_2$. From (4.89), (4.90), and the fact $S_1 \leq S_2$, we have

$$0 \geq \int_{S_1}^{S_2} \mathrm{d}W_2(u) \quad \text{and} \quad \int_{s_1}^{S_1} \mathrm{d}W_1(u) \geq \int_{s_1}^{S_1} \mathrm{d}W_2(u). \tag{4.99}$$

It follows from (4.98) with $\alpha = s_2$ and (4.99) that

$$\begin{aligned} \int_{s_1}^{S_1} \mathrm{d}W_1(u) &> \int_{s_2}^{s_1} \mathrm{d}W_2(u) + \int_{s_1}^{S_1} \mathrm{d}W_2(u) + \int_{S_1}^{S_2} \mathrm{d}W_2(u) \\ &= \int_{s_2}^{S_2} \mathrm{d}W_2(u). \end{aligned} \tag{4.100}$$

On the other hand, by (4.91) we have

$$-\int_{s_1}^{S_1} \mathrm{d}W_1(u) = K = -\int_{s_2}^{S_2} \mathrm{d}W_2(u),$$

which contradicts with (4.100). This proves $s_1 \leq s_2$ in part 2. □

Proof of Theorem 4.7 We consider only parts (i) and (ii)—that is, when $g_k(I_k^1, I_k^2, v_k)$ is conditionally stochastically increasing with respect to I_k^1. Parts (iii) and (iv) can be similarly treated.

When $n = N$, from (4.84) of Lemma 4.3,

$$\begin{aligned} &\frac{\partial\left(c_N^f u + \mathsf{E}\left[H_{N+1}(u - g_N(i_N^1, I_N^2, v_N))\right]\right)}{\partial u} \\ &\quad \geq \frac{\partial\left(c_N^f u + \mathsf{E}\left[H_{N+1}(u - g_N(\hat{i}_N^1, I_N^2, v_N))\right]\right)}{\partial u}. \end{aligned}$$

Using this and Lemma 4.4, we have part (i) of the theorem.

Now we prove part (ii). Since set-up costs are zero in period n and subsequent periods, we know that $\tilde{V}_{n+1}(w, i_{n+1}^1)$ is convex, $\Phi_n(i_n^1) = \phi_n(i_n^1)$, and $\Sigma_n(i_n^1) = \sigma_n(i_n^1)$. It follows from (4.84) of Lemma 4.3 that

$$\begin{aligned} &c_n^s + \frac{\partial \mathsf{E}\left[\tilde{V}_{n+1}(w - g_n(i_n^1, I_n^2, v_n), I_{n+1}^1)\right]}{\partial w} \\ &\quad \geq c_n^s + \frac{\partial \mathsf{E}\left[\tilde{V}_{n+1}(w - g_n(\hat{i}_n^1, I_n^2, v_n), I_{n+1}^1)\right]}{\partial w}. \end{aligned} \tag{4.101}$$

From Lemma 4.4, therefore, $\Sigma_n(i_n^1) \le \Sigma_n(\hat{i}_n^1)$. Let

$$L_n(u, i_n^1) = \inf_{w \ge u} \left\{ c_n^s w + \mathsf{E}\left[\tilde{V}_{n+1}(w - g_n(i_n^1, I_n^2, v_n), I_{n+1}^1) \right] \right\}. \quad (4.102)$$

From (4.101), $\Sigma_n(i_n^1) \le \Sigma_n(\hat{i}_n^1)$, and the convexity of

$$c_n^s w + \mathsf{E}\left[\tilde{V}_{n+1}(w - g_n(i_n^1, I_n^2, v_n), I_{n+1}^1) \right],$$

we have that

$$\frac{dL_n(u, i_n^1)}{du} = \frac{dL_n(u, \hat{i}_n^1)}{du} = 0 \quad \text{when } u \in (-\infty, \ \Sigma_n(i_n^1)), \quad (4.103)$$

$$\frac{dL_n(u, i_n^1)}{du} \ge 0 = \frac{dL_n(u, \hat{i}_n^1)}{du} \quad \text{when } u \in (\Sigma_n(i_n^1), \ \Sigma_n(\hat{i}_n^1)), \quad (4.104)$$

and

$$\begin{aligned} \frac{dL_n(u, i_n^1)}{du} &= c_n^s + \frac{\partial \mathsf{E}\left[\tilde{V}_{n+1}(w - g_n(i_n^1, I_n^2, v_n), I_{n+1}^1) \right]}{\partial w} \\ &\ge c_n^s + \frac{\partial \mathsf{E}\left[\tilde{V}_{n+1}(w - g_n(\tilde{i}_n^1, I_n^2, v_n), I_{n+1}^1) \right]}{\partial w} \\ &= \frac{dL_n(u, \hat{i}_n^1)}{du}, \quad \text{for } u \in (\Sigma_n(\hat{i}_n^1), \ +\infty). \end{aligned} \quad (4.105)$$

Using (4.84) of Lemma 4.3 and (4.103)–(4.105), we have

$$\begin{aligned} &\frac{\partial \left((c_n^f - c_n^s)u + \mathsf{E}\left[H_{n+1}(u - g_n(i_n^1, I_n^2, v_n)) \right] + L_n(u, i_n^1) \right)}{\partial u} \\ &\ge \frac{\partial \left((c_n^f - c_n^s)u + \mathsf{E}\left[H_{n+1}(u - g_n(\hat{i}_n^1, I_n^2, v_n)) \right] + L_n(u, \hat{i}_n^1) \right)}{\partial u}. \end{aligned}$$

Then it follows from Lemma 4.4 that $\Phi_n(i_n^1) \le \Phi_n(\hat{i}_n^1)$. □

REMARK 4.6 Result (ii) of Theorem 4.7 can be obtained for positive fixed costs, provided we specialize $\Psi_n(\cdot|i_n^1)$, the conditional distribution of the demand D_n given $I_n^1 = i_n^1$, as in the following. If I_n^1 is the location parameter of $g_n(I_n^1, I_n^2, v_n)$, then it is clear that for $\hat{i}_n^1 \ge i_n^1$,

$$\Psi_n(x|\hat{i}_n^1) = \Psi_n(x - \hat{i}_n^1 + i_n^1|i_n^1) \quad (4.106)$$

(see Chapter 8, Huang, Sethi, and Yan [10], or Law and Kelton [11] for a detailed discussion on the location parameter distribution). That is, $g_n(I_n^1, I_n^2, v_n)$ is conditionally stochastically increasing in I_n^1. With $K_k^s > 0$, (4.102) should have been

$$\begin{aligned} L_n(u, i_n^1) &= \inf_{w \geq u} \Big\{ K_k^s \cdot \delta(w - u) + c_n^s w \\ &\qquad + \mathsf{E}\left[\tilde{V}_{n+1}(w - g_n(i_n^1, I_n^2, v_n), I_{n+1}^1) \right] \Big\}. \end{aligned}$$

From this and (4.106), we can conclude that $\Sigma_n(\hat{i}_n^1) = \Sigma_n(i_n^1) + (\hat{i}_n^1 - i_n^1)$ and $\sigma_n(\hat{i}_n^1) = \sigma_n(i_n^1) + (\hat{i}_n^1 - i_n^1)$. Similarly, it is possible to show that

$$\Phi_n(\hat{i}_n^1) = \Phi_n(i_n^1) + (\hat{i}_n^1 - i_n^1) \text{ and } \phi_n(\hat{i}_n^1) = \phi_n(i_n^1) + (\hat{i}_n^1 - i_n^1).$$

REMARK 4.7 From the proof of Theorem 4.7, it is easy to see that in part (ii) the property $\Sigma_n(i_n^1) \leq \Sigma_n(\hat{i}_n^1)$ does not require K_n^f and K_n^s to be zero, and the property $\Phi_n(i_n^1) \leq \Phi_n(\hat{i}_n^1)$ does not require K_n^f to be zero. Similar statements hold true for part (iv).

4.6. The Nonstationary Infinite-Horizon Problem

We now consider an infinite-horizon version of the problem formulated in Section 4.3. By letting $N = \infty$ and $(\boldsymbol{F}, \boldsymbol{S}) = ((F_n, S_n), (F_{n+1}, S_{n+1}), \ldots)$, the extended real-valued objective function of the problem is

$$\begin{aligned} &J_n^{\infty}(x_n, s_{n-1}, i_n^1, (\boldsymbol{F}, \boldsymbol{S})) \\ &\quad = H_n(x_n) + \sum_{k=n}^{\infty} \alpha^{k-n} \mathsf{E}\Big[K_k^f \cdot \delta(F_k) + C_k^f(F_k) \\ &\qquad + K_k^s \cdot \delta(S_k) + C_k^s(S_k) + \alpha H_{k+1}(X_{k+1}) \Big], \end{aligned} \tag{4.107}$$

where α is a given discount factor, $0 < \alpha < 1$,

$$X_{n+1} = x_n + s_{n-1} + F_n - g_n(i_n^1, I_n^2, v_n),$$

and X_k $(k > n + 1)$ are defined by (4.8). We make the following assumptions on the costs $C_n^f(\cdot)$, $C_n^s(\cdot)$, and $H_n(\cdot)$ and demands $g_n(I_n^1, I_n^2, v_n)$: there exist constants $c > 0$ and $M > 0$ such that for all $n \geq 1$,

$$\left|C_n^f(x_1) - C_n^f(x_2)\right| \leq c \cdot |x_1 - x_2|, \tag{4.108}$$

$$|C_n^s(x_1) - C_n^s(x_2)| \leq c \cdot |x_1 - x_2|, \tag{4.109}$$

$$|H_n(x_1) - H_n(x_2)| \leq c \cdot |x_1 - x_2|, \tag{4.110}$$

$$\mathsf{E}[g_n(I_n^1, I_n^2, v_n)] < M < \infty. \tag{4.111}$$

Furthermore, we assume that

$$C_n^f(t) + \mathsf{E}\left[H_{n+1}(t - g_n(I_n^1, I_n^2, v_n))\right] \to \infty \text{ as } t \to \infty, \tag{4.112}$$

$$C_n^s(t) + \mathsf{E}\left[H_{n+1}(t - g_n(I_n^1, I_n^2, v_n))\right] \to \infty \text{ as } t \to \infty, \tag{4.113}$$

uniformly hold with respect to n.

Similar to (4.14), the dynamic programming equations for the infinite-horizon problem are

$$\begin{aligned} & U_n^\infty(x_n, s_{n-1}, i_n^1) \\ &= H_n(x_n) + \inf_{\substack{\phi \geq 0 \\ \sigma \geq 0}} \Big\{ K_n^f \cdot \delta(\phi) + C_n^f(\phi) + K_n^s \cdot \delta(\sigma) + C_n^s(\sigma) \\ &\quad + \alpha \mathsf{E}\left[U_{n+1}^\infty(x_n + s_{n-1} + \phi - g_n(i_n^1, I_n^2, v_n), \sigma, I_{n+1}^1)\right] \Big\}, \\ &\quad n = 1, 2, ... \end{aligned} \tag{4.114}$$

Similar to Chapter 3, let us first examine the finite-horizon approximation $J_{n,k}(x_n, s_{n-1}, i_n^1)$ of (4.107), which is obtained by the first k-period truncation of the infinite-horizon problem. The objective function for this problem is to minimize

$$\begin{aligned} & J_{n,k}(x_n, s_{n-1}, i_n^1, (\boldsymbol{F}, \boldsymbol{S})) \\ &= H_n(x_n) + \sum_{\ell=n}^{n+k-1} \alpha^{\ell-n} \mathsf{E}\Big[K_\ell^f \cdot \delta(F_\ell) + C_\ell^f(F_\ell) \\ &\qquad + K_\ell^s \cdot \delta(S_\ell) + C_\ell^s(S_\ell) + \alpha H_{\ell+1}(X_{\ell+1}) \Big] \\ &\quad + \alpha^k \mathsf{E}\left[K_{n+k}^f \cdot \delta(F_{n+k}) + C_{n+k}^f(F_{n+k}) + \alpha H_{n+k+1}(X_{n+k+1})\right]. \end{aligned} \tag{4.115}$$

Let $V_{n,k}(x_n, s_{n-1}, i_n^1)$ be the value function of the truncated problem—that is,

$$V_{n,k}(x_n, s_{n-1}, i_n^1) = \inf_{(\boldsymbol{F}, \boldsymbol{S}) \in \mathcal{A}_{n,k}} \left\{ J_{n,k}(x_n, s_{n-1}, i_n^1, (\boldsymbol{F}, \boldsymbol{S})) \right\}. \tag{4.116}$$

Since (4.115) is a finite-horizon problem on the interval $\langle n, n+k\rangle$, we can apply Theorem 4.1 to prove that $V_{n,k}(x_n, s_{n-1}, i_n^1)$ satisfies the dynamic programming equations,

$$\begin{aligned} &U_{n+\ell,k-\ell}(x_{n+\ell}, s_{n+\ell-1}, i_{n+\ell}^1) \\ &= H_{n+\ell}(x_{n+\ell}) + \inf_{\substack{\phi\geq 0\\ \sigma\geq 0}} \Big\{ K_{n+\ell}^f \cdot \delta(\phi) + C_{n+\ell}^f(\phi) \\ &\quad + K_{n+\ell}^s \cdot \delta(\sigma) + C_{n+\ell}^s(\sigma) \\ &\quad + \alpha \mathsf{E}\left[U_{n+\ell+1,k-\ell-1}(Z_{n+\ell+1}(x_{n+\ell}+\phi), \sigma, I_{n+\ell+1}^1)\right] \Big\}, \\ &\ell = 0, ..., k-1, \end{aligned} \tag{4.117}$$

$$\begin{aligned} &U_{n+k,0}(x_{n+k}, s_{n+k-1}, i_{n+k}^1) \\ &= H_{n+k}(x_{n+k}) + \inf_{\phi\geq 0} \Big\{ K_{n+k}^f \cdot \delta(\phi) + C_{n+k}^f(\phi) \\ &\quad + \alpha \mathsf{E}\left[H_{n+k+1}(Z_{n+k+1}(x_{n+k}+\phi))\right] \Big\}, \end{aligned} \tag{4.118}$$

where

$$Z_{n+\ell+1}(t) = t + s_{n+\ell-1} - g_{n+\ell}(i_{n+\ell}^1, I_{n+\ell}^2, v_{n+\ell}).$$

The following results can be proved by the method of successive approximations of the infinite-horizon problem by longer and longer finite-horizon problems as in Chapter 3. The proof is similar to Theorem 3.6, and is therefore omitted.

THEOREM 4.8 *Assume that* (4.1), (4.3)–(4.4), *and* (4.108)–(4.113) *hold, and*

$$\min\{K_k^f, K_k^s\} \geq \max\{K_{k+1}^f, K_{k+1}^s\}, \;\; k = 1, 2, ... \tag{4.119}$$

Then the limit of $V_{n,k}(x_n, s_{n-1}, i_n^1)$ *exists as* $k \to \infty$. *Let the limit be denoted by* $V_n^\infty(x_n, s_{n-1}, i_n^1)$. *Then*

(i) $V_n^\infty(x_n, s_{n-1}, i_n^1)$ *is Lipschitz continuous in* (x_n, s_{n-1}) *on* $(-\infty, +\infty)\times[0, +\infty)$;

(ii) $V_n^\infty(x_n, s_{n-1}, i_n^1)$ *is a solution of* (4.114);

(iii) *there exist functions* $\bar{F}_n(x_n, s_{n-1}, i_n^1)$ *and* $\bar{S}_n(x_n, s_{n-1}, i_n^1)$ *which provide the infima in* (4.114) *with* $U_n^\infty(x_n, s_{n-1}, i_n^1) = V_n^\infty(x_n, s_{n-1}, i_n^1)$, *and*

$$(\bar{\boldsymbol{F}}, \bar{\boldsymbol{S}}) = \left\{(\bar{F}_n(x_n, s_{n-1}, i_n^1), \bar{S}_n(x_n, s_{n-1}, i_n^1)),\; n \geq 1\right\}$$

is an optimal nonanticipative policy—that is,

$$\begin{aligned} V_1^\infty(x_1, s_0, i_1^1) &= J_1^\infty(x_1, s_0, i_1^1, (\bar{\boldsymbol{F}}, \bar{\boldsymbol{S}})) \\ &= \inf_{(\boldsymbol{F},\boldsymbol{S})\in\mathcal{A}} \left\{ J_1^\infty(x_1, s_0, i_1^1, (\boldsymbol{F}, \boldsymbol{S})) \right\}. \end{aligned}$$

With Theorem 4.8 in hand, the optimal policy in the infinite-horizon case corresponding to Theorem 4.5 can be proved similarly as in the finite-horizon case. Here we present it without the proof.

THEOREM 4.9 *Assume that* (4.1), (4.3)–(4.4), (4.30)–(4.31), *and* (4.110)–(4.111) *hold. Furthermore, let* (4.119) *hold. For each period k, the initial inventory level at the beginning of period k is x_k, the slow-order quantity in period $(k-1)$ is s_{k-1}, and the observed value of I_k^1 is i_k^1 in period $(k-1)$. Then there exist two sequence of pair numbers $\{(\phi_k, \Phi_k), k \geq 1\}$ and $\{(\sigma_k, \Sigma_k), k \geq 1\}$ with $\phi_k \leq \Phi_k$ and $\sigma_k \leq \Sigma_k$, which do not depend on the inventory position $x_k + s_{k-1}$, such that the optimal fast-order quantity $\tilde{F}_k(x_k, s_{k-1}, i_k^1)$ and the optimal slow-order quantity $\tilde{S}_k(x_k, s_{k-1}, i_k^1)$ in period k can be determined by the following expressions*:

$$\begin{aligned} &\tilde{F}_k(x_k, s_{k-1}, i_k^1) \\ &= \begin{cases} \Phi_k - (x_k + s_{k-1}), & \textit{if } \ x_k + s_{k-1} \leq \phi_k, \\ 0, & \textit{if } \ x_k + s_{k-1} > \phi_k, \end{cases} \end{aligned}$$

$$\begin{aligned} &\tilde{S}_k(x_k, s_{k-1}, i_k^1) \\ &= \begin{cases} \Sigma_k - (x_k + s_{k-1} + \tilde{F}_k(\cdot)), & \textit{if } \ x_k + s_{k-1} + \tilde{F}_k(\cdot) \leq \sigma_k, \\ 0, & \textit{if } \ x_k + s_{k-1} + \tilde{F}_k(\cdot) > \sigma_k. \end{cases} \end{aligned}$$

4.7. Concluding Remarks

In this chapter, we have studied a periodic-review inventory model with fixed order costs, dual supply modes, and demand-forecast updates. We show that the optimal policies for both fast and slow orders are of (s, S)-type. For fast orders, the initial inventory position in any given period includes the slow order issued in the previous period. For slow orders, the (s, S) policy is based on a slow-order inventory position, which includes also the fast order issued during the period. We show that the policy parameters behave in an intuitive fashion with respect to changes in the cost parameters. We show also that the policy parameters depend on the most recent forecast update but not on the inventory position. We show further that the policy parameters exhibit monotonic behavior with respect to the forecast update. This behavior is consistent with our intuition, in the sense that an update indicating an "increased" demand implies higher order-up-to levels and reorder points. The chapter generalizes several existing results in the literature. One potential extension is to develop the algorithm to compute the optimal policies. For a classical inventory model with a fixed cost, see Zheng and Federgruen [21] and Feng and Xiao [6]. Another future extension could incorporate Markovian demand as in Song and Zipkin [18] and Sethi and Cheng [15] and promotion policies influencing the demand as in Sethi and Cheng [15]. In the latter extension, the optimal promotion policy would depend on the current forecast updates, while at the same time it could influence future forecast updates and future demands. It is also of interest to consider more than two delivery modes. Research dealing with three delivery modes with no fixed order costs is discussed in the next chapter.

4.8. Notes

This chapter is based on Sethi, Yan, and Zhang [16].

Results in this chapter differ from the dual delivery source models of Hausmann, Lee, and Zhang [9] and Scheller-Wolf and Tayur [14] in the sense that we make use of demand-forecast updates in making decisions. In contrast with two-stage or two-period models of Yan, Liu, and Hsu [20], Barnes-Schuster, Bassok, and Anupindi [1], Donohue [5], and Gurnani and Tang [8], we consider N-periods, $1 \leq N \leq \infty$. The demand process in our model is nonstationary, whereas Toktay and Wein [19] consider a stationary demand. Our forecast-updating process covers as a special case, the additive demand-information updates employed in the single-delivery-source model of Gallego and Özer [7].

For the inventory models involving multiple delivery modes and fixed costs, Bensoussan, Crouhy, and Proth [2] consider an inventory model with two supply modes—one instantaneous and the other with a one-period lead time. They obtain an (s, S)-type optimal policy. Huang, Sethi, and Yan [10] consider a single-period, two-stage supply-contract model and a fixed cost of ordering via the fast mode at the second stage. For a uniformly distributed demand, they are able to provide an explicit solution of the (s, S)-type. The explicit nature of their solution enables them to obtain some important insights into a better supply-contract management.

References

[1] D. Barnes-Schuster, Y. Bassok, and R. Anupindi. Coordination and flexibility in supply contracts with options. *Manufacturing and Service Operations Management*, 4:171–207, 2002.

[2] A. Bensoussan, M. Crouhy, and J.M. Proth. *Mathematical Theory of Production Planning*. North-Holland, Amsterdam, the Netherlands, 1983.

[3] D. Beyer, S. Sethi, and M. Taksar. Inventory models with Markovian demands and cost functions of polynomial growth. *Journal of Optimization Theory and Applications*, 98:281–323, 1998.

[4] A.O. Brown and H.L. Lee. Optimal pay-to-delay capacity reservation with application to the semiconductor industry. Working Paper, Stanford University, Stanford, CA, 1997.

[5] K.L. Donohue. Efficient supply contracts for fashion goods with forecast updating and two production modes. *Management Science*, 46:1397–1411, 2000.

[6] Y. Feng and B. Xiao. A new algorithm for computing optimal (s, S) policies in a stochastic single item/location inventory system. *IIE Transactions*, 32:1081–1090, 2000.

[7] G. Gallego and Ö. Özer. Integrating replenishment decisions with advance-demand information. *Management Science*, 47:1344–1360, 2001.

[8] H. Gurnani and C.S. Tang. Note: optimal ordering decisions with uncertain cost and demand forecast updating. *Management Science*, 45:1456–1462, 1999.

[9] W.H. Hausman, H.L. Lee, and V.L. Zhang. Optimal ordering for an inventory system with dual lead times. Working Paper, Stanford University, Stanford, CA, 1993.

[10] H.Y. Huang, S.P. Sethi, and H. Yan. Purchase contract management with demand forecast updates. *IIE Transactions*, to appear.

[11] A.M. Law and W.D. Kelton. *Simulation Modeling and Analysis*. McGraw-Hill, New York, 2000.

[12] S. Ross. *Stochastic Processes*. John Wiley, New York, 1983.

[13] H. Scarf. The optimality of (s, S) policies in the dynamic inventory problem, Chapter 13 in Arrow, K.J., Karlin S. and P. Suppes (editors), *Mathematical Methods in Social Sciences*. Stanford University Press, Stanford, CA, 1960.

[14] A. Scheller-Wolf and S. Tayur. A Markovian dual-source production-inventory model with order bands. Working Paper, Carnegie Mellon University, Pittsburgh, PA, 1998.

[15] S.P. Sethi and F. Cheng. Optimality of (s, S) policies in inventory models with Markovian demand. *Operations Research*, 45:931–939, 1997.

[16] S.P. Sethi, H. Yan, and H. Zhang. Inventory models with fixed costs, forecast updates, and two delivery modes. *Operations Research*, 51:321–328, 2003.

[17] M. Shaked and J.G. Shanthikumar. *Stochastic Orders and Their Applications*, Academic Press, New York, 1994.

[18] J. Song and P. Zipkin. Inventory control in a fluctuating demand environment. *Operations Research*, 41:351–370, 1993.

[19] L.B. Toktay and L. Wein. Analysis of a forecasting-production-inventory system with stationary demand. *Management Science*, 47:1268–1281, 2001.

[20] H. Yan, K. Liu, and A. Hsu. Optimal ordering in a dual-supplier system with demand forecast updates. *Production and Operations Management*, 12:30–45, 2003.

[21] Y.S. Zheng and A. Federgruen. Finding optimal (s, S) policies is about as simple as evaluating a single policy. *Operations Research*, 39:654–665, 1991.

Chapter 5

INVENTORY MODELS WITH THREE CONSECUTIVE DELIVERY MODES

5.1. Introduction

In this chapter, we consider a periodic-review inventory system with three delivery modes and two demand-forecast updates before demand is realized. We denote the three delivery modes as fast, medium, and slow. Fast, medium, and slow orders made at the beginning of a period are delivered at the end of the current period, at the end of the next period, and at the end of the second next period, respectively. In other words, fast, medium, and slow orders have lead times of one, two, and three periods, respectively. Fast orders are more expensive than medium orders, which, in turn, are more expensive than slow ones.

The sequence of events is as follows. At the beginning of each period, the inventory or backlog level is reviewed, and forecasts are updated for the demands to be realized at the end of the next three periods, counting the current period as the first period. In addition, the size of the slow order issued two periods ago and the size of the medium order issued in the previous period are also known. With these data in hand, decisions regarding the amounts to be ordered by slow, medium, and fast modes are made. At the end of the current period, the slow order issued two periods ago, the medium order issued in the previous period, and the fast order issued at the beginning of the current period are delivered. Then the demand for the current period materializes, which determines the inventory or backlog level at the beginning of the next period. Quantities ordered by slow, medium, and fast modes in each period determine the total cost of ordering, inventory holding, and backlogging. The objective is to make ordering decisions that minimize the total cost over the problem horizon.

In this chapter, we prove the existence of an optimal policy for the model that allows three delivery modes as well as forecast updates. We show this for finite-horizon and discounted-cost infinite-horizon problems. We show that there exist optimal base-stock levels for the fast and medium delivery modes. These levels are independent of the inventory level and the outstanding slow and medium orders to be delivered at the end of the current period. But these levels depend in general on the outstanding slow order issued in the previous period to be delivered in two periods hence and on the observed forecast updates. Moreover, under the optimal policy, the slow mode does not follow a base-stock policy in general.

Given the inventory position (relevant for the fast mode), the fast-mode level acts like the traditional base stock. Once the fast order is decided, it is added to the inventory position along with the slow order issued in the previous period to come up with the "inventory position relevant for the medium mode." Given this position and the medium-mode base-stock level, one can easily obtain the medium-order decision. This decision is added to the medium inventory position to obtain the "inventory position relevant for the slow mode." With that, we can obtain the slow-order decision.

The dependence of the optimal base-stock levels, as mentioned above, on the outstanding slow order issued in the previous period is a critical departure from the results obtained in single- and two-delivery-mode systems. Because of the presence of the inventory position and an outstanding order as two of the states of the system, there is a priori every reason to expect that any policy expressed in terms of order-up-to levels (which, it should be noted, can always be done) would have these levels depend on those two states. Such levels cannot be considered base stocks, and therefore, there is no a priori reason to expect that there is an optimal base-stock policy. Thus, obtaining a structure of the optimal policy in the three mode case represents a contribution to the inventory literature. This discussion is further elaborated in Section 5.4.

The remainder of this chapter is organized as follows. In Section 5.2, we provide the required notation and formulate the model. In Section 5.3, we develop dynamic programming equations, and prove that an optimal Markov policy exists for the problem. In Section 5.4, we examine the structure of the optimal policy. Section 5.5 is devoted to extending the results to the infinite-horizon case. The chapter is concluded in Sections 5.6 and 5.7.

5.2. Notation and Model Formulation

We consider a discrete-time, single-product, periodic-review inventory system. The dynamics of the system consists of two parts: the material flows and the information flows. The inbound material flows come from three supply sources (fast, medium, and slow), and the outbound material flows go to customers. The information flows include the initial forecast of demand for each given period, its first forecast update, its second forecast update, and its realization at the end of the given period. When the demand realizes, it is satisfied if there is sufficient available inventory on hand, and the excess is carried over to the next period. Otherwise, the demand is partially satisfied, and the remainder is backlogged.

In what follows, we use the word *inventory* to mean inventory when positive and backlog when negative. The decision variables are the quantities ordered from fast, medium, and slow sources at the beginning of each period. Decisions in a period are based on the inventory level, all outstanding orders, and all observed forecast update parameters.

We introduce the following notation and precisely formulate the model under consideration:

$\langle 1, N\rangle$ = $\{1, 2, ..., N\}$, the time horizon;

F_k = the nonnegative fast-order quantity in period k, $1 \leq k \leq N$;

M_k = the nonnegative medium-order quantity in period k, $1 \leq k \leq N-1$;

S_k = the nonnegative slow-order quantity in period k, $1 \leq k \leq N-2$;

$C_k^f(u)$ = the cost of fast order $u \geq 0$ units in period k;

$C_k^m(u)$ = the cost of medium order $u \geq 0$ units in period k;

$C_k^s(u)$ = the cost of slow order $u \geq 0$ units in period k;

I_k^1 = the first determinant (a random variable) of the demand in period k;

I_k^2 = the second determinant (a random variable) of the demand in period k;

I_k^3 = the third determinant (a random variable) of the demand in period k;

D_k = the nonnegative demand in period k modeled as a function $g_k(I_k^1, I_k^2, I_k^3)$;

X_k = the inventory level at the beginning of period k;

Y_k = $X_k + S_{k-2} + M_{k-1}$ = the inventory position at the beginning of period k;

X_{N+1} = the inventory level at the end of the last period N;

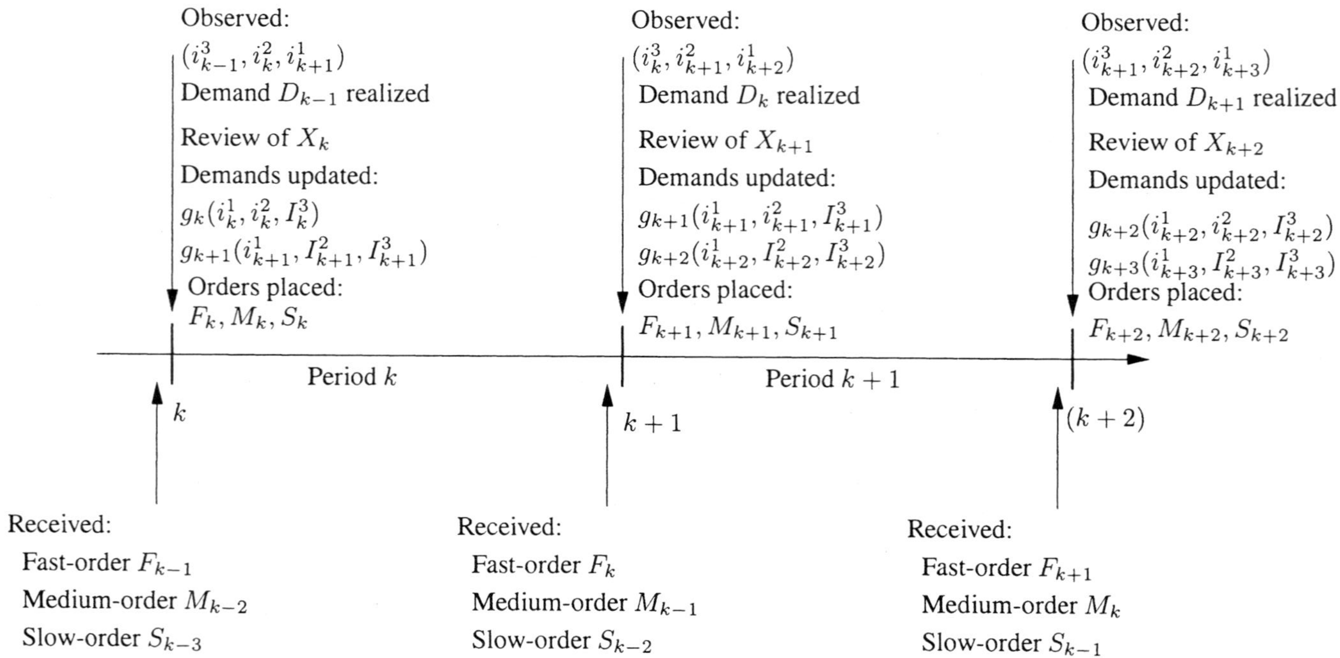

Figure 5.1. A time line of the inventory-system dynamics and ordering decisions

$$\begin{aligned} H_k(x) &= \text{the inventory cost when } X_k = x; \\ H_{N+1}(x) &= \text{the inventory cost when } X_{N+1} = x \geq 0 \\ &\quad \text{or penalty cost when } X_{N+1} = x < 0. \end{aligned}$$

REMARK 5.1 The demand at period k is given in Chapters 3 and 4, but for the sake of notation convenience, we write it here as

$$g_k(I_k^1, I_k^2, I_k^3).$$

We impose the following assumptions on I_k^1, I_k^2, and I_k^3:

$$\left\{(I_k^1, I_k^2, I_k^3), 1 \leq k \leq N\right\} \text{ is a sequence of independent random vectors.} \tag{5.1}$$

Without loss of generality, we may assume that $I_1^1 = i_1^1$, $I_1^2 = i_1^2$, and $I_2^1 = i_2^1$ are given constants. Let us define $\mathcal{F}_{k+1}, k \geq 1$, to be the sigma algebra or σ-field generated by the random variables $\{(I_\ell^1, I_\ell^2, I_\ell^3), 1 \leq \ell \leq k\}$, (I_{k+1}^1, I_{k+1}^2), and I_{k+2}^1—that is, for $1 \leq k \leq N-2$,

$$\mathcal{F}_{k+1} = \sigma\left\{(I_\ell^1, I_\ell^2, I_\ell^3), 1 \leq \ell \leq k, (I_{k+1}^1, I_{k+1}^2), I_{k+2}^1\right\}, \tag{5.2}$$

and

$$\mathcal{F}_N = \sigma\left\{(I_\ell^1, I_\ell^2, I_\ell^3), 1 \leq \ell \leq N-1, (I_N^1, I_N^2)\right\}. \tag{5.3}$$

Let $\mathcal{F}_0 = \mathcal{F}_1 = \{\emptyset, \Omega\}$ and $\mathcal{F}_{N+1} = \mathcal{F}$. It is clear that the demand D_k is an $\mathcal{F}_{k+1}$-adapted random variable. We assume further that

$$\mathsf{E}[D_k] = \mathsf{E}\left[g_k(I_k^1, I_k^2, I_k^3)\right] < \infty, \quad 1 \leq k \leq N. \tag{5.4}$$

We also suppose that for each k,

$$C_k^f(u),\ C_k^m(u) \text{ and } C_k^s(u) \text{ are increasing, nonnegative and convex.} \tag{5.5}$$

Furthermore, the inventory-cost function $H_k(x)$ satisfies

$$\begin{cases} H_k(x) \text{ is convex and} \\ |H_k(x) - H_k(\hat{x})| \leq c_H \cdot |x - \hat{x}|, \quad 1 \leq k \leq N+1, \end{cases} \tag{5.6}$$

for some $c_H > 0$. Similar to (3.6) and (3.7), we assume that

$$\begin{cases} C_k^f(t) + \mathsf{E}[H_{k+1}(t - g_k(I_k^1, I_k^2, I_k^3))] \to \infty \text{ as } t \to \infty, \\ C_k^m(t) + \mathsf{E}[H_{k+1}(t - g_k(I_k^1, I_k^2, I_k^3))] \to \infty \text{ as } t \to \infty, \\ C_k^s(t) + \mathsf{E}[H_{k+1}(t - g_k(I_k^1, I_k^2, I_k^3))] \to \infty \text{ as } t \to \infty. \end{cases} \tag{5.7}$$

Throughout this chapter we assume that (5.1) and (5.4)–(5.7) *hold.*

The inventory-balance equations are defined as

$$X_{k+1} = X_k + S_{k-2} + M_{k-1} + F_k - g_k(I_k^1, I_k^2, I_k^3), \quad 1 \le k \le N, \tag{5.8}$$

where $X_1 = x_1$ is the initial inventory level, and

$$M_0 = m_0, \; S_{-1} = s_{-1}, \text{ and } S_0 = s_0 \tag{5.9}$$

are inherited orders at the beginning of period 1 that are still outstanding. Specifically, m_0 and s_{-1} will be delivered at the end of period 1, and s_0 will be delivered at the end of period 2.

Furthermore, since the decision F_k is adapted to the σ-field $\mathcal{F}_k$, the decision M_{k-1} is adapted to the σ-field $\mathcal{F}_{k-1}$, and the decision S_{k-2} is adapted to the σ-field $\mathcal{F}_{k-2}$, we can see from (5.8) that X_k is an $\mathcal{F}_k$-adapted random variable and that X_{k+1} is an $\mathcal{F}_{k+1}$-adapted random variable.

Let us explain the dynamics (5.8) with the help of Figure 5.1. At the beginning of period k, we observe the value x_k of the inventory level X_k, the value i_k^2 of the second determinant I_k^2 of D_k, and the value i_{k+1}^1 of the first determinant I_{k+1}^2 of D_{k+1}. These observations provide updated forecasts $g_k(i_k^1, i_k^2, I_k^3)$ and $g_{k+1}(i_{k+1}^1, I_{k+1}^2, I_{k+1}^3)$ of D_k and D_{k+1}, respectively. We know the inventory level X_k and outstanding orders S_{k-2}, M_{k-1}, and S_{k-1}. The amount $Y_k = X_k + S_{k-2} + M_{k-1}$ is the inventory position available to meet the demand in period k, where S_{k-2} is the amount delivered in period k as a result of the slow-order decision made in period $(k-2)$ and M_{k-1} is the amount delivered in period $(k+1)$ as a result of the medium-order decision made in period $(k-1)$. In addition, we know S_{k-1}, the amount to be delivered in period $k+1$ as a result of the slow-order decision made in period $(k-1)$. Given these, we can decide on the slow-order S_k, the medium-order M_k, and the fast-order F_k. Since F_k is to be delivered at the end of the period, the total quantity available to meet the kth period demand D_k is $(X_k + S_{k-2} + M_{k-1} + F_k)$. At the end of period k, the value i_k^3 of the random variable I_k^3 is observed, which is tantamount to observing the demand $D_k = g_k(i_k^1, i_k^2, i_k^3)$. The difference of $(X_k + S_{k-2} + M_{k-1} + F_k)$ and D_k is the inventory level X_{k+1} at the beginning of period $(k+1)$. This last statement represents a sample path of the dynamics (5.8).

The objective is to choose a sequence of orders from the fast, medium, and slow sources over time to minimize the total expected value of all the costs incurred in periods $\langle 1, N \rangle$. Thus, the objective function is

$$\begin{aligned} &J_1\left(x_1, s_{-1}, m_0, s_0, i_1^1, i_1^2, i_2^1, (\boldsymbol{F}, \boldsymbol{M}, \boldsymbol{S})\right) \\ &= \mathsf{E}\left\{ \sum_{\ell=1}^{N} \left[C_\ell^f(F_\ell) + C_\ell^m(M_\ell) + C_\ell^s(S_\ell) + H_{\ell+1}(X_{\ell+1}) \right] \right\} \\ &+ H_1(x_1), \end{aligned} \tag{5.10}$$

where

$$(\boldsymbol{F}, \boldsymbol{M}, \boldsymbol{S}) = ((F_1, ..., F_N), (M_1, ..., M_N), (S_1, ..., S_N)) \tag{5.11}$$

is a sequence of *history-dependent* or *nonanticipative admissible decisions.* That is, (F_k, M_k, S_k) is adapted to the σ-field $\mathcal{F}_k$, $1 \le k \le N$. In other words, each of F_k, M_k, and S_k with $1 \le k \le N-1$ is a nonnegative real-valued function of the history of the demand information given by $\{(I_\ell^1, I_\ell^2, I_\ell^3), 0 \le \ell \le k-1\}$, (I_k^1, I_k^2), and I_{k+1}^1, and (F_N, M_N, S_N) is a nonnegative real-valued function of the history of the demand information given by $\{(I_\ell^1, I_\ell^2, I_\ell^3), 0 \le \ell \le N-1\}$ and (I_N^1, I_N^2).

Let $\mathcal{A}_1$ denote the class of all history-dependent admissible decisions, and define the value function for the problem over $\langle 1, N \rangle$ with the initial inventory level x_1 to be

$$\begin{aligned} &V_1\left(x_1, s_{-1}, m_0, s_0, i_1^1, i_1^2, i_2^1\right) \\ &= \inf_{(\boldsymbol{F},\boldsymbol{M},\boldsymbol{S}) \in \mathcal{A}_1} \left\{ J_1\left(x_1, s_{-1}, m_0, s_0, i_1^1, i_1^2, i_2^1, (\boldsymbol{F}, \boldsymbol{M}, \boldsymbol{S})\right) \right\}. \end{aligned} \tag{5.12}$$

Note that the existence of an optimal policy is not required to define the value function. Of course, once the existence is established, the "inf" in (5.12) can be replaced by "min."

It is important to note here that since the horizon length is N periods and the orders M_N, S_{N-1}, and S_N will not be delivered during the problem horizon, it is obvious in view of the ordering costs given in (5.5) that

$$M_N = S_{N-1} = S_N = 0 \tag{5.13}$$

is any optimal solution. So in what follows, we still allow these orders in any feasible solution, but we set them to zero in any optimal solution. This is equivalent to a situation in which these orders are not altogether issued in practice.

5.3. Dynamic Programming and Optimal Nonanticipative Policies

In this section, we use dynamic programming to study the problem. We verify whether the cost of a nonanticipative policy obtained from the solution of the dynamic programming equations equals the value function of the problem over $\langle 1, N\rangle$. First, as usual, we define the problem over $\langle n, N\rangle$. Let

$$\begin{aligned} &J_n\left(x_n, s_{n-2}, m_{n-1}, s_{n-1}, i_n^1, i_n^2, i_{n+1}^1, (\boldsymbol{F}, \boldsymbol{M}, \boldsymbol{S})\right) \\ &\quad = \mathsf{E}\Big\{\sum_{\ell=n}^{N}\Big[C_\ell^f(F_\ell) + C_\ell^m(M_\ell) + C_\ell^s(S_\ell) + H_{\ell+1}(X_{\ell+1})\Big]\Big\} \\ &\quad\quad + H_n(x_n). \end{aligned} \tag{5.14}$$

where, with a slight abuse of notation,

$$(\boldsymbol{F}, \boldsymbol{M}, \boldsymbol{S}) = ((F_n, ..., F_N), (M_n, ..., M_N), (S_n, ..., S_N))$$

the *history-dependent* or *nonanticipative admissible decisions* for the problem defined over periods $\langle n, N\rangle$. That is, given x_n, s_{n-2}, m_{n-1}, s_{n-1}, i_n^1, i_n^2, and i_{n+1}^1 as constants, (F_n, M_n, S_n) is a vector of nonnegative constants, (F_k, M_k, S_k) $(n < k < N)$ are nonnegative real-valued functions of the history of the demand information from period n to period k, given by

$$\left\{(I_\ell^1, I_\ell^2, I_\ell^3), n \le \ell \le k-1, (I_k^1, I_k^2), I_{k+1}^1\right\}, \tag{5.15}$$

(F_N, M_N, S_N) is a positive real-valued function of the history of the demand information from period n to period $(N-1)$ given by $\{(I_\ell^1, I_\ell^2, I_\ell^3), n \le \ell \le N-1\}$ and (I_N^1, I_N^2). Define the value function associated with the problem over $\langle n, N\rangle$ as follows:

$$\begin{aligned} &V_n\left(x_n, s_{n-2}, m_{n-1}, s_{n-1}, i_n^1, i_n^2, i_{n+1}^1\right) \\ &\quad = \inf_{(\boldsymbol{F},\boldsymbol{M},\boldsymbol{S})\in\mathcal{A}_n}\Big\{J_n\big(x_n, s_{n-2}, m_{n-1}, s_{n-1}, \\ &\quad\quad\quad i_n^1, i_n^2, i_{n+1}^1, (\boldsymbol{F}, \boldsymbol{M}, \boldsymbol{S})\big)\Big\}, \end{aligned} \tag{5.16}$$

where $\mathcal{A}_n$ denotes the class of all history-dependent admissible decisions for the problem over $\langle n, N\rangle$.

In view of (5.14), we can write the dynamic programming equations corresponding to the problem as follows:

$$
\begin{aligned}
&U_\ell\left(x_\ell, s_{\ell-2}, m_{\ell-1}, s_{\ell-1}, i_\ell^1, i_\ell^2, i_{\ell+1}^1\right) \\
&= H_\ell(x_\ell) + \inf_{\substack{F\geq 0\\ M\geq 0\\ S\geq 0}} \Big\{C_\ell^f(F) + C_\ell^m(M) + C_\ell^s(S) \\
&\quad + \mathsf{E}\Big[U_{\ell+1}\left(Z_{\ell+1}(F), s_{\ell-1}, M, S, i_{\ell+1}^1, I_{\ell+1}^2, I_{\ell+2}^1\right)\Big]\Big\}, \\
&\ell = 1, ..., N-1, \qquad (5.17)
\end{aligned}
$$

$$
\begin{aligned}
&U_N\left(x_N, s_{N-2}, m_{N-1}, s_{N-1}, i_N^1, i_N^2, i_{N+1}^1\right) \\
&= H_N(x_N) + \inf_{\substack{F\geq 0\\ M\geq 0\\ S\geq 0}} \Big\{C_N^f(F) + C_N^m(M) + C_N^s(M) \\
&\quad + \mathsf{E}\Big[H_{N+1}\left(Z_{N+1}(F)\right)\Big]\Big\}, \qquad (5.18)
\end{aligned}
$$

where the notation $Z_{\ell+1}(\cdot)$ is defined as

$$
Z_{\ell+1}(F) = x_\ell + s_{\ell-2} + m_{\ell-1} + F - g_\ell(i_\ell^1, i_\ell^2, I_\ell^3), \ \ell = 1, ..., N, \qquad (5.19)
$$

and F, M, and S are arguments for minimization in (5.17)–(5.18).

REMARK 5.2 In the dynamic programming equations (5.17)–(5.18), the inventory cost is also charged for the initial inventory level. As in Chapter 3, this charge is of no consequence.

Based on the dynamic programming equations, we state the following theorem.

THEOREM 5.1 *Assume that* (5.1) *and* (5.4)–(5.7) *hold. Then the value functions*

$$
V_k\left(x_k, s_{k-2}, m_{k-1}, s_{k-1}, i_k^1, i_k^2, i_{k+1}^1\right), \ 1 \leq k \leq N,
$$

defined by (5.16), *satisfy the dynamic programming equations* (5.17)–(5.18).

Proof It follows from the definition of $V_N(x_N, s_{N-2}, m_{N-1}, i_N^1, i_N^2, i_{N+1}^1)$ that it satisfies the last equation in (5.18). Now we use induction. Suppose that $V_\ell(x_\ell, s_{\ell-2}, m_{\ell-1}, s_{\ell-1}, i_\ell^1, i_\ell^2, i_{\ell+1}^1)$, $\ell = k+1, ..., N$, for any given $k \leq N-1$, satisfy (5.17) and (5.18) for $\ell \leq N-1$ and $\ell = N$, respectively. Then we show that $V_k(x_k, s_{k-2}, m_{k-1}, s_{k-1}, i_k^1, i_k^2, i_{k+1}^1)$ satisfies (5.17). It suffices to show that

$$\begin{aligned} &V_k\left(x_k, s_{k-2}, m_{k-1}, s_{k-1}, i_k^1, i_k^2, i_{k+1}^1\right) \\ &\quad = H_k(x_k) + \inf_{\substack{F\geq 0\\ M\geq 0\\ S\geq 0}} \Big\{ C_k^f(F) + C_k^m(M) + C_k^s(S) \\ &\qquad + \mathsf{E}\left[V_{k+1}\left(Z_{k+1}(F), s_{k-1}, M, S, i_{k+1}^1, I_{k+1}^2, I_{k+2}^1\right)\right] \Big\}. \end{aligned} \tag{5.20}$$

From the definition of $V_k(x_k, s_{k-2}, m_{k-1}, s_{k-1}, i_k^1, i_k^2, i_{k+1}^1)$ for any given $k \leq N-1$ (see (5.16)) and the history-dependence of decisions $(\boldsymbol{F}, \boldsymbol{M}, \boldsymbol{S})$, we have

$$\begin{aligned} &V_k\left(x_k, s_{k-2}, m_{k-1}, s_{k-1}, i_k^1, i_k^2, i_{k+1}^1\right) \\ &\quad = H_k(x_k) + \inf_{(\boldsymbol{F},\boldsymbol{M},\boldsymbol{S})\in\mathcal{A}_k} \Big\{ \mathsf{E}\Big[\sum_{\ell=k}^{N} \Big(C_\ell^f(F_\ell) + C_\ell^m(M_\ell) \\ &\qquad\qquad + C_\ell^s(S_\ell) + H_{\ell+1}(X_{\ell+1})\Big)\Big]\Big\} \\ &\quad = H_k(x_k) + \inf_{(\boldsymbol{F},\boldsymbol{M},\boldsymbol{S})\in\mathcal{A}_k} \Big\{ C_k^f(F_k) + C_k^m(M_k) + C_k^s(S_k) \\ &\qquad + \mathsf{E}\Big[H_{k+1}(X_{k+1}) \\ &\qquad + \sum_{\ell=k+1}^{N} \Big(C_\ell^f(F_\ell) + C_\ell^m(M_\ell) + C_\ell^s(S_\ell) + H_{\ell+1}(X_{\ell+1})\Big)\Big]\Big\}. \end{aligned} \tag{5.21}$$

It follows from the definition of the σ-field $\mathcal{F}_{k+1}$ that

$$\begin{aligned} &\mathsf{E}\Big[H_{k+1}(X_{k+1}) + \sum_{\ell=k+1}^{N} \Big(C_\ell^f(F_\ell) + C_\ell^m(M_\ell) \\ &\qquad\qquad + C_\ell^s(S_\ell) + H_{\ell+1}(X_{\ell+1})\Big)\Big] \end{aligned}$$

$$
\begin{aligned}
&= \mathsf{E}\Big\{\mathsf{E}\Big[H_{k+1}(X_{k+1}) \\
&\quad + \sum_{\ell=k+1}^{N} \Big(C_\ell^f(F_\ell) + C_\ell^m(M_\ell) + C_\ell^s(S_\ell) + H_{\ell+1}(X_{\ell+1})\Big) \Big| \mathcal{F}_{k+1}\Big]\Big\} \\
&= \mathsf{E}\Big\{H_{k+1}(X_{k+1}) + C_{k+1}^f(F_{k+1}) + C_{k+1}^m(M_{k+1}) + C_{k+1}^s(S_{k+1}) \\
&\quad + \mathsf{E}\Big[H_{k+2}(X_{k+2}) + \sum_{\ell=k+2}^{N} \Big(C_\ell^f(F_\ell) + C_\ell^m(M_\ell) \\
&\qquad + C_\ell^s(S_\ell) + H_{\ell+1}(X_{\ell+1})\Big)\Big| \mathcal{F}_{k+1}\Big]\Big\}.
\end{aligned}
\tag{5.22}
$$

By induction on the index $(k+1)$,

$$
\begin{aligned}
&\inf_{(\boldsymbol{F},\boldsymbol{M},\boldsymbol{S})\in\mathcal{A}_{k+1}} \Big\{\mathsf{E}\Big[H_{k+1}(X_{k+1}) \\
&\quad + \sum_{\ell=k+1}^{N} \Big(C_\ell^f(F_\ell) + C_\ell^m(M_\ell) + C_\ell^s(S_\ell) + H_{\ell+1}(X_{\ell+1})\Big)\Big]\Big\} \\
&\quad = \mathsf{E}\left[V_{k+1}\left(X_{k+1}, S_{k-1}, M_k, S_k, i_{k+1}^1, I_{k+1}^1, I_{k+2}^1\right)\right].
\end{aligned}
\tag{5.23}
$$

Then (5.21)–(5.23) complete the proof. ☐

Next we discuss how we can obtain an optimal solution for our inventory model. It follows from (5.7) that there exists an upper bound order quantity $Q > 0$ such that

$$
\begin{aligned}
&\inf_{\substack{F\geq 0\\ M\geq 0\\ S\geq 0}} \Big\{C_\ell^f(F) + C_\ell^m(M) + C_\ell^s(S) \\
&\qquad + \mathsf{E}\Big[V_{\ell+1}\left(Z_{\ell+1}(F), s_{\ell-1}, M, S, i_{\ell+1}^1, I_{\ell+1}^2, I_{\ell+2}^1\right)\Big]\Big\} \\
&= \inf_{0\leq F,M,S\leq Q} \Big\{C_\ell^f(F) + C_\ell^m(M) + C_\ell^s(S) \\
&\quad + \mathsf{E}\Big[V_{\ell+1}\left(Z_{\ell+1}(F), s_{\ell-1}, M, S, i_{\ell+1}^1, I_{\ell+1}^2, I_{\ell+2}^1\right)\Big]\Big\}, \\
&\quad \ell = 1, ..., N-1,
\end{aligned}
$$

and

$$\begin{aligned}
& V_N\left(x_N, s_{N-2}, m_{N-1}, s_{N-1}, i_N^1, i_N^2, i_{N+1}^1\right) \\
&= \inf_{\substack{F\geq 0\\ M\geq 0\\ S\geq 0}} \left\{C_N^f(F) + C_N^m(M) + C_N^s(S) + \mathsf{E}\left[H_{N+1}(Z_{N+1}(F))\right]\right\} \\
&= \inf_{0\leq F,M,S\leq Q} \Big\{C_N^f(F) + C_N^m(M) + C_N^s(S) \\
&\qquad\qquad + \mathsf{E}\left[H_{N+1}(Z_{N+1}(F))\right]\Big\}.
\end{aligned}$$

By a well-known selection theorem (see Theorem 3.9), there exist Borel-measurable functions

$$\begin{cases}
\bar{\phi}_\ell = \bar{\phi}_\ell(x_\ell, s_{\ell-2}, m_{\ell-1}, s_{\ell-1}, i_\ell^1, i_\ell^2, i_{\ell+1}^1), & 1\leq \ell\leq N, \\
\bar{\mu}_\ell = \bar{\mu}_\ell(x_\ell, s_{\ell-2}, m_{\ell-1}, s_{\ell-1}, i_\ell^1, i_\ell^2, i_{\ell+1}^1), & 1\leq \ell\leq N, \\
\bar{\sigma}_\ell = \bar{\sigma}_\ell(x_\ell, s_{\ell-2}, m_{\ell-1}, s_{\ell-1}, i_\ell^1, i_\ell^2, i_{\ell+1}^1), & 1\leq \ell\leq N,
\end{cases} \tag{5.24}$$

such that

$$\begin{aligned}
& C_\ell^f(\bar{\phi}_\ell) + C_\ell^m(\bar{\mu}_\ell) + C_\ell^s(\bar{\sigma}_\ell) \\
& + \mathsf{E}\Bigg[V_{\ell+1}\Bigg(x_\ell + s_{\ell-2} + m_{\ell-1} + \bar{\phi}_\ell - g_\ell(i_\ell^1, i_\ell^2, I_\ell^3), s_{\ell-1}, \bar{\mu}_\ell, \bar{\sigma}_\ell, \\
&\qquad i_{\ell+1}^1, I_{\ell+1}^2, I_{\ell+2}^1\Bigg)\Bigg] \\
&= \inf_{0\leq F,M,S\leq Q} \Big\{C_\ell^f(F) + C_\ell^m(M) + C_\ell^s(S) \\
&\qquad + \mathsf{E}\left[V_{\ell+1}\left(Z_{\ell+1}(F), s_{\ell-1}, M, S, i_{\ell+1}^1, I_{\ell+1}^2, I_{\ell+2}^1\right)\right]\Big\}, \\
&\qquad \ell = 1, \ldots, N-1,
\end{aligned} \tag{5.25}$$

and

$$\begin{aligned}
& V_N\left(x_N, s_{N-2}, m_{N-1}, s_{N-1}, i_N^1, i_N^2, i_{N+1}^1\right) \\
&= C_N^f(\bar{\phi}_N) + C_N^m(\bar{\mu}_N) + C_N^s(\bar{\sigma}_N) \\
&\quad + \mathsf{E}\left[H_{N+1}\left(x_N + s_{N-2} + m_{N-1} + \bar{\phi}_N - g_N(i_N^1, i_N^2, I_N^3)\right)\right]
\end{aligned}$$

$$= \inf_{0 \le F,M,S \le Q} \Big\{ C_N^f(F) + C_N^m(M) + C_N^s(S) + \mathsf{E}\left[H_{N+1}(Z_{N+1}(F))\right] \Big\}. \tag{5.26}$$

Note also that in view of the discussion leading to (5.13), in (5.24) we have

$$\bar{\mu}_N(\cdot) = \bar{\sigma}_{N-1}(\cdot) = \bar{\sigma}_N(\cdot) = 0. \tag{5.27}$$

Next we show that the minimizers (5.24) of the dynamic programming equations give rise to optimal solution. Define

$$\begin{cases} \bar{X}_1 = x_1, \\ \bar{F}_1 = \bar{\phi}_1(\bar{X}_1, s_{-1}, m_0, s_0, i_1^1, i_1^2, i_2^1), \\ \bar{M}_1 = \bar{\mu}_1(\bar{X}_1, s_{-1}, m_0, s_0, i_1^1, i_1^2, i_2^1), \\ \bar{S}_1 = \bar{\sigma}_1(\bar{X}_1, s_{-1}, m_0, s_0, i_1^1, i_1^2, i_2^1), \end{cases} \tag{5.28}$$

and for $\ell = 2, \cdots, N$,

$$\begin{cases} \bar{X}_\ell = \bar{X}_{\ell-1} + \bar{S}_{\ell-3} + \bar{M}_{\ell-2} + \bar{F}_{\ell-1} - g_{\ell-1}(I_{\ell-1}^1, I_{\ell-1}^2, I_{\ell-1}^3), \\ \bar{F}_\ell = \bar{\phi}_\ell\left(\bar{X}_\ell, \bar{S}_{\ell-2}, \bar{M}_{\ell-1}, \bar{S}_{\ell-1}, I_\ell^1, I_\ell^2, I_{\ell+1}^1\right), \\ \bar{M}_\ell = \bar{\mu}_\ell\left(\bar{X}_\ell, \bar{S}_{\ell-2}, \bar{M}_{\ell-1}, \bar{S}_{\ell-1}, I_\ell^1, I_\ell^2, I_{\ell+1}^1\right), \\ \bar{S}_\ell = \bar{\sigma}_\ell\left(\bar{X}_\ell, \bar{S}_{\ell-2}, \bar{M}_{\ell-1}, \bar{S}_{\ell-1}, I_\ell^1, I_\ell^2, I_{\ell+1}^1\right), \end{cases} \tag{5.29}$$

where $\bar{S}_{-1} = s_{-1}$, $\bar{S}_0 = s_0$, and $\bar{M}_0 = m_0$. Using the dynamic programming equations (5.17)–(5.18), we can prove the following result.

THEOREM 5.2 (VERIFICATION THEOREM) *Assume that* (5.1) *and* (5.4)–(5.7) *hold. Then*

$$(\bar{\boldsymbol{F}}, \bar{\boldsymbol{M}}, \bar{\boldsymbol{S}}) = \left(\left(\bar{F}_1, ..., \bar{F}_N\right), \left(\bar{M}_1, ..., \bar{M}_N\right), \left(\bar{S}_1, ..., \bar{S}_N\right)\right)$$

given in (5.28)–(5.29) *are optimal decisions to the problem. That is,*

$$H_1(x_1) + \mathsf{E}\left[\sum_{\ell=1}^{N} \left(C_\ell^f(\bar{F}_\ell) + C_\ell^m(\bar{M}_\ell) + C_\ell^s(\bar{S}_\ell) + H_{\ell+1}(\bar{X}_{\ell+1})\right)\right]$$
$$= V_1\left(x_1, s_{-1}, m_0, s_0, i_1^1, i_1^2, i_2^1\right). \tag{5.30}$$

REMARK 5.3 Theorems 5.1 and 5.2 establish the existence of an optimal nonanticipative policy. That is, there exists a policy in the class of all history-dependent policies whose objective function value equals the value function defined in (5.12), and there, in turn, exists a nonanticipative policy defined by (5.28)–(5.29), which provides the same value for the objective function.

Proof of Theorem 5.2 By (5.25)–(5.26), we know that

$$((\bar{F}_1, ..., \bar{F}_N), (\bar{M}_1, ..., \bar{M}_N), (\bar{S}_1, ..., \bar{S}_N)) \in \mathcal{A}_1$$

—that is, it is a history-dependent policy. Next we show that equation (5.30) holds. It suffices to show that for any

$$((F_1, ..., F_N), (M_1, ..., M_N)), (S_1, ..., S_N)) \in \mathcal{A}_1,$$

with the corresponding X_ℓ $(1 \le \ell \le N)$ obtained from (5.8), we have

$$\begin{aligned} H_1(x_1) + \mathsf{E}\Bigg[\sum_{\ell=1}^{N}\Big(C_\ell^f(F_\ell) + C_\ell^m(M_\ell) + C_\ell^s(S_\ell) + H_{\ell+1}(X_{\ell+1})\Big)\Bigg] \\ \ge \mathsf{E}\Bigg[\sum_{\ell=1}^{N}\Big(C_\ell^f(\bar{F}_\ell) + C_\ell^m(\bar{M}_\ell) + C_\ell^s(\bar{S}_\ell) + H_{\ell+1}(\bar{X}_{\ell+1})\Big)\Bigg] \\ +H_1(x_1). \end{aligned} \tag{5.31}$$

By the definition of $(\bar{F}_1, \bar{M}_1, \bar{S}_1)$ and (5.25) with $\ell = 1$, it is possible to obtain

$$\begin{aligned} & C_1^f(\bar{F}_1) + C_1^m(\bar{M}_1) + C_1^s(\bar{S}_1) \\ & +\mathsf{E}\left[V_2\left(\bar{X}_2, \bar{S}_0, \bar{M}_1, \bar{S}_1, I_2^1, I_2^2, I_3^1\right)\right] \\ & \quad \le C_1^f(F_1) + C_1^m(M_1) + C_1^s(S_1) \\ & \quad\quad +\mathsf{E}\left[V_2\left(X_2, S_0, M_1, S_1, I_2^1, I_2^2, I_3^1\right)\right]. \end{aligned} \tag{5.32}$$

Furthermore, from the history-dependent property of the decisions, we know that $(\bar{F}_1, \bar{M}_1, \bar{S}_1)$ and (F_1, M_1, S_1) are constant vectors and that $(\bar{F}_2, \bar{M}_2, \bar{S}_2)$ and (F_2, M_2, S_2) are dependent on only $\{I_1^1, I_1^2, I_1^3, I_2^1, I_2^2, I_3^1\}$ $(= \mathcal{F}_2)$. Thus, by (5.25) and (5.29),

$$\begin{aligned}
&V_2\left(X_2, S_0, M_1, S_1, I_2^1, I_2^2, I_3^1\right) \\
&\quad = H_2(X_2) + \inf_{0 \le F, M, S \le Q} \Big\{ C_2^f(F) + C_2^m(M) + C_2^s(S) \\
&\quad +\mathsf{E}\Big[V_3\left(X_2 + S_0 + M_1 + F - g_2(I_2^1, I_2^2, I_2^3),\right. \\
&\qquad\qquad \left. S_1, M, S, I_3^1, I_3^2, I_4^1\right) \Big| \mathcal{F}_2\Big]\Big\} \\
&\quad \le H_2(X_2) + C_2^f(F_2) + C_2^m(M_2) + C_2^s(S_2) \\
&\quad +\mathsf{E}\Big[V_3\left(X_3, S_1, M_2, S_2, I_3^1, I_3^2, I_4^1\right) \Big| \mathcal{F}_2\Big],
\end{aligned} \tag{5.33}$$

and

$$\begin{aligned}
&V_2\left(\bar{X}_2, \bar{S}_0, \bar{M}_1, \bar{S}_1, I_2^1, I_2^2, I_3^1\right) \\
&\quad = H_2(\bar{X}_2) + C_2^f(\bar{F}_2) + C_2^m(\bar{M}_2) + C_2^s(\bar{S}_2) \\
&\quad +\mathsf{E}\Big[V_3\left(\bar{X}_3, \bar{S}_1, \bar{M}_2, \bar{S}_2, I_3^1, I_3^2, I_4^1\right) \Big| \mathcal{F}_2\Big].
\end{aligned} \tag{5.34}$$

Therefore, it follows from (5.34) that

$$\begin{aligned}
&\mathsf{E}\left[V_2\left(\bar{X}_2, \bar{S}_0, \bar{M}_1, \bar{S}_1, I_2^1, I_2^2, I_3^1\right)\right] \\
&\quad = \mathsf{E}\Big\{H_2(\bar{X}_2) + C_2^f(\bar{F}_2) + C_2^m(\bar{M}_2) + C_2^s(\bar{S}_2) \\
&\quad +\mathsf{E}\Big[V_3(\bar{X}_3, \bar{S}_1, \bar{M}_2, \bar{S}_2, I_3^1, I_3^2, I_4^1)\Big| \mathcal{F}_2\Big]\Big\},
\end{aligned} \tag{5.35}$$

and from (5.33) that

$$\begin{aligned}
&\mathsf{E}\left[V_2\left(X_2, S_0, M_1, S_1, I_2^1, I_2^2, I_3^1\right)\right] \\
&\quad \le \mathsf{E}\Big\{H_2(X_2) + C_2^f(F_2) + C_2^m(M_2) + C_2^s(S_2) \\
&\quad +\mathsf{E}\Big[V_3\left(X_3, S_1, M_2, S_2, I_3^1, I_3^2, I_4^1\right) \Big| \mathcal{F}_2\Big]\Big\}.
\end{aligned} \tag{5.36}$$

Combining (5.32)–(5.36) yields

$$
\begin{aligned}
&H_1(x_1) + \mathsf{E}\left[\sum_{\ell=1}^{2}\left(C_\ell^f(\bar{F}_\ell) + C_\ell^m(\bar{M}_\ell) + C_\ell^s(\bar{S}_\ell)\right) + H_2(\bar{X}_2)\right] \\
&+\mathsf{E}\left[V_3\left(\bar{X}_3, \bar{S}_1, \bar{M}_2, \bar{S}_2, I_3^1, I_3^2, I_4^1\right)\right] \\
&\le H_1(x_1) + \mathsf{E}\left[\sum_{\ell=1}^{2}\left(C_\ell^f(F_\ell) + C_\ell^m(M_\ell) + C_\ell^s(S_\ell)\right) + H_2(X_2)\right] \\
&+\mathsf{E}\left[V_3\left(X_3, S_1, M_2, S_2, I_3^1, I_3^2, I_4^1\right)\right].
\end{aligned}
\tag{5.37}
$$

Repeating (5.35) and (5.36), we finally prove that (5.31) holds. □

5.4. Optimality of Base-Stock Type Policies

For a further analysis of the problem, the dynamic programming equations (5.17)–(5.18) involving order quantities F, M, and S as decision variables traditionally are recast as variables involving respective inventory positions that would be attained after the respective orders are delivered. Thus, we replace F by $\phi - y_\ell$, M by $\mu - (\phi + s_{\ell-1})$, and S by $\sigma - \mu$ in (5.17)–(5.18), so that ϕ, μ, and σ are the *postorder inventory positions* after the delivery of fast, medium, and slow orders, respectively. When there are only two delivery modes, this transformation of variables changes the problem into a standard one-delivery-mode problem. As a result, such a transformation has been used widely to analyze problems with two delivery modes (see, e.g., Chapters 3 and 4; Scheller-Wolf and Tayur [8]; and Sethi, Yan, and Zhang [10]).

However, when there are more than two delivery modes, the transformation does not reduce the problem to a single-delivery mode problem. Thus, the methodology developed in Chapter 3 or in Sethi, Yan, and Zhang [10, 11] does not work. However, to get the optimal policy, it is possible to directly analyze natural constraints that are required on the minimizers of the convex cost functions resulting from fast, medium, and slow orders.

Before we write the dynamic programming equations in terms of ϕ, μ, and σ, we make another simplification. From (5.17) and (5.19), one can see that it is possible to replace $(x_\ell+s_{\ell-2}+m_{\ell-1})$ by y_ℓ, called the *inventory position.* This is because in the infimand (the expression inside the inf operation) of (5.17), the terms x_ℓ, $s_{\ell-2}$, and $m_{\ell-1}$ appear only as their sum. Thus, the decisions F, M, and S depend on x_ℓ, $s_{\ell-2}$, and $m_{\ell-1}$ only through their sum $(x_\ell+s_{\ell-2}+m_{\ell-1})$. The same simplification holds for (5.18). With these observations, the dynamic programming equations (5.17)–(5.18) can be modified as follows:

$$\begin{aligned}
&\hat{U}_\ell\left(y_\ell, s_{\ell-1}, i_\ell^1, i_\ell^2, i_{\ell+1}^1\right)\\
&= \inf_{\substack{\phi \geq y_\ell\\ \mu \geq \phi + s_{\ell-1}\\ \sigma \geq \mu}} \Big\{ C_\ell^f(\phi - y_\ell) + C_\ell^m(\mu - (\phi + s_{\ell-1}))\\
&\quad + C_\ell^s(\sigma - \mu) + \mathsf{E}\left[H_{\ell+1}(\phi - g_\ell(i_\ell^1, i_\ell^2, I_\ell^3))\right]\\
&\quad + \mathsf{E}\left[\hat{U}_{\ell+1}\left(\mu - g_\ell(i_\ell^1, i_\ell^2, I_\ell^3), \sigma - \mu, i_{\ell+1}^1, I_{\ell+1}^2, I_{\ell+2}^1\right)\right]\Big\},\\
&\ell = 1, ..., N-1,
\end{aligned} \tag{5.38}$$

$$\begin{aligned}
&\hat{U}_N\left(y_N, s_{N-1}, i_N^1, i_N^2, i_{N+1}^1\right)\\
&= \inf_{\substack{\phi \geq y_N\\ \mu \geq \phi + s_{N-1}\\ \sigma \geq \mu}} \Big\{ C_N^f(\phi - y_N) + C_N^m(\mu - (\phi + s_{N-1}))\\
&\quad + C_N^s(\sigma - \mu) + \mathsf{E}\left[H_{N+1}\left(\phi - g_N(i_N^1, i_N^2, I_N^3)\right)\right]\Big\}.
\end{aligned} \tag{5.39}$$

Let $\hat{V}_\ell(y_\ell, s_{\ell-1}, i_\ell^1, i_\ell^2, i_{\ell+1}^1)$ be the solution of (5.38)–(5.39). Similar to the discussion preceding Theorem 5.2, there exist feasible minimizing functions

$$\left\{\begin{array}{ll}
\hat{\phi}_\ell\left(y_\ell, s_{\ell-1}, i_\ell^1, i_\ell^2, i_{\ell+1}^1\right) & (1 \leq \ell \leq N),\\
\hat{\mu}_\ell\left(y_\ell, s_{\ell-1}, i_\ell^1, i_\ell^2, i_{\ell+1}^1\right) & (1 \leq \ell \leq N),\\
\hat{\sigma}_\ell\left(y_\ell, s_{\ell-1}, i_\ell^1, i_\ell^2, i_{\ell+1}^1\right) & (1 \leq \ell \leq N),
\end{array}\right. \tag{5.40}$$

such that

$$\begin{aligned}
&\hat{V}_\ell\left(y_\ell, s_{\ell-1}, i_\ell^1, i_\ell^2, i_{\ell+1}^1\right)\\
&= C_\ell^f(\hat{\phi}_\ell - y_\ell) + C_\ell^m(\hat{\mu}_\ell - (\hat{\phi}_\ell + s_{\ell-1})) + C_\ell^s(\hat{\sigma}_\ell - \hat{\mu}_\ell)\\
&\quad + \mathsf{E}\left[H_{\ell+1}\left(\hat{\phi}_\ell - g_\ell(i_\ell^1, i_\ell^2, I_\ell^3)\right)\right]\\
&\quad + \mathsf{E}\left[\hat{V}_{\ell+1}\left(\hat{\mu}_\ell - g_\ell(i_\ell^1, i_\ell^2, I_\ell^3), \hat{\sigma}_\ell - \hat{\mu}_\ell, i_{\ell+1}^1, I_{\ell+1}^2, I_{\ell+2}^1\right)\right],\\
&\quad \ell = 1, ..., N-1,
\end{aligned}$$

$$\begin{aligned}
&\hat{V}_N\left(y_N, s_{N-1}, i_N^1, i_N^2, i_{N+1}^1\right)\\
&= C_N^f\left(\hat{\phi}_N - y_N\right) + C_N^m\left(\hat{\mu}_N - (\hat{\phi}_N + s_{N-1})\right)\\
&\quad + C_N^s\left(\hat{\sigma}_N - \hat{\mu}_N\right) + \mathsf{E}\left[H_{N+1}\left(\hat{\phi}_N - g_N(i_N^1, i_N^2, I_N^3)\right)\right].
\end{aligned}$$

Note that just as in (5.27), in (5.40) we have

$$\hat{\mu}_N(\cdot) = \hat{\phi}_N(\cdot), \ \hat{\sigma}_{N-1}(\cdot) = \hat{\mu}_{N-1}(\cdot), \text{ and } \hat{\sigma}_N(\cdot) = \hat{\phi}_N(\cdot). \tag{5.41}$$

Let $\hat{\sigma}_0 = m_0 + s_{-1}$ and $\hat{\mu}_0 = m_0 + s_{-1} + s_0$, and let

$$\begin{cases} \hat{Y}_1 = y_1, \\ \hat{Y}_2 = \hat{\mu}_1 \left(\hat{Y}_1, \hat{\sigma}_0 - \hat{\mu}_0, i_1^1, i_1^2, i_2^1\right) - g_1(i_1^1, i_1^2, I_1^3\right), \\ \hat{Y}_\ell = \hat{\mu}_{\ell-1} \left(\hat{Y}_{\ell-1}, \hat{\sigma}_{\ell-2} - \hat{\mu}_{\ell-2}, I_{\ell-1}^1, I_{\ell-1}^2, I_\ell^1\right) \\ \quad -g_{\ell-1}(I_{\ell-1}^1, I_{\ell-1}^2, I_{\ell-1}^3), \\ \quad \ell = 3, ..., N. \end{cases}$$

Define with a slight abuse of notation, for $\ell = 1, \cdots, N$,

$$\begin{cases} \hat{F}_\ell = \hat{\phi}_\ell \left(\hat{Y}_\ell, \hat{\sigma}_{\ell-1} - \hat{\mu}_{\ell-1}, I_\ell^1, I_\ell^2, I_{\ell+1}^1\right) - \hat{Y}_\ell, \\ \hat{M}_\ell = \hat{\mu}_\ell \left(\hat{Y}_\ell, \hat{\sigma}_{\ell-1} - \hat{\mu}_{\ell-1}, I_\ell^1, I_\ell^2, I_{\ell+1}^1\right) \\ \qquad - \left(\hat{Y}_\ell + \hat{\sigma}_{\ell-1} - \hat{\mu}_{\ell-1} + \hat{F}_\ell\right), \\ \hat{S}_\ell = \hat{\sigma}_\ell \left(\hat{Y}_\ell, \hat{\sigma}_{\ell-1} - \hat{\mu}_{\ell-1}, I_\ell^1, I_\ell^2, I_{\ell+1}^1\right) \\ \qquad - \left(\hat{Y}_\ell + \hat{\sigma}_{\ell-1} - \mu_{\ell-1} + \hat{F}_\ell + \hat{M}_\ell\right). \end{cases} \tag{5.42}$$

THEOREM 5.3 *Assume that* (5.1) *and* (5.4)–(5.7) *hold. Then*

$$\left(\hat{\boldsymbol{F}}, \hat{\boldsymbol{M}}, \hat{\boldsymbol{S}}\right) = \left(\left(\hat{F}_1, ..., \hat{F}_N\right), \left(\hat{M}_1, ..., \hat{M}_N\right), \left(\hat{S}_1, ..., \hat{S}_N\right)\right) \tag{5.43}$$

defined in (5.42) *are optimal decisions for the problem over* $\langle 1, N \rangle$.

Proof The proof follows the dynamic programming equations (5.38)–(5.39) in the same way as the proof of Theorem 5.2 follows from the dynamic programming equations (5.17)–(5.18). □

To obtain the optimality of a base-stock type policy, we assume that the order-cost functions are linear—that is,

$$\begin{cases} C_k^f(t) = c_k^f \cdot t, & c_k^f > 0, \\ C_k^m(t) = c_k^m \cdot t, & c_k^m > 0, \\ C_k^s(t) = c_k^s \cdot t, & c_k^s > 0, \\ c_k^s \le c_{k+1}^m \le c_{k+2}^f, & k = 1, \cdots, N-2, \\ c_{N-1}^m \le c_N^f. \end{cases} \tag{5.44}$$

Then (5.38)–(5.39) can be written as

$$\begin{aligned}
&\tilde{U}_\ell \left(y_\ell, s_{\ell-1}, i^1_\ell, i^2_\ell, i^1_{\ell+1}\right) \\
&= \inf_{\substack{\phi \geq y_\ell \\ \mu \geq \phi + s_{\ell-1} \\ \sigma \geq \mu}} \Big\{ - c^f_\ell \cdot y_\ell - c^m_\ell \cdot s_{\ell-1} + [c^f_\ell - c^m_\ell] \cdot \phi \\
&\quad + [c^m_\ell - c^s_\ell] \cdot \mu + c^s_\ell \sigma + \mathsf{E}\left[H_{\ell+1}\left(\phi - g_\ell\left(i^1_\ell, i^2_\ell, I^3_\ell\right)\right)\right] \\
&\quad + \mathsf{E}\left[\tilde{U}_{\ell+1}\left(\mu - g_\ell\left(i^1_\ell, i^2_\ell, I^3_\ell\right), \sigma - \mu, i^1_{\ell+1}, I^2_{\ell+1}, I^1_{\ell+2}\right)\right] \Big\}, \\
&\qquad \ell = 1, ..., N-1,
\end{aligned} \tag{5.45}$$

and

$$\begin{aligned}
&\tilde{U}_N \left(y_N, s_{N-1}, i^1_N, i^2_N, i^1_{N+1}\right) \\
&= \inf_{\substack{\phi \geq y_N \\ \mu \geq \phi + s_{N-1} \\ \sigma \geq \mu}} \Big\{ - c^f_N \cdot y_N - c^m_N \cdot s_{N-1} + [c^f_N - c^m_N] \cdot \phi \\
&\quad + [c^m_N - c^s_N] \cdot \mu + c^s_N \cdot \sigma + \mathsf{E}\left[H_{N+1}\left(\phi - g_N\left(i^1_N, i^2_N, I^3_N\right)\right)\right] \Big\}.
\end{aligned} \tag{5.46}$$

Let $\tilde{V}_\ell(y_\ell, s_{\ell-1}, i^1_\ell, i^2_\ell, i^1_{\ell+1})$ be the solution of (5.45)–(5.46). We have the following result on the optimality of a base-stock type policy.

THEOREM 5.4 *Assume that* (5.1), (5.4), (5.6) *and* (5.44) *hold. At the beginning of period k, $k = 1, \ldots, N$, suppose that the observed values of I^1_k, I^2_k, and I^1_{k+1} are i^1_k, i^2_k, and i^1_{k+1}, respectively, the initial inventory position is y_k, and the slow-order quantity ordered in period $(k-1)$ is s_{k-1}. Then there are base-stock levels $\bar{F}_k$ (independent of y_k but dependent on s_{k-1}, i^1_k, i^2_k, and i^1_{k+1}) and $\bar{M}_k$ (independent of y_k but dependent on s_{k-1}, i^1_k, i^2_k, and i^1_{k+1}) such that the optimal fast-order quantity F^*_k and the optimal medium-order quantity M^*_k in period k are as follows:*

$$\begin{cases} F^*_k = (\bar{F}_k - y_k)^+, \ k = 1, \ldots, N, \\ M^*_k = (\bar{M}_k - y_k - s_{k-1} - F^*_k)^+, \ \ k = 1, \ldots N-1, \ \ M^*_N = 0. \end{cases} \tag{5.47}$$

REMARK 5.4 In view of the fact that the base-stock levels $\bar{F}_k$ and $\bar{M}_k$ depend on s_{k-1}, i^1_k, i^2_k, and i^1_{k+1}, from the definitions of $\hat{\phi}_k(y_k, s_{k-1}, i^1_k, i^2_k, i^1_{k+1})$ and $\hat{\mu}_k(y_k, s_{k-1}, i^1_k, i^2_k, i^1_{k+1})$ given by (5.40),

$$\hat{\phi}_k(y_k, s_{k-1}, i^1_k, i^2_k, i^1_{k+1}) = \bar{F}_k \vee y_k, \ k = 1, \ldots, N,$$

$$\hat{\mu}_k(y_k, s_{k-1}, i^1_k, i^2_k, i^1_{k+1}) = \bar{M}_k \vee \left(\bar{F}_k \vee y_k + s_{k-1}\right), \ k = 1, \ldots, N.$$

These, with a slight abuse of notation, are the same as those given in (5.42).

The proof of Theorem 5.4 needs an important lemma, which we prove after some general discussion on base-stock policies.

A fundamental characteristic of the optimal base-stock level in the classical single-delivery-mode inventory problem is that the level is independent of the inventory position. Since any ordering policy can be converted to an order-up-to policy simply by adding the order quantity to the inventory position, the proof of the optimality of a base-stock policy requires, therefore, that we can find order-up-to levels that are independent of the inventory position.

While the base-stock level must be independent of the inventory position, it may depend on time if the problem is nonstationary or one with a finite horizon and on the states other than the inventory position if the system behavior is influenced by these, usually exogenous, states (see Sethi and Cheng [9] and Song and Zipkin [12]). In such cases, the system is sometimes referred to as world-driven and the optimal policy as the state-dependent base-stock policy, even though it must be independent of the state called the inventory position.

Since the inventory position is also a state variable, the terminology *state-dependent* base-stock levels is not quite correct. Our preference is, therefore, to use the term *order-up-to levels* if the levels can depend on time or any of the state variables including the inventory position and use the term *base-stock levels* if the levels are independent of the current inventory position.

A number of papers cited earlier prove also that the base-stock policy remains optimal with two consecutive delivery modes, provided that the ordering cost is linear (see Sethi, Yan, and Zhang [10], for example).

When we move from two modes to three modes, the issues become substantially more complicated because now there is an additional endogenous state variable—namely, the slow order placed in the previous period. From (5.40), we see that the order-up-to levels or postorder inventory positions ϕ_ℓ^*, μ_ℓ^*, and σ_ℓ^* depend, in general, on $y_\ell, s_{\ell-1}$ and the observed demand signals. However, in Theorem 5.4 we show that there are levels $\bar{F}_\ell$ and $\bar{M}_\ell$, such that $\bar{F}_\ell$ is independent of the inventory position y_ℓ and that $\bar{M}_\ell$ is independent of $y_\ell + s_{\ell-1}$. It is easily seen from (5.47) that $\bar{F}_\ell$ acts like the base-stock level in the single-mode case. Once the fast-order F_ℓ^* is placed, the "inventory position relevant for the medium order" is $y_\ell + s_{\ell-1} + F_\ell^*$. If it is less than the level $\bar{\mu}_\ell$, we order M_ℓ^*. Otherwise, we do not (that is, $M_\ell^* = 0$). This behavior is the natural generalization of the single-mode base-stock policy to the case of three modes. And since these levels are independent of their corresponding relevant inventory positions, $\bar{F}_\ell$ and $\bar{M}_\ell$ can be called the base-stock levels for the first and the second modes, respectively.

LEMMA 5.1 *Let $g(\cdot)$ and $h(\cdot)$ be convex functions with $\tilde{x}$ and $\tilde{z}$ as their respective unconstrained minima—that is, $g(\tilde{x}) = \min_x g(x)$ and $h(\tilde{z}) = \min_z h(z)$.*

For given $b \geq 0$, let $\hat{a}$ minimize $g(x) + h(x+b)$—that is,

$$g(\hat{a}) + h(\hat{a}+b) = \min_x [g(x) + h(x+b)].$$

Then for any a,

$$\min_{\substack{x \geq a \\ z \geq x+b}} [g(x) + h(z)]$$

$$= \begin{cases} g(\tilde{x}) + h(\tilde{z}), & \textit{if } \tilde{x} \geq a,\ \tilde{z} \geq \tilde{x}+b, \ \textit{Case } (i) \\ g(a) + h(\tilde{z}), & \textit{if } \tilde{x} < a,\ \tilde{z} \geq a+b, \ \textit{Case } (ii) \\ g(a) + h(a+b), & \textit{if } \tilde{x} < a,\ \tilde{z} < a+b, \ \textit{Case } (iii) \\ g(\hat{a} \vee a) + h((\hat{a} \vee a)+b), & \textit{if } \tilde{x} \geq a,\ \tilde{z} < \tilde{x}+b, \ \textit{Case } (iv) \end{cases}$$

$$= \begin{cases} g(a \vee \tilde{x}) + h(\tilde{z} \vee (a+b)), & \textit{if } \tilde{z} \geq \tilde{x}+b, \ \textit{Case I} \\ g(\hat{a} \vee a) + h((\hat{a} \vee a)+b), & \textit{if } \tilde{z} < \tilde{x}+b, \ \textit{Case II}, \end{cases} \tag{5.48}$$

where in Case II, we can always choose $\hat{a}$ so that $\tilde{z} - b \leq \hat{a} \leq \tilde{x}$. In other words, $x^* = (a \vee \tilde{x})$ and $z^* = ((a+b) \vee \tilde{z})$ in Case I and $x^* = (\hat{a} \vee a)$ and $z^* = ((\hat{a} \vee a) + b)$ in Case II minimize $g(x) + h(z)$ subject to the constraints $x \geq a$ and $z \geq x + b$.

Furthermore, if we define

$$(\bar{x}, \bar{z}) = \begin{cases} (\tilde{x}, \tilde{z}) & \textit{in Case I}, \\ (\hat{a}, \hat{a}+b) \quad \textit{or} \quad (\hat{a}, \tilde{z}) & \textit{in Case II}, \end{cases} \tag{5.49}$$

then x^* and z^* can be expressed as

$$x^* = a + (\bar{x} - a)^+, \tag{5.50}$$

$$\begin{aligned} z^* &= (x^* + b) + [\bar{z} - x^* - b]^+ \\ &= a + b + (\bar{x} - a)^+ + (\bar{z} - a - b - (\bar{x} - a)^+)^+. \end{aligned} \tag{5.51}$$

Finally, $(\bar{x}, \bar{z})$ is independent of a.

Proof Let us denote the feasible set for minimization as

$$\mathcal{D} = \{(x, z) | x \geq a,\ z \geq x + b\}.$$

We prove the results for each of the four cases, shown also in Figure 5.2. Note that a is not restricted to be positive.

Case (i): $[\tilde{x} \geq a$ and $\tilde{z} \geq \tilde{x} + b]$

Since $(\tilde{x}, \tilde{z}) \in \mathcal{D}$, the result holds trivially.

Case (ii): $[\tilde{x} < a$ and $\tilde{z} \geq a + b]$

For any $(x, z) \in \mathcal{D}$, we have $x \geq a > \tilde{x}$. By convexity of $g(\cdot)$ and the definition of $\tilde{z}$, it is obvious that

$$g(a) + h(\tilde{z}) \leq g(a) + h(z) \leq g(x) + h(z).$$

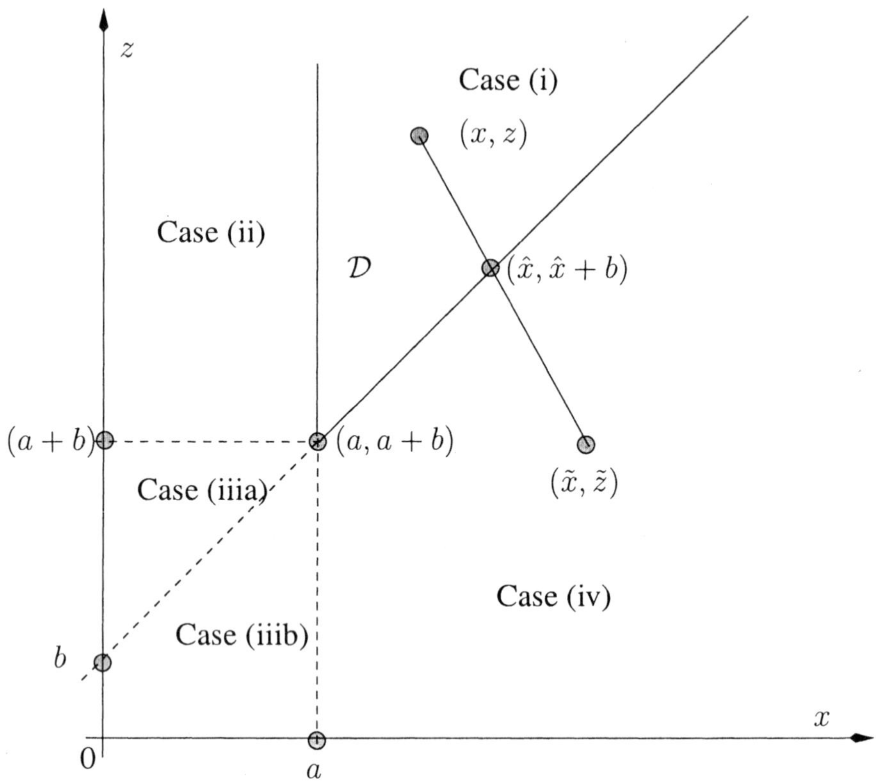

Figure 5.2. Cases (i)–(iv) and details of Case (iv)

Case (iii): $[\tilde{x} < a \text{ and } \tilde{z} < \tilde{x} + b]$
For any $(x, z) \in \mathcal{D}$, we have $x \geq a \geq \tilde{x}$ and $z \geq x + b \geq a + b > \tilde{z}$. Then

$$g(a) + h(a + b) \leq g(x) + h(z).$$

Case (iv): $[\tilde{x} \geq a \text{ and } \tilde{z} < a + b]$
It is easy to see from Figure 5.2 that a line joining any $(x, z) \in \mathcal{D}$ and $(\tilde{x}, \tilde{z})$ will intersect the line $z = x + b$, which is the 45 degree line passing through the point $(a, a + b)$. Let $(\hat{x}, \hat{x} + b)$ denote the point of intersection. Certainly, $(\hat{x}, \hat{x} + b) \in \mathcal{D}$. Moreover, depending on the location (see Figure 5.2) of (x, z), either $x \geq \hat{x} \geq \tilde{x}$ or $x \leq \hat{x} \leq \tilde{x}$, and either $z \geq \hat{x} + b \geq \tilde{z}$ or $z \leq \hat{x} + b \leq \tilde{z}$. That is, $\hat{x}$ is in the middle of x and $\tilde{x}$, and $\hat{x} + b$ is in the middle of z and $\tilde{z}$. Therefore, we have an $(\hat{x}, \hat{x} + b)$ for each $(x, z) \in \mathcal{D}$ such that

$$g(\hat{x}) + h(\hat{x} + b) \leq g(x) + h(z). \tag{5.52}$$

There are two cases to consider. When $\hat{a} \geq a$, then $(\hat{a}, \hat{a} + b)$ is feasible and

$$g(\hat{a}) + h(\hat{a} + b) \leq g(\hat{x}) + h(\hat{x} + b). \tag{5.53}$$

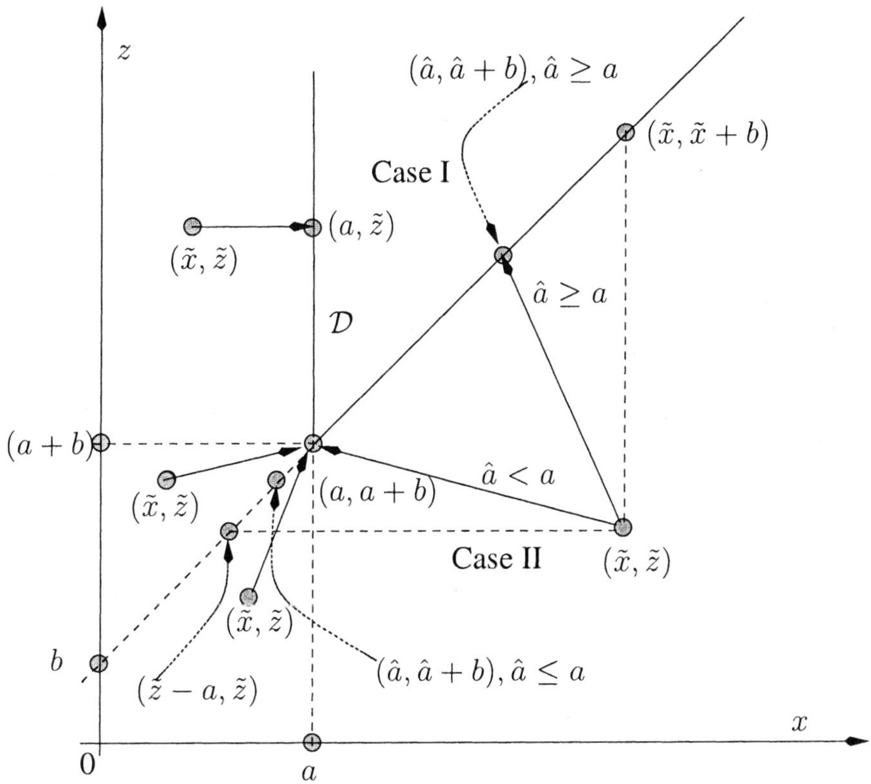

Figure 5.3. Solutions in Cases I and II

When $\hat{a} < a$, then in view of the convexity of $g(x) + h(x+b)$ in x and the fact that $(a, a+b)$ is in the middle of $(\hat{a}, \hat{a}+b)$ and $(\hat{x}, \hat{x}+b)$, we have

$$g(a) + h(a+b) \leq g(\hat{x}) + h(\hat{x}+b). \tag{5.54}$$

Case (iv) follows from (5.52)–(5.54).

We now derive the second equality in (5.48). For this, we observe that Case I consists of Cases (i), (ii), and (iiia) and that Case II consists of Cases (iiib)and (iv), where Case (iiia) is the part of Case (iii) above and including the line $z = x + b$ and Case (iiib) is the remaining part of Case (iii). We then need to show that $\tilde{z} - b \leq \hat{a} \leq \tilde{x}$ in Case II (see Figure 5.3).

To show that $\tilde{z} - b \leq \hat{a} \leq \tilde{x}$, note that for any $(x, x+b)$ with $x \geq \tilde{x}$, we have $\tilde{z} \leq \tilde{x} + b \leq x + b$. Thus, $h(\tilde{x}+b) \leq h(x+b)$, and therefore

$$g(\tilde{x}) + h(\tilde{x}+b) \leq g(x) + h(x+b). \tag{5.55}$$

Likewise, for any $(x, x+b)$ with $x \leq \tilde{z} - a$, we have $x \leq \tilde{z} - b \leq \tilde{x}$. Thus, $g(\tilde{z} - a) \leq g(x)$, so that

$$g(\tilde{z} - a) + h(\tilde{z}) \leq g(x) + h(x + b). \tag{5.56}$$

Inequalities (5.55) and (5.56) show that we can always choose an $\hat{a}$ such that $\tilde{z} - b \leq \hat{a} \leq \tilde{x}$.

Next, we show (5.49), (5.50),and (5.51).

Case I: $[\tilde{z} \geq \tilde{x} + b]$

We have

$$\begin{aligned} x^* &= \tilde{x} \vee a = a + (\tilde{x} - a)^+ = a + (\bar{x} - a)^+, \\ z^* &= \tilde{z} \vee (a + b) = \tilde{z} \vee (\tilde{x} + b) \vee (a + b) = \tilde{z} \vee ((\tilde{x} \vee a) + b) \\ &= \tilde{z} \vee (x^* + b) = (x^* + b) + [\tilde{z} - x^* - b]^+. \end{aligned}$$

Case II: $[\,\tilde{z} < \tilde{x} + b]$

We have

$$\begin{aligned} x^* &= \hat{a} \vee a = a + (\hat{a} - a)^+ = a + (\bar{x} - a)^+, \\ z^* &= \hat{a} \vee a + b = (\hat{a} + b) \vee (a + b). \end{aligned}$$

If we take $\bar{z} = \hat{a} + b$, then $z^* = \bar{z} \vee (a + b) = \bar{z} \vee ((\hat{a} \vee a) + b) = \bar{z}(x^* + b)$, and (5.51) follows from the derivation in Case I. If we take $\bar{z} = \tilde{z}$, we know from previous discussions that $\tilde{z} \leq \hat{a} + b$ in this case. Thus, $z^* = \tilde{z} \vee (\hat{a} + b) \vee (a + b) = \bar{z} \vee ((\hat{a} \vee a) + b) = \bar{z} \vee (x^* + b)$.

Finally, for either Case I or Case II, it is obvious that $\tilde{x}$ and $\hat{a}$ do not depend on a, and therefore, $\bar{x}$ as defined in (5.49) is independent of a. □

LEMMA 5.2 *Let $g(x)$ and $h(z, w)$ be two convex functions with $\tilde{x}$ and $(\tilde{z}, \tilde{w})$ as their respective unconstrained minima. There exist reals $\bar{x}$, $\bar{z}$, and $\bar{w}$ (independent of a) such that the solution to*

$$\min\{g(x) + h(z, w) |\ x \geq a, z \geq x + b, w \geq z\}$$

is given by

$$\begin{aligned} x^* &= a \vee \bar{x}, & (5.57) \\ z^* &= (x^* + b) \vee \bar{z}, & (5.58) \\ w^* &= z^* \vee \bar{w}, & (5.59) \end{aligned}$$

if $\tilde{w} \leq \tilde{z}$.

Proof Define $\tilde{w}^c(z) = \arg\min_w\{h(z, w)|w \geq z\}$. Then

$$h(z, \tilde{w}^c(z)) = \min_{w \geq z} h(z, w) \text{ is a convex function in } z. \tag{5.60}$$

This will be proved at the end.

We take $(\bar{x}, \bar{z})$ as suggested in Lemma 5.1 equation (5.49). Then the optimal solution (x^*, z^*) satisfies (5.57) and (5.58). Moreover, from the proof of Lemma 5.1, we have $\bar{z} \geq \tilde{z}$. Together with the fact that $z^* \geq \bar{z}$ and $\tilde{z} \geq \tilde{w}$, we deduce that $z^* \geq \tilde{w}$. Note that for each fixed $z^* \geq \tilde{w}$, $h(z^*, w)$ constrained on $\{w \geq z^*\}$ is convex in w with constrained minimizer $w^c(z^*) = z^*$. If we take $\bar{w} = \tilde{w}$, then (5.59) holds.

Finally we show (5.60). For each $\delta \in [0, 1]$, we have

$$\begin{aligned}
&\delta \cdot h(z_1, w^c(z_1)) + (1-\delta) \cdot h(z_2, w^c(z_2)) \\
&\quad = \delta \cdot \inf_{w \geq z_1} h(z_1, w) + (1-\delta) \cdot \inf_{w \geq z_2} h(z_2, w) \\
&\quad \geq h\left(\delta z_1 + (1-\delta) z_2, \delta w^c(z_1) + (1-\delta) w^c(z_2)\right) \\
&\quad \geq \inf_{w \geq \delta z_1 + (1-\delta) z_2} h\left(\delta z_1 + (1-\delta) z_2, w\right) \\
&\quad = h\left(\delta z_1 + (1-\delta) z_2, w^c(\delta z_1 + (1-\delta) z_2)\right).
\end{aligned}$$

□

REMARK 5.5 Going along the same lines of the proof, we can prove that if $b < 0$, Lemma 5.2 still holds.

Proof of Theorem 5.4 First, we show (5.47) for period N. We know from (5.6) that

$$c_N^f \phi + \mathsf{E}\left[H_{N+1}(\phi - g_N(i_N^1, i_N^2, I_N^3))\right] \text{ is convex in } \phi \tag{5.61}$$

and that it attains its unconstrained minimum. Let this be attained at $\bar{F}_N$, which is clearly independent of y_N. Then the minimizer of (5.61) on the region $[y_N, +\infty)$ is given by

$$\phi_N^*(y_N, i_N^1, i_N^2) = \begin{cases} \bar{F}_N, & \text{if } y_N \leq \bar{F}_N, \\ y_N, & \text{if } y_N > \bar{F}_N. \end{cases} \tag{5.62}$$

In view of this and (5.39), (5.47) for period N follows from (5.46).

Next, we prove (5.47) for period $(N-1)$. It follows from (5.46) and the convexity of $H_N(\cdot)$ and $H_{N+1}(\cdot)$ that

$$\begin{aligned}
g_{N-1}(\phi) &= [c_{N-1}^f - c_{N-1}^m] \cdot \phi \\
&\quad + \mathsf{E}\left[H_N(\phi - g_{N-1}(i_{N-1}^1, i_{N-1}^2, I_{N-1}^3))\right] \text{ is convex in } \phi,
\end{aligned} \tag{5.63}$$

and

$$\begin{aligned} h_{N-1}(\mu) &= c^m_{N-1}\mu \\ &+\mathsf{E}\left[\tilde{V}_N(\mu - g_{N-1}(i^1_{N-1}, i^2_{N-1}, I^3_{N-1}), i^1_N, I^2_N)\right] \text{ is convex in } \mu. \end{aligned} \tag{5.64}$$

Let $\tilde{F}_{N-1}$ and $\tilde{M}_{N-1}$ be unconstrained minimizers of $g_{N-1}(\phi)$ and $h_{N-1}(\mu)$, respectively. Consider, as in Lemma 5.1, two cases:

$$\begin{aligned} &\text{Case I.} \quad \tilde{M}_{N-1} \geq \tilde{F}_{N-1} + s_{N-2}, \\ &\text{Case II.} \quad \tilde{M}_{N-1} < \tilde{F}_{N-1} + s_{N-2}. \end{aligned}$$

Then by Lemma 5.1,

$$\begin{cases} \phi^*_{N-1}(y_{N-1}, s_{N-2}, i^1_{N-1}, i^2_{N-1}, i^1_N) = y_{N-1} \vee \tilde{F}_{N-1}, \\ \mu^*_{N-1}(y_{N-1}, s_{N-2}, i^1_{N-1}, i^2_{N-1}, i^1_N) = (y_{N-1} + s_{N-2}) \vee \tilde{M}_{N-1} \end{cases} \tag{5.65}$$

minimize $g_{N-1}(\phi) + h_{N-1}(\mu)$ on the region $\{(\phi, \mu) : \phi \geq y_{N-1} \text{ and } \mu \geq \phi + s_{N-2}\}$ in Case I. Therefore, by Theorem 5.3 and (5.45), we have the result for period $(N-1)$ with

$$\bar{F}_{N-1} = \tilde{F}_{N-1} \text{ and } \bar{M}_{N-1} = \tilde{M}_{N-1} \text{ in Case I.} \tag{5.66}$$

Now consider Case II. Let $\hat{F}_{N-1}$ minimize $g_{N-1}(\phi) + h_{N-1}(\phi + s_{N-2})$. Then by Case II of Lemma 5.1,

$$\begin{cases} \phi^*_{N-1}(y_{N-1}, s_{N-2}, i^1_{N-1}, i^2_{N-1}, i^1_N) = \hat{F}_{N-1} \vee y_{N-1}, \\ \mu^*_{N-1}(y_{N-1}, s_{N-2}, i^1_{N-1}, i^2_{N-1}, i^1_N) = (\hat{F}_{N-1} \vee y_{N-1}) + s_{N-2} \end{cases} \tag{5.67}$$

minimize $g_{N-1}(\phi) + h_{N-1}(\mu)$ on the region $\{(\phi, \mu) : \phi \geq y_{N-1} \text{ and } \mu \geq \phi + s_{N-2}\}$ in Case II. Consequently, by Theorem 5.3 and (5.45), we have the result for period $(N-1)$ with

$$\bar{F}_{N-1} = \hat{F}_{N-1} \text{ and } \bar{M}_{N-1} = \hat{F}_{N-1} + s_{N-2} \text{ in Case II.} \tag{5.68}$$

Combining Cases I and II, we get (5.47) for period $(N-1)$.

Next, we prove (5.47) for period $(N-2)$. We rewrite (5.45) as

$$
\begin{aligned}
&\tilde{V}_{N-2}\left(y_{N-2}, s_{N-3}, i^1_{N-2}, i^2_{N-2}, i^1_{N-1}\right) \\
&= \inf_{\substack{\phi \geq y_{N-2} \\ \mu \geq \phi + s_{N-3} \\ \sigma \geq \mu}} \Big\{ - c^f_{N-2} \cdot y_{N-2} - c^m_{N-2} \cdot s_{N-3} \\
&\quad + \mathsf{E}\left[H_{N-1}\left(\phi - g_{N-2}(i^1_{N-2}, i^2_{N-2}, I^3_{N-2})\right)\right] \\
&\quad + [c^f_{N-2} - c^m_{N-2}] \cdot \phi + [c^m_{N-2} - c^s_{N-2}] \cdot \mu + c^s_{N-2} \cdot \sigma \\
&\quad + \mathsf{E}\Big[\tilde{V}_{N-1}\left(\mu - g_{N-2}(i^1_{N-2}, i^2_{N-2}, I^3_{N-2}), \sigma - \mu,\right. \\
&\qquad\qquad \left. i^1_{N-1}, I^2_{N-1}, I^1_N\right)\Big]\Big\} \\
&= -c^f_{N-2} \cdot y_{N-2} - c^m_{N-2} \cdot s_{N-3} \\
&\quad + \inf_{\substack{\phi \geq y_{N-2} \\ \mu \geq \phi + s_{N-3}}} \Big\{ [c^f_{N-2} - c^m_{N-2}] \cdot \phi \\
&\quad + \mathsf{E}\left[H_{N-1}(\phi - g_{N-2}(i^1_{N-2}, i^2_{N-2}, I^3_{N-2}))\right] \\
&\quad + [c^m_{N-2} - c^s_{N-2}] \cdot \mu \\
&\quad + \inf_{\sigma \geq \mu} \Big(c^s_{N-2}\sigma + \mathsf{E}\Big[\tilde{V}_{N-1}\left(\mu - g_{N-2}(i^1_{N-2}, i^2_{N-2}, I^3_{N-2}),\right. \\
&\qquad\qquad \left. \sigma - \mu, i^1_{N-1}, I^2_{N-1}, I^1_N\right)\Big]\Big)\Big\}.
\end{aligned} \tag{5.69}
$$

Let $\tilde{S}_{N-2}(\mu)$ (dependent on μ, write $\tilde{S}_{N-2}(\mu)$) be the minimizer of

$$
\begin{aligned}
&\mathsf{E}\left[\tilde{V}_{N-1}\left(\mu - g_{N-2}(i^1_{N-2}, i^2_{N-2}, I^3_{N-2}), \sigma - \mu, i^1_{N-1}, I^2_{N-1}, I^1_N\right)\right] \\
&+ c^s_{N-2} \cdot \sigma
\end{aligned}
$$

with respect to σ. Then

$$
\begin{aligned}
&\inf_{\sigma \geq \mu} \Big\{ c^s_{N-2}\sigma + \mathsf{E}\Big[\tilde{V}_{N-1}\left(\mu - g_{N-2}(i^1_{N-2}, i^2_{N-2}, I^3_{N-2}),\right. \\
&\qquad\qquad \left. \sigma - \mu, i^1_{N-1}, I^2_{N-1}, I^1_N\right)\Big]\Big\}
\end{aligned}
$$

$$
\begin{aligned}
&= c^s_{N-2} \cdot (\tilde{S}_{N-2}(\mu) \vee \mu) \\
&\quad + \mathsf{E}\Big[\tilde{V}_{N-1}\Big(\mu - g_{N-2}(i^1_{N-2}, i^2_{N-2}, I^3_{N-2}), (\tilde{S}_{N-2}(\mu) - \mu)^+, \\
&\qquad\qquad i^1_{N-1}, I^2_{N-1}, I^1_N\Big)\Big].
\end{aligned}
\tag{5.70}
$$

Since the infimand is jointly convex in (σ, μ), it is easy to show that the right-hand side of (5.70) is convex in μ.

Now let

$$
\begin{aligned}
g_{N-2}(\phi) &= [c^f_{N-2} - c^m_{N-2}] \cdot \phi \\
&\quad + \mathsf{E}\left[H_{N-1}\left(\phi - g_{N-2}(i^1_{N-2}, i^2_{N-2}, I^3_{N-2})\right)\right],
\end{aligned}
\tag{5.71}
$$

and

$$
\begin{aligned}
h_{N-2}(\mu) &= [c^m_{N-2} - c^s_{N-2}] \cdot \mu + c^s_{N-2} \cdot (\tilde{S}_{N-2} \vee \mu) \\
&\quad + \mathsf{E}\Big[\tilde{V}_{N-1}\Big(\mu - g_{N-2}(i^1_{N-2}, i^2_{N-2}, I^3_{N-2}), (\tilde{S}_{N-2} - \mu)^+, \\
&\qquad\qquad i^1_{N-1}, I^2_{N-1}, I^1_N\Big)\Big].
\end{aligned}
\tag{5.72}
$$

Since $g_{N-2}(\cdot)$ and $h_{N-2}(\cdot)$ are convex, let $\tilde{F}_{N-2}$, $\tilde{M}_{N-2}$, and $\hat{F}_{N-2}$ be the minimizers of $g_{N-2}(\phi)$, $h_{N-2}(\mu)$, and $g_{N-2}(\phi) + h_{N-2}(\phi + s_{N-3})$, respectively. Note that these minimizers are independent of y_{N-2}. If

$$
\tilde{M}_{N-2} \geq \tilde{F}_{N-2} + s_{N-3}, \tag{5.73}
$$

then by Case I of Lemma 5.1, we know that

$$
\left\{
\begin{aligned}
\phi^*_{N-2} &= \phi^*_{N-2}(y_{N-2}, s_{N-3}, i^1_{N-2}, i^2_{N-2}, i^1_{N-1}) \\
&= y_{N-2} \vee \tilde{F}_{N-2}, \\
\mu^*_{N-2} &= \mu^*_{N-2}(y_{N-2}, s_{N-3}, i^1_{N-2}, i^2_{N-2}, i^1_{N-1}) \\
&= (y_{N-2} + s_{N-3}) \vee \tilde{M}_{N-2}
\end{aligned}
\right.
\tag{5.74}
$$

minimize $g_{N-2}(\phi) + h_{N-2}(\mu)$ on the region $\{(\phi, \mu) : \phi \geq y_{N-2}$ and $\mu \geq \phi + s_{N-3}\}$. Consequently, it follows from (5.70) that $(\phi^*_{N-2}, \mu^*_{N-2}, \sigma^*_{N-2})$, with

$$\sigma^*_{N-2} = (y_{N-2} + s_{N-3}) \vee \tilde{M}_{N-2} + \left(\tilde{S}_{N-2}(\tilde{M}_{N-2}) - \left[\mu^*_{N-2} \vee \tilde{S}_{N-2}(y_{N-2} + s_{N-3})\right]\right)^+,$$

is a solution of (5.69). Therefore, by Theorem 5.3 and Lemma 5.1, we know that if (5.73) holds, then

$$\bar{F}_{N-2} = \tilde{F}_{N-2},\ \bar{M}_{N-2} = \tilde{M}_{N-2},\ \bar{S}_{N-1}(\cdot) = \tilde{S}_{N-2}(\cdot). \tag{5.75}$$

If

$$\tilde{M}_{N-2} < \tilde{F}_{N-2} + s_{N-3}, \tag{5.76}$$

then, by Case II of Lemma 5.1, we know that

$$\begin{cases} \phi^*_{N-2} = \phi^*_{N-2}(y_{N-2}, s_{N-3}, i^1_{N-2}, i^2_{N-2}, i^1_{N-1}) \\ \qquad = y_{N-2} \vee \hat{F}_{N-2}, \\ \mu^*_{N-2} = \mu^*_{N-2}(y_{N-2}, s_{N-3}, i^1_{N-2}, i^2_{N-2}, i^1_{N-1}) \\ \qquad = (\hat{F}_{N-2} \vee y_{N-2}) + s_{N-3} \end{cases} \tag{5.77}$$

minimize $g_{N-2}(\phi) + h_{N-2}(\mu)$ on the region $\{(\phi, \mu) : \phi \geq y_{N-2} \text{ and } \mu \geq \phi + s_{N-3}\}$. Consequently, it follows from (5.70) that $(\phi^*_{N-2}, \mu^*_{N-2}, \sigma^*_{N-2})$ with

$$\sigma^*_{N-2} = \mu^*_{N-2} + \left(\tilde{S}_{N-2}\left(\hat{F}_{N-2} + s_{N-3}\right) - \left[\mu^*_{N-2} \vee \tilde{S}_{N-2}\left(y_{N-2} + s_{N-3}\right)\right]\right)^+,$$

is a solution of (5.69). Once again, by Theorem 5.3 and Lemma 5.1, we know that if (5.76) holds, then

$$\bar{F}_{N-2} = \hat{F}_{N-2},\ \bar{M}_{N-2} = \hat{F}_{N-2} + s_{N-3},\ \bar{S}_{N-2}(\cdot) = \tilde{S}_{N-2}(\cdot). \tag{5.78}$$

Combining (5.75) and (5.78), we have the result (5.47) for $k = N-2$. Repeating this procedure, we can prove the theorem for any period ℓ $(1 \leq \ell \leq N-3)$.□

REMARK 5.6 Thus, we have found a structural form of the optimal inventory-replenishment policy with three delivery modes and demand-forecast updates—that is, the optimal ordering decisions for fast and medium delivery modes are characterized by critical numbers known as the base stocks. The base stocks for these modes are independent of the inventory position. However, for period k, the base-stock level for slow mode is a function of the slow-delivery decision made in period $(k-1)$. In general, the optimal order policy for the slow mode is not a base-stock policy (see Feng *et al.* [4] for details).

5.5. The Nonstationary Infinite-Horizon Problem

We now consider an infinite-horizon version of the problem formulated in Section 5.2. By letting $N = \infty$ and

$$(\boldsymbol{F}, \boldsymbol{M}, \boldsymbol{S}) = ((F_n, M_n, S_n), (F_{n+1}, M_{n+1}, S_{n+1}), ...),$$

the extended real-valued objective function of the problem is

$$J_n^\infty(x_n, s_{n-2}, m_{n-1}, s_{n-1}, i_n^1, i_n^2, i_{n+1}^1, (\boldsymbol{F}, \boldsymbol{M}, \boldsymbol{S}))$$
$$= H_n(x_n) + \sum_{k=n}^{\infty} \alpha^{k-n} \mathsf{E}\Big[C_k^f(F_k) + C_k^m(M_k)$$
$$+ C_k^s(S_k) + \alpha H_{k+1}(X_{k+1})\Big], \quad (5.79)$$

where α is a given discount factor, $0 < \alpha < 1$,

$$X_{n+1} = x_n + s_{n-2} + m_{n-1} + F_n - g_n(i_n^1, i_n^2, I_n^3),$$

and X_k $(k > n+1)$ are defined by (5.8). Similar to (5.17)–(5.18), the dynamic programming equations for the problem are

$$U_n^\infty(x_n, s_{n-2}, m_{n-1}, s_{n-1}, i_n^1, i_n^2, i_{n+1}^1)$$
$$= H_n(x_n) + \inf_{\substack{F \geq 0 \\ M \geq 0 \\ S \geq 0}} \Big\{ C_n^f(F) + C_n^m(M) + C_n^s(S)$$
$$+ \alpha \mathsf{E}\Big[U_{n+1}^\infty \big(x_n + s_{n-2} + m_{n-1} + F - g_n(i_n^1, i_n^2, I_n^3), s_{n-1}, M, S,$$
$$i_{n+1}^1, I_{n+1}^2, I_{n+2}^1\big)\Big]\Big\},$$
$$n = 1, 2, ... \quad (5.80)$$

In what follows, we shall show that there exists a solution of the dynamic programming equation (5.80). Our method is that of successive approximation of the infinite-horizon problem by longer and longer finite-horizon problems. Let us therefore examine the finite-horizon approximation

$$J_n^\infty(x_n, s_{n-2}, m_{n-1}, s_{n-1}, i_n^1, i_n^2, i_{n+1}^1 (\boldsymbol{F}, \boldsymbol{M}, \boldsymbol{S}))$$

of (5.79), which is obtained by the first k-period truncation of the infinite-horizon problem. The objective function for this truncated problem is to minimize

$$J_{n,k}(x_n, s_{n-2}, m_{n-1}, s_{n-1}, i_n^1, i_n^2, i_{n+1}^1, (\boldsymbol{F}, \boldsymbol{M}, \boldsymbol{S}))$$

$$= H_n(x_n) + \sum_{k=n}^{n+k} \alpha^{k-n} \mathsf{E}\Big[C_k^f(F_k) + C_k^m(M_k) + C_k^s(S_k)$$

$$+\alpha H_{k+1}(X_{k+1})\Big]. \tag{5.81}$$

Let $V_{n,k}(x_n, s_{n-2}, m_{n-1}, s_{n-1}, i_n^1, i_n^2, i_{n+1}^1)$ be the value function of the truncated problem—that is,

$$V_{n,k}(x_n, s_{n-2}, m_{n-1}, s_{n-1}, i_n^1, i_n^2, i_{n+1}^1)$$

$$= \inf_{(\boldsymbol{F},\boldsymbol{M},\boldsymbol{S})\in\mathcal{A}_n} \Big\{ J_{n,k}\left(x_n, s_{n-2}, m_{n-1}, s_{n-1},\right.$$

$$\left. i_n^1, i_n^2, i_{n+1}^1, (\boldsymbol{F}, \boldsymbol{M}, \boldsymbol{S})\right) \Big\}.$$

Since (5.81) is a finite-horizon problem on the interval $\langle n, n+k\rangle$, we can apply Theorem 5.1 to prove that $V_{n,k}(x_n, s_{n-2}, m_{n-1}, s_{n-1}, i_n^1, i_n^2, i_{n+1}^1)$ satisfies the dynamic programming equations

$$\begin{cases} U_{n+i,k-i}\left(x_{n+i}, s_{n+i-2}, m_{n+i-1}, s_{n+i-1}, i_{n+i}^1, i_{n+i}^2, i_{n+i+1}^1\right) \\ = H_{n+i}(x_{n+i}) + \inf_{\substack{F\geq 0\\ M\geq 0\\ S\geq 0}} \Big\{C_{n+i}^f(F) + C_{n+i}^m(M) + C_{n+i}^s(S) \\ \qquad +\alpha \mathsf{E}\Big[U_{n+i+1,k-i-1}\left(Z_{n+i+1}(F), s_{n+i-1}, M, S,\right. \\ \qquad\qquad \left. i_{n+i+1}^1, I_{n+i+1}^2, I_{n+i+2}^1\right)\Big]\Big\}, \\ \qquad i = 0, ..., k-1, \\ U_{n+k,0}\left(x_{n+k}, s_{n+k-2}, m_{n+k-1}, i_{n+k}^1, i_{n+k}^2\right) \\ = H_{n+k}(x_{n+k}) + \inf_{\substack{F\geq 0\\ M\geq 0\\ S\geq 0}} \Big\{C_{n+k}^f(F) + C_{n+k}^m(M) + C_{n+k}^s(S) \\ \qquad +\alpha \mathsf{E}\Big[H_{n+k+1}(Z_{n+k+1}(F))\Big]\Big\}, \end{cases} \tag{5.82}$$

where $Z_\ell(F)$ $(n+1 \leq \ell \leq n+k)$ is as defined in (5.19). Similar to the discussion of Section 3.5, we assume that there exist constants $c > 0$ and

$M > 0$ such that for all $k \geq 1$,

$$\left|C_k^f(x_1) - C_k^f(x_2)\right| \leq c \cdot |x_1 - x_2|, \tag{5.83}$$

$$|C_k^m(x_1) - C_k^m(x_2)| \leq c \cdot |x_1 - x_2|, \tag{5.84}$$

$$|C_k^s(x_1) - C_k^s(x_2)| \leq c \cdot |x_1 - x_2|, \tag{5.85}$$

$$|H_k(x_1) - H_k(x_2)| \leq c \cdot |x_1 - x_2|, \tag{5.86}$$

$$\mathsf{E}[g_k(I_k^1, I_k^2, I_k^3)] < M < \infty. \tag{5.87}$$

Furthermore, we assume that

$$C_k^f(t) + \mathsf{E}\left[H_{k+1}(t - g_k(I_k^1, I_k^2, I_k^3))\right] \to \infty \text{ as } t \to \infty, \tag{5.88}$$

$$C_k^m(t) + \mathsf{E}\left[H_{k+1}(t - g_k(I_k^1, I_k^2, I_k^3))\right] \to \infty \text{ as } t \to \infty, \tag{5.89}$$

$$C_k^s(t) + \mathsf{E}\left[H_{k+1}(t - g_k(I_k^1, I_k^2, I_k^3))\right] \to \infty \text{ as } t \to \infty, \tag{5.90}$$

uniformly hold with respect to k.

We state the following result for the infinite-horizon problem; its proof is similar to Theorem 3.6. Here we omit it.

THEOREM 5.5 *Assume that* (5.1), (5.5)–(5.6), *and* (5.83)–(5.90) *hold. Then the limit of* $V_{n,k}(x_n, s_{n-2}, m_{n-1}, s_{n-1}, i_n^1, i_n^2, i_{n+1}^1)$ *exists as* $k \to \infty$. *Let the limit be denoted by* $V_n^\infty(x_n, s_{n-2}, m_{n-1}, s_{n-1}, i_n^1, i_n^2, i_{n+1}^1)$, *we have that*

$$V_n^\infty(x_n, s_{n-2}, m_{n-1}, s_{n-1}, i_n^1, i_n^2, i_{n+1}^1)$$

is a solution of (5.80). *Furthermore, There exist functions*

$$\bar{F}_n\left(x_n, s_{n-2}, m_{n-1}, s_{n-1}, i_n^1, i_n^2, i_{n+1}^1\right),$$
$$\bar{M}_n\left(x_n, s_{n-2}, m_{n-1}, s_{n-1}, i_n^1, i_n^2, i_{n+1}^1\right),$$

and

$$\bar{S}_n\left(x_n, s_{n-2}, m_{n-1}, s_{n-1}, i_n^1, i_n^2, i_{n+1}^1\right),$$

which provide the infima in (5.80) *with*

$$\begin{aligned} &U_n^\infty(x_n, s_{n-2}, m_{n-1}, s_{n-1}, i_n^1, i_n^2, i_{n+1}^1) \\ &= V_n^\infty(x_n, s_{n-2}, m_{n-1}, s_{n-1}, i_n^1, i_n^2, i_{n+1}^1), \end{aligned}$$

and

$$\begin{aligned} (\bar{\boldsymbol{F}}, \bar{\boldsymbol{M}}, \bar{\boldsymbol{S}}) \;=\; & \big\{\big(\bar{F}_n(x_n, s_{n-2}, m_{n-1}, s_{n-1}, i_n^1, i_n^2, i_{n+1}^1), \\ & \bar{M}_n(x_n, s_{n-2}, m_{n-1}, s_{n-1}, i_n^1, i_n^2, i_{n+1}^1), \\ & \bar{S}_n(x_n, s_{n-2}, m_{n-1}, s_{n-1}, i_n^1, i_n^2, i_{n+1}^1)\big),\ n \geq 1\big\} \end{aligned}$$

is an optimal nonanticipative policy—that is,

$$\begin{aligned} & V_1^{\infty}\left(x_1, s_{-1}, m_0, s_0, i_1^1, i_1^2, i_2^1\right) \\ & = J_1^{\infty}\left(x_1, s_{-1}, m_0, s_0, i_1^1, i_1^2, i_2^1, (\bar{\boldsymbol{F}}, \bar{\boldsymbol{M}}, \bar{\boldsymbol{S}})\right) \\ & = \inf_{(\boldsymbol{F}, \boldsymbol{M}, \boldsymbol{S}) \in \mathcal{A}} \left\{ J_1^{\infty}\left(x_1, s_{-1}, m_0, s_0, i_1^1, i_1^2, i_2^1, (\boldsymbol{F}, \boldsymbol{M}, \boldsymbol{S})\right) \right\}. \end{aligned}$$

REMARK 5.7 Theorem 5.5 does not imply that there is a unique solution of the dynamic programming equations (5.80). Moreover, it is possible to show that the value function is the minimal positive solution of (5.80). Furthermore, it is also possible to obtain a uniqueness proof provided that the cost functions $C_n^f(\cdot)$, $C_n^m(\cdot)$, $C_n^s(\cdot)$, and $H_n(\cdot)$ are subject to some additional conditions.

Next, we establish the optimality of a base-stock type policy in the same way as in Section 5.4.

THEOREM 5.6 *Assume that* (5.1), (5.6), *and* (5.44) *hold. Furthermore, let* (5.86)–(5.87) *hold. There are base-stock levels* $\bar{F}_n$ *(independent of* $y_n = x_n + s_{n-2} + m_{n-1}$*) and* $\bar{M}_n$ *(independent of* y_n*) such that if the initial inventory position at the beginning period* n *is* y_n*, and the slow-order quantity ordered in period* $(n-1)$ *is denoted by* s_{n-1}*, then the optimal fast-order quantity* F_n^* *and the optimal medium-order quantity* M_n^* *in period* n*,* $n = 1, 2, \ldots$*, are as follows:*

$$\left\{ \begin{array}{l} F_n^* = (\bar{F}_n - y_n)^+ \\ M_n^* = (\bar{M}_n - y_n - s_{n-1} - F_n^*)^+. \end{array} \right. \tag{5.91}$$

Proof The proof is a standard extension of the proof of Theorem 5.4, and is therefore omitted. □

5.6. Concluding Remarks

In this chapter, we consider a discrete-time, periodic-review inventory system with three delivery modes and demand-information updates. We show that only the fastest two modes have optimal base stocks, and provide a simple counterexample to show that the remaining one does not.

Our model generalizes several special cases in the literature. Extension of our model to include fixed order cost as in Chapter 4 for the case of the dual delivery modes, would be an interesting problem for future research. Feng, Gallego, Sethi, Yan, and Zhang [4] also generalize the notion of the base-stock policy to an inventory system with multiple delivery modes. For multiple consecutive delivery modes, they show that only the fastest two modes have optimal base stocks and that the remaining ones do not, in general.

5.7. Notes

The chapter is based on Feng, Gallego, Sethi, Yan, and Zhang [3].

For the same sets of goods, companies commonly provide their customers with a choice between different lead times or delivery alternatives. For examples, Hewlett-Packard's MOD0 boxes are assembled in its Singapore factory, but the factory allows HP's distribution centers in Roseville, CA, Grenoble, Guadalajara, and Singapore to choose between ocean and air shipments (Beyer and Ward [2]). Inventory models with more than two delivery alternatives have not received much attention in the literature. To our knowledge, Fukuda [5] and Zhang [14] are the only ones who address three-supply-mode problems. Fukuda [5] investigates the problem under an artificial assumption that the orders could be placed only in every other period. Under this assumption, he shows that the problem is equivalent to a two-supply-mode problem. Zhang [14] extends Fukuda's work to allow for three consecutive delivery modes. Zhang [14] takes unconstrained minimizers of the cost function as the base-stock levels for the three delivery modes. She uses a heuristic procedure to estimate their values. This method does not yield an optimal policy in general.

Allowing for three delivery modes extends Chapter 3 and also represents an extension of Hausmann, Lee, and Zhang [7], Scheller-Wolf and Tayur [8], Yan, Liu, and Hsu [13], Barnes-Schuster, Bassok, and Anupindi [1], and Gurnani and Tang [6], all dealing with two delivery modes.

Feng, Gallego, Sethi, Yan and Zhang [4] show that for problems with three or more consecutive modes, the base stock policies are not optimal for all but the fastest two modes. For problems with non-consecutive modes, the base-stock policy is optimal for the fastest mode, and also for the second fastest mode if it is consecutive to the fastest one.

References

[1] D. Barnes-Schuster, Y. Bassok, and R. Anupindi. Coordination and flexibility in supply contracts with options. *Manufacturing and Service Operations Management*, 4:171–207, 2002.

[2] D. Beyer and J. Ward. Network server supply chain at HP: A case study. In *Suupply Chain Structures: Coordination, Information and Optimization*, S. Song and D.D. Yao (editors), pp. 257–282, Kluwer Academic Publishers, Boston, 2001.

[3] Q. Feng, G. Gallego, S. Sethi, H. Yan, and H. Zhang. Periodic review inventory model with three consecutive delivery modes and forecast updates. *Journal of Optimization Theory and Applications*, 124:137–155, 2005.

[4] Q. Feng, G. Gallego, S. Sethi, H. Yan, and H. Zhang. Optimality and nonoptimality of the base-stock policy in inventory problems with multiple delivery modes. Working Paper, University of Texas at Dallas, Richardson, TX, 2004; a short version of the paper to appear in *Operations Research*.

[5] Y. Fukuda. Optimal policies for the inventory problem with negotiable lead time. *Management Science*, 10:690–708, 1964.

[6] H. Gurnani and C.S. Tang. Note: optimal ordering decisions with uncertain cost and demand forecast updating. *Management Science*, 45:1456–1462, 1999.

[7] W.H. Hausman, H.L. Lee, and V.L. Zhang. Optimal ordering for an inventory system with dual lead times. Working Paper, Stanford University, Stanford, CA, 1993.

[8] A. Scheller-Wolf and S. Tayur. A Markovian dual-source production-inventory model with order bands. Working Paper, Carnegie Mellon University, Pittsburgh, PA, 1998.

[9] S.P. Sethi and F. Cheng. Optimality of (s, S) policies in inventory models with Markovian demand. *Operations Research*, 45:931–939, 1997.

[10] S.P. Sethi, H. Yan, and H. Zhang. Peeling layers of an onion: An inventory model with multiple delivery modes and forecast updates. *Journal of Optimization Theory and Applications*, 108:253–281, 2001.

[11] S. Sethi, H. Yan, and H. Zhang. Inventory models with fixed costs, forecast updates and two delivery modes. *Operations Research*, 51:321–328, 2003.

[12] J. Song and P. Zipkin. Inventory control in a fluctuating demand environment. *Operations Research*, 41:351–370, 1993.

[13] H. Yan, K. Liu, and A. Hsu. Optimal ordering in a dual-supplier system with demand forecast updates. *Production and Operations Management*, 12:30–45, 2003.

[14] V.L. Zhang. Ordering policies for an inventory system with three supply modes. *Naval Research Logistics*, 43:691–708, 1996.

Chapter 6

MULTIPERIOD QUANTITY-FLEXIBILITY CONTRACTS

6.1. Introduction

As economic globalization, product proliferation and technology progression continue, customer demand and market price have become highly uncertain across many industry sectors. Improving their ability to forecast demand and price have become a major challenge for many companies. At the same time, various supply chain management tools and instruments have emerged to help companies streamline their supply chain operations. Quantity-flexibility contracts are one of these widely used supply chain management tools. The quantity-flexibility contract accommodates the lead time requirement of production and procurement and allows a timely response to changing demand.

In a stochastic production-planning environment where production and procurement decisions are made based on a rolling-horizon demand forecasting, a quantity-flexibility contract is an apparatus that can resolve clashes between suppliers and buyers. For each planning iteration, a flexible bound limits the upside and downside changes and provides a smooth production requirement for the suppliers. On the other hand, the contract allows an order to be increased or reduced with updated demand information and provides a cushion against demand uncertainty for the buyer. Specifically, a quantity-flexibility contract specifies that the supplier charges a fixed unit purchase price but gives the retailer a partial or full refund on the first ςq units returned, where q is the number of units purchased and $\varsigma \in (0, 1]$ is the flexibility factor. Alternatively, the supplier allows the retailer to add an additional purchase up to ςq at the same or a premium price.

In this chapter, we develop a model that analyzes a quantity-flexibility contract involving multiple periods, rolling-horizon demand, and forecast updates.

The contract permits the buyer to order at two distinct time stages—one at the beginning of a period and another at the time before the demand realizes at the end of the period. At the first stage, the buyer purchases q units of a product at price p. This gives him an option to purchase up to ςq units of the same product at price $p_c > p$ at the second stage, where $0 < \varsigma \leq 1$ is known as the flexibility bound. In addition, the buyer can purchase any amount in the spot market at the prevailing market price. The contract provides the buyer with both price and quantity protection against the demand and price uncertainties and, at the same time, ensures minimum production level for the supplier.

Our model differs from most of the existing models of quantity-flexibility contracts in the following ways: (i) we provide a model that allows spot-market purchases in addition to contract purchases; (ii) the contract has a flexibility bound that specifies the degree of the flexibility; (iii) we model both speculative and reactive decisions—in particular, how both speculative and reactive decision are related to the information revisions, such as demand- and price-information updates; (iv) with stochastic comparison theory, we characterize the impacts on the optimal policy and profit induced by the quality of forecast updates; and (v) we extend our results to the multiple-period case.

The rest of the chapter is organized as follows. In Section 6.2, we model a single-period contract and give some fundamental structural results. In Section 6.3, we provide explicit optimal solutions for every possible observation of the signal and the market price at stage 2. For the worthless and perfect information updates, respectively, we obtain closed-form solutions at stage 1 in Section 6.4. In Section 6.5, we use the stochastic comparison theory to establish results relating to the quality of information revisions. The model is extended to allow for a finite number of periods in Section 6.6. Section 6.7 is devoted to a numerical example. The chapter is concluded in Sections 6.8 and 6.9.

6.2. Model and Problem Formulation

In this section, we design a one-period, two-stage quantity-flexibility supply contract between a buyer and a supplier. The contract is an agreement between a buyer and a supplier. The contract makes it possible for the buyer to have an option to increase a certain percentage of its initial orders in a later stage. Specially, with limited information about its customer demand and market price, the buyer signs a quantity-flexibility contract with the supplier that details the terms of supply: the purchase quantity q and the unit price p. The contract allows the buyer to argument the initial purchase quantity by up to an amount ςq in a later stage at a price p_c such that $p_c \geq p$. In addition to the contract, the buyer has an option to purchase the same product from a spot market at the market price. The decision and information dynamics are illustrated in Figure 6.1.

At stage 1, with the knowledge of unit price p, the contract-unit price p_c of the future optional purchase, the distribution of the spot-market price, and the customer demand, the buyer makes a decision of initial purchase quantity q. The buyer is also aware that the information of the customer demand and the spot-market price will be updated at stage 2. At that time, the uncertainty of customer demand is reduced.

At stage 2, it is possible for the buyer to make a final adjustment in responding to the new information obtained between stage 1 and stage 2. The buyer can purchase additional product q_c, such that $q_c \leq \varsigma q$, at the contract price p_c. Moreover, the buyer can purchase the same product from a spot market at the market price. We further assume that the spot-market price can be modeled as a random variable P_s taking value in the interval $[p_{sl},\ p_{sh}]$ with $p_{sh} \geq p_{sl} > 0$. The decision at stage 2 is to choose the purchase quantity q_s from the spot market at the prevailing market price p_s and $q_c(q_c \leq \varsigma q)$ on-contract at price p_c. Note that the degree of quantity flexibility is determined by the flexibility bound ς and the initial-purchase quantity q jointly.

Finally, after stage 2, the customer demand realizes. The buyer is assumed to lose revenue r for each unit of unsatisfied demand, and excess inventory is assumed to have a salvage value of s. To avoid trivial cases, we assume throughout this chapter that

$$r > \max\{p_{sh}, p_c\} \ \text{ and } \ s < \min\{p_{sl}, p\}. \tag{6.1}$$

The above sequence of events is displayed in Figure 6.1

We use D to denote customer demand and I to represent the information observed between stage 1 and stage 2. We assume that D and I are random variables, not necessarily independent. Let

$$\begin{aligned}
\Theta(\cdot,\cdot) &= \text{the joint distribution function of } D \text{ and } I;\\
\theta(\cdot,\cdot) &= \text{the joint density function of } D \text{ and } I;\\
\Lambda(\cdot) &= \text{the marginal distribution function of } I;\\
\lambda(\cdot) &= \text{the marginal density function of } I;\\
\psi(\cdot|i) &= \text{the conditional density function of } D \text{ given } I = i;\\
\Psi(\cdot|i) &= \text{the conditional distribution function of } D \text{ given } I = i.
\end{aligned}$$

The optimal profit is defined as

$$\begin{aligned}
\pi_1^* &= \max_{q\geq 0} \Pi_1(q)\\
&= \max_{q\geq 0}\left\{-pq + \mathsf{E}\left(\max_{\substack{0\leq q_s<\infty\\ 0\leq q_c\leq\varsigma q}} \Pi_2(q, q_s, q_c, I, P_s)\right)\right\}, \qquad (6.2)
\end{aligned}$$

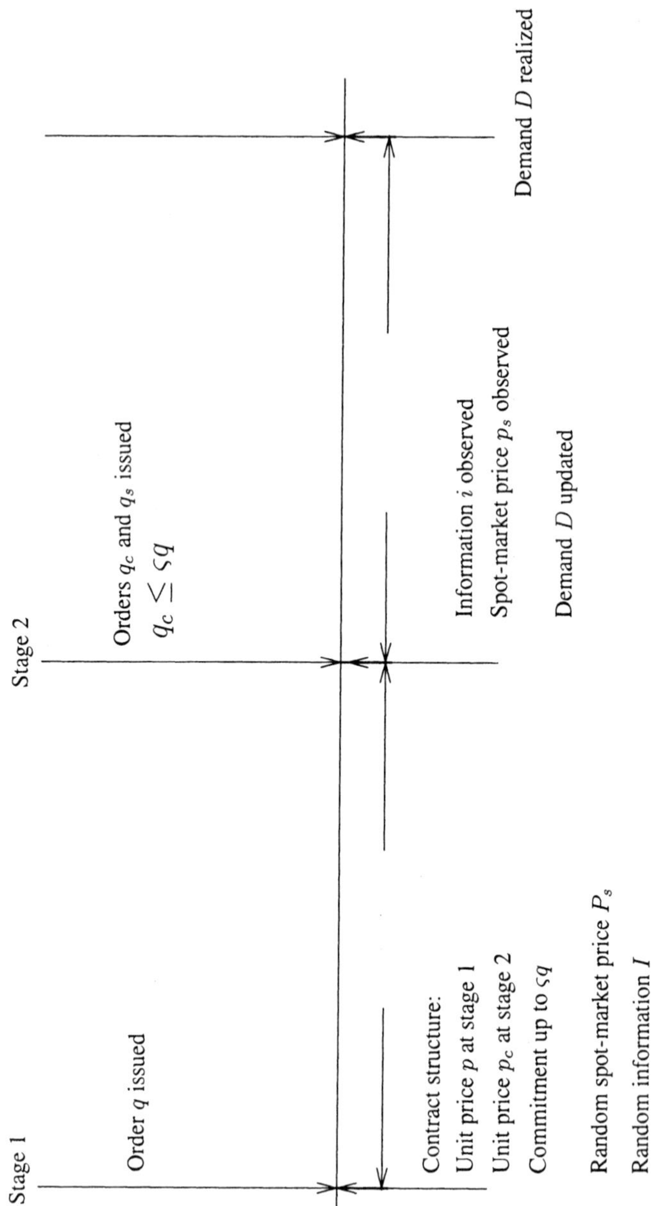

Figure 6.1. A time line of the decision and information dynamics

where

$$
\begin{aligned}
&\Pi_2(q, q_s, q_c, I, P_s) \\
&= \mathsf{E}\Bigg[\Bigg(r \cdot (D \wedge (q + q_s + q_c)) \\
&\qquad + s \cdot (q + q_s + q_c - D)^+ - p_c q_c - P_s q_s\Bigg)\Bigg|(I, P_s)\Bigg].
\end{aligned} \tag{6.3}
$$

In (6.2), pq, represents the ordering cost incurred at stage 1. The second term of (6.2), $\Pi_2(q, q_s, q_c, I, P_s)$, corresponds to the random profit received by the buyer at stage 2 given I and P_s. Therefore, the buyer's problem is to determine the optimal purchase decisions, denoted by (q^*, q_s^*, q_c^*), for maximizing the total expected profit. Clearly, q_s^* and q_c^* depend on q, I, and P_s. To highlight the above dependence, we sometimes write these contingent decisions as $q_s^*(q, I, P_s)$ and $q_c^*(q, I, P_s)$, respectively. To solve the problem, we first determine the optimal $q_s^*(q, i, p_s)$ and $q_c^*(q, i, p_s)$ for given q, $I = i$ and $P_s = p_s$—that is, first solve

$$
\max_{\substack{0 \le q_s < \infty \\ 0 \le q_c \le \varsigma q}} \Pi_2(q, q_s, q_c, i, p_s). \tag{6.4}
$$

With the notation defined above, given $(I, P_s) = (i, p_s)$, equation (6.3) can be written as

$$
\begin{aligned}
&\Pi_2(q, q_s, q_c, i, p_s) \\
&= r \int_0^{q+q_s+q_c} z \cdot \psi(z|i) \mathrm{d}z + r \cdot (q + q_s + q_c) \int_{q+q_s+q_c}^{\infty} \psi(z|i) \mathrm{d}z \\
&\quad + s \int_0^{q+q_s+q_c} [(q + q_s + q_c) - z] \cdot \psi(z|i) \mathrm{d}z - p_c q_c - p_s q_s. \\
&= -(r - s) \int_0^{q+q_s+q_c} [(q + q_s + q_c) - z] \cdot \psi(z|i) \mathrm{d}z \\
&\quad + r \cdot (q + q_c + q_s) - p_c q_c - p_s q_s.
\end{aligned} \tag{6.5}
$$

If the unit-order cost at stage 1 and the contractual unit-order cost are larger than the unit-order costs of the spot market at stage 2—that is,

$$
p_{sl} \le p_{sh} \le p \le p_c,
$$

then for any observed market price, the best strategy is to purchase all required product from the spot market—that is, $q^* = 0$ and $q_c^* = 0$. To find out $q_s^*(0, i, p_s)$, in view of (6.5), it suffices to find the value of q_s that maximizes the function

$$
r \int_0^{q_s} z \cdot \psi(z|i) \mathrm{d}z + r q_s \int_{q_s}^{\infty} \psi(z|i) \mathrm{d}z + s \int_0^{q_s} [q_s - z] \cdot \psi(z|i) \mathrm{d}z - p_s q_s.
$$

This is a newsvendor problem, and its solution is

$$q_s^*(0, i, p_s) = \Psi^{-1}\left(\frac{r - p_s}{r - s}\bigg|\, i\right).$$

As a result, the model described above reduces to a classic newsvendor model. If $p_{sl} < p < p_{sh} < p_c$, then for any observed market price, $q_c^* = 0$. Consequently, this case is the same as the case $p_{sl} < p < p_c < p_{sh}$ with $\varsigma = 0$. Similarly, if $p < p_{sl} < p_{sh} < p_c$, then $q_c^* = 0$, and it is the case $p < p_{sl} < p_c < p_{sh}$ with $\varsigma = 0$. In summary, based on $p < p_c$, it suffices to consider the following cases:

$$p < p_c \le p_{sl} \le p_{sh}; \;\; p \le p_{sl} \le p_c \le p_{sh}; \;\; p_{sl} \le p < p_c \le p_{sh}. \qquad (6.6)$$

REMARK 6.1 Note that if the spot-market price is very large—that is, $p_{sl} \to \infty$ and $p_{sh} \to \infty$—then the spot market is prohibitively expensive or nonexistent. Thus, the model reduces to a pure contract model.

REMARK 6.2 Here the spot-market price is realized at stage 2. If we were to use information I to update both the demand D and the spot-marker price P_s, an extension of the following analysis could be easily carried out.

In the next section, we take up the buyer's problem at stage 2.

6.3. Contingent Order Quantity at Stage 2

In this section, we solve for the contingent order quantities for every possible realization of the signal I and the market price P_s at stage 2. We also characterize monotonicity properties of the solutions with respect to these realizations.

THEOREM 6.1 *For any observed value* (i, p_s) *of* (I, P_s), *we have the following solutions*:

(i) *if the market price turns out to be low—that is,* $p_s \le p_c$*—then the optimal reaction at stage 2 is to order all additional required product from the spot market. That is,*

$$q_c^*(q, i, p_s) = 0, \;\; q_s^*(q, i, p_s) = \left[\Psi^{-1}\left(\frac{r - p_s}{r - s}\bigg|\, i\right) - q\right]^+;$$

(ii) *if the market price turns out to be high—that is,* $p_s > p_c$*—then the optimal reaction at stage 2 is to order additional product on the contract and to order from the spot market only when the required product exceeds the quantity-*

flexible bound. That is,

$$q_c^*(q,i,p_s) = (\varsigma q) \wedge \left[\Psi^{-1}\left(\frac{r-p_c}{r-s}\bigg| i\right) - q\right]^+,$$

$$q_s^*(q,i,p_s) = \left[\Psi^{-1}\left(\frac{r-p_s}{r-s}\bigg| i\right) - (1+\varsigma)q\right]^+.$$

Before giving the proof, let us explain the theorem in words. Statement (i) says that when the contract price p_c is higher than the prevailing market price p_s, then the buyer purchases nothing on the contract at stage 2. Instead, the buyer purchases the product from the spot market. The purchase quantity is determined by the difference of the critical fractile of the updated demand distribution and the amount purchased at stage 1. The critical fractile is determined by the demand distribution, the sales price r, the salvage value s, and the spot-market price p_s. When the market price p_s is higher than the contractual price p_c, then the buyer purchases on the contract first and considers purchasing from the spot market only after exhausting the quantity flexibility provided in the contract. Note that the buyer can purchase ςq at most. Therefore, the marginal purchase price can be the contract price p_c or the spot-market price p_s. The buyer first exhausts its option to purchase on the contract with the contract price p_c as the marginal purchasing price in the critical fractile calculation. Otherwise, in addition to exhausting the purchase option in the contract, the buyer purchases a desired additional amount from the spot market with the spot-market price p_s as the marginal price in the critical fractile calculation.

REMARK 6.3 When $\varsigma = 0$—that is, when there is no flexibility at stage 2—the contract price p_c does not impact the decision maker. So the optimal order quantity at the spot market is given by

$$q_s^*(q,i,p_s) = \left[\Psi^{-1}\left(\frac{r-p_s}{r-s}\bigg| i\right) - q\right]^+. \tag{6.7}$$

Note that, for this special case with the assumption in which P_s has a geometric distribution, Gurnani and Tang [12] also obtain (6.7).

Proof of Theorem 6.1 Let us first consider (i). Note that

$$\begin{aligned}
&\max_{\substack{0\le q_s<\infty \\ 0\le q_c\le \varsigma q}} \Pi_2(q,q_s,q_c,i,p_s) \\
&= \max_{\substack{0\le q_s<\infty \\ 0\le q_c\le \varsigma q}} \left\{ r\int_0^{q+q_s+q_c} z\cdot\psi(z|i)\mathrm{d}z \right. \\
&\qquad\qquad + r\cdot(q+q_s+q_c)\int_{q+q_s+q_c}^{\infty}\psi(z|i)\mathrm{d}z
\end{aligned}$$

$$+s\int_0^{q+q_s+q_c}[q+q_s+q_c-z]\cdot\psi(z|i)\mathrm{d}z-p_cq_c-p_sq_s\Bigg\}. \tag{6.8}$$

It follows from simple calculations that $q_s^*(q,i,p_s)$ given by (i) of the theorem maximizes

$$-(r-s)\int_0^t(t-z)\cdot\psi(z|i)\mathrm{d}z+(r-p_s)t+p_sq$$

on the interval $[q,\ +\infty)$. If $p_s\le p_c$, then for any $q_s\ge 0$ and $q_c\ge 0$,

$$\begin{aligned}
&r\int_0^{q+q_s+q_c}z\cdot\psi(z|i)\mathrm{d}z+r\cdot(q+q_s+q_c)\int_{q+q_s+q_c}^{\infty}\psi(z|i)\mathrm{d}z\\
&+s\int_0^{q+q_s+q_c}[q+q_s+q_c-z]\cdot\psi(z|i)\mathrm{d}z-p_cq_c-p_sq_s\\
&\le r\int_0^{q+q_s+q_c}z\cdot\psi(z|i)\mathrm{d}z+r\cdot(q+q_s+q_c)\int_{q+q_s+q_c}^{\infty}\psi(z|i)\mathrm{d}z\\
&+s\int_0^{q+q_s+q_c}[q+q_s+q_c-z]\cdot\psi(z|i)\mathrm{d}z-p_s\cdot(q_s+q_c)\\
&\le\max_{q\le t<\infty}\left\{-(r-s)\int_0^t(t-z)\cdot\psi(z|i)\mathrm{d}z+(r-p_s)t+p_sq\right\}.
\end{aligned}$$

Consequently, $(0,q_s^*(q,i,p_s))$ also maximizes the following function in (q_c,q_s),

$$\begin{aligned}
&r\int_0^{q+q_s+q_c}z\cdot\psi(z|i)\mathrm{d}z+r\cdot(q+q_s+q_c)\int_{q+q_s+q_c}^{\infty}\psi(z|i)\mathrm{d}z\\
&+s\int_0^{q+q_s+q_c}(q+q_s+q_c-z)\cdot\psi(z|i)\mathrm{d}z-p_cq_c-p_sq_s
\end{aligned}$$

on the region $[0,\ \varsigma q]\times[0,\ \infty)$. Therefore, the proof of (i) is completed.

Now we consider (ii). Using (6.8), it follows from simple calculations that $[(1+\varsigma)q]\wedge\left[\Psi^{-1}\left(\frac{r-p_c}{r-s}\middle|\,i\right)\vee q\right]$ maximizes

$$-(r-s)\int_0^t(t-z)\cdot\psi(z|i)\mathrm{d}z+(r-p_c)t+p_cq$$

on the interval $[q,\ (1+\varsigma)q]$, and $[(1+\varsigma)q]\vee\Psi^{-1}\left(\frac{r-p_s}{r-s}\middle|\,i\right)$ maximizes

$$\begin{aligned}
&-(r-s)\int_0^t(t-z)\cdot\psi(z|i)\mathrm{d}z+(r-p_s)t\\
&+p_s\cdot\left([(1+\varsigma)q]\wedge\left[\Psi^{-1}\left(\frac{r-p_s}{r-s}\middle|\,i\right)\vee q\right]\right)
\end{aligned}$$

on the interval $((1+\varsigma)q,\ \infty)$. If $p_s > p_c$, then for any given $q_s \geq 0$, $q_c \geq 0$, $\varepsilon > 0$, we have

$$p_s q_s + p_c q_c < p_s \cdot (q_s + \varepsilon) + p_c \cdot (q_c - \varepsilon).$$

This implies that

$$\begin{aligned} & r \int_0^{q+q_s+q_c} z\psi(z|i)\mathrm{d}z + r \cdot (q + q_s + q_c) \int_{q+q_s+q_c}^{\infty} \psi(z|i)\mathrm{d}z \\ & + s \int_0^{q+q_s+q_c} (q + q_s + q_c - z) \cdot \psi(z|i)\mathrm{d}z - p_c(q_c - \varepsilon) - p_s(q_s + \varepsilon) \\ & < r \int_0^{q+q_s+q_c} z \cdot \psi(z|i)\mathrm{d}z + r \cdot (q + q_s + q_c) \int_{q+q_s+q_c}^{\infty} \psi(z|i)\mathrm{d}z \\ & + s \int_0^{q+q_s+q_c} [q + q_s + q_c - z] \cdot \psi(z|i)\mathrm{d}z - p_s \cdot (q_s + q_c). \end{aligned}$$

Consequently, $(q_c^*(q, i, p_s), q_s^*(q, i, p_s))$ also maximizes the function

$$\begin{aligned} & r \int_0^{q+q_s+q_c} z \cdot \psi(z|i)\mathrm{d}z + r \cdot (q + q_s + q_c) \int_{q+q_s+q_c}^{\infty} \psi(z|i)\mathrm{d}z \\ & + s \int_0^{q+q_s+q_c} (q + q_s + q_c - z) \cdot \psi(z|i)\mathrm{d}z - p_c q_c - p_s q_s \end{aligned}$$

of (q_c, q_s) on the region $[0,\ \varsigma q] \times [0,\ \infty)$. Therefore, the proof of (ii) is completed. □

With an assumption that the demand D is conditionally stochastically monotone with respect to signal I, we provide an explicit expression of the optimal purchase quantity with respect to i. Without loss of generality, we assume D to be conditionally stochastically increasing with respect to I. For the case of a conditionally stochastically decreasing with respect to I, it is possible for us to redefine the signal I so that the case of the conditionally stochastically decreasing can be translated to the case of a conditionally stochastically increasing.

THEOREM 6.2 *Let the demand D be conditionally stochastically increasing with respect to I. Then for an observed market price* p_s, *there exist* $\bar{i}(q, p_c)$, $\hat{i}(q, p_c)$, $\bar{i}(q, p_s)$, *and* $\hat{i}(q, p_s)$ *defined by the relations*

$$\Psi^{-1}\left(\frac{r - p_c}{r - s}\middle|\, \bar{i}(q, p_c)\right) = q,\ \Psi^{-1}\left(\frac{r - p_c}{r - s}\middle|\, \hat{i}(q, p_c)\right) = (1 + \varsigma)q,$$

$$\Psi^{-1}\left(\frac{r - p_s}{r - s}\middle|\, \bar{i}(q, p_s)\right) = q,\ \Psi^{-1}\left(\frac{r - p_s}{r - s}\middle|\, \hat{i}(q, p_s)\right) = (1 + \varsigma)q,$$

such that

(i) *if* $p_s \leq p_c$, *then*

$$q_c^*(q, i, p_s) = 0,$$

$$q_s^*(q, i, p_s) = \begin{cases} 0, & \text{if } i \leq \bar{i}(q, p_s), \\ \Psi^{-1}\left(\frac{r-p_s}{r-s}\Big| i\right) - q, & \text{if } i > \bar{i}(q, p_s); \end{cases}$$

(ii) *if* $p_c < p_s$, *then*

$$q_c^*(q, i, p_s) = \begin{cases} 0, & \text{if } i \leq \bar{i}(q, p_c), \\ \Psi^{-1}\left(\frac{r-p_c}{r-s}\Big| i\right) - q, & \text{if } \bar{i}(q, p_c) < i \leq \hat{i}(q, p_c), \\ \varsigma q, & \text{if } i > \hat{i}(q, p_c), \end{cases}$$

$$q_s^*(q, i, p_s) = \begin{cases} 0, & \text{if } i \leq \hat{i}(q, p_s), \\ \Psi^{-1}\left(\frac{r-p_s}{r-s}\Big| i\right) - (1+\varsigma)q, & \text{if } i > \hat{i}(q, p_s). \end{cases}$$

Proof Statements (i) and (ii) follow directly from the corresponding results (i) and (ii) in Theorem 6.1, respectively, when D is conditionally stochastically increasing with respect to I. □

REMARK 6.4 Statements (i) and (ii) indicate that when the conditional demand distribution D given $I = i$ has a monotonicity structure, the optimal purchase quantity at stage 2 has the same monotone structure with respect to the observed information i.

REMARK 6.5 When $\varsigma = 0$—that is, when there is no flexibility at stage 2, then, for any observed market price, if the conditional distribution of D given $I = i$ is increasing in i, the optimal spot-market purchase is

$$q_s^*(q, i, p_s) = \begin{cases} 0, & \text{if } i \leq \bar{i}(q, p_s), \\ \Psi^{-1}\left(\frac{r-p_s}{r-s}\Big| i\right) - q, & \text{if } i > \bar{i}(q, p_s). \end{cases} \tag{6.9}$$

Note that for this special case with the assumption in which P_s has a geometric distribution, Gurnani and Tang [12] also obtain (6.9).

REMARK 6.6 Note that (6.1) implies

$$r > \max\{\mathsf{E}[P_s], p_c\} \text{ and } s < \min\{\mathsf{E}[P_s], p\}. \tag{6.10}$$

Regarding Theorem 6.1, since its proof is based on the classical newsboy problem, it can be easily shown that if $p_c < r \leq p_s$, then $q_s^*(q, i, p_s) = 0$, and if $p_s \leq s < p$, then $q_s^*(q, i, p_s) = 0$. These are the cases that do not occur under (6.1), but occur under (6.10). Going along the lines of the proof of Theorem 6.2, we can show that Theorem 6.2 holds also for these cases.

6.4. Optimal Purchase Quantity at Stage 1

With the knowledge of the optimal reaction plan at stage 2 derived in the previous section, it is possible to determine the purchase quantity q at stage 1. This is done by substituting in (6.2) for q_s and q_c by their optimal quantities $q_c^*(q, I, P_s)$ and $q_s^*(q, I, P_s)$ and solving the optimization problem

$$\pi_1^* = \max_{q \geq 0} \Big\{ -pq + \mathsf{E}\Big[\Pi_2\left(q, q_s^*(q, I, P_s), q_c^*(q, I, P_s), I, P_s\right)\Big]\Big\}. \tag{6.11}$$

This is a problem of maximizing an objective function with a single variable q. For given values of the problem parameters and observations i and p_s, it can be easily solved numerically. One could also use the Kuhn-Tucker theory to derive the first-order conditions for a maximum. Such an approach was used by Brown and Lee [5] on a related problem.

For a further mathematical analysis of the problem, we need to simplify the distributions of the random variables involved. To begin with, we assume that the market price is geometrically distributed. Specifically, we make the following assumptions:

ASSUMPTION 6.1 *The market price P_s has the value p_{sl} with probability β and the value p_{sh} with probability $(1 - \beta)$.*

ASSUMPTION 6.2 *D is conditionally stochastically increasing with respect to I.*

It is clear from (6.11) that the initial order quantity q^* depends on several factors, including ς. It is also easy to see that the "level" of flexibility is jointly determined by ς and q^*. The flexibility level increases as ς increases and as q^* increases. It is therefore important to know how q^* relates to ς. This is the subject of the following theorem.

THEOREM 6.3 *Under Assumptions* 6.1 *and* 6.2, *for a given set of purchase and contract prices, we have*

(i) *the initial optimal order quantity q^* is nonincreasing in ς;*

(ii) *the optimal expected profit is nondecreasing in ς.*

REMARK 6.7 In the contractual framework, when the initial purchase quantity is made at stage 1, the buyer must consider not only the unit-order costs at stage 2 but also the level of flexibility. Recall that the flexibility is jointly determined by flexibility factor ς and the initial purchase quantity. To maintain a same level of flexibility, it is possible either to increase ς and to reduce the initial purchase quantity or to increase the initial purchase quantity and to reduce ς.

Proof of Theorem 6.3 First, we consider the case $p_{sl} < p_c \leq p_{sh}$. It follows from Theorem 6.2 that

$$
\begin{aligned}
&-pq + \mathsf{E}\left(\max_{\substack{0\leq q_s<\infty \\ 0\leq q_c\leq\varsigma q}} \Pi_2(q, q_s, q_c, I, P_s)\right) \\
&= -pq + \beta \int_{-\infty}^{\bar{i}(q,p_{sl})} \left[(s-r)\int_0^q (q-z)\cdot\psi(z|i)\mathrm{d}z + rq\right] \mathrm{d}\Lambda(i) \\
&\quad +\beta \int_{\bar{i}(q,p_{sl})}^{+\infty} \Bigg\{(s-r)\cdot \\
&\quad \cdot \int_0^{\Psi^{-1}((r-p_{sl})/(r-s)|i)} \left[\Psi^{-1}\left(\frac{r-p_{sl}}{r-s}\middle| i\right) - z\right]\cdot\psi(z|i)\mathrm{d}z \\
&\quad +(r-p_{sl})\cdot\Psi^{-1}\left(\frac{r-p_{sl}}{r-s}\middle| i\right) + p_{sl}q\Bigg\}\mathrm{d}\Lambda(i) \\
&\quad +(1-\beta) \int_{-\infty}^{\bar{i}(q,p_c)} \left[(s-r)\int_0^q (q-z)\cdot\psi(z|i)\mathrm{d}z + rq\right] \mathrm{d}\Lambda(i) \\
&\quad +(1-\beta) \int_{\bar{i}(q,p_c)}^{\hat{i}(q,p_c)} \Bigg\{(s-r)\cdot \\
&\quad \cdot \int_0^{\Psi^{-1}((r-p_c)/(r-s)|i)} \left[\Psi^{-1}\left(\frac{r-p_c}{r-s}\middle| i\right) - z\right]\cdot\psi(z|i)\mathrm{d}z \\
&\quad +(r-p_c)\cdot\Psi^{-1}\left(\frac{r-p_c}{r-s}\middle| i\right) + p_c q\Bigg\}\mathrm{d}\Lambda(i) \\
&\quad +(1-\beta) \int_{\hat{i}(q,p_{sh})}^{+\infty} \Bigg\{(s-r)\cdot \\
&\quad \cdot \int_0^{\Psi^{-1}((r-p_{sh})/(r-s)|i)} \left[\Psi^{-1}\left(\frac{r-p_{sh}}{r-s}\middle| i\right) - z\right]\cdot\psi(z|i)\mathrm{d}z \\
&\quad +(r-p_{sh})\cdot\Psi^{-1}\left(\frac{r-p_{sh}}{r-s}\middle| i\right) - p_c\varsigma q + p_{sh}(1+\varsigma)q\Bigg\}\mathrm{d}\Lambda(i).
\end{aligned}
$$

$$+(1-\beta)\int_{\hat{i}(q,p_c)}^{\hat{i}(q,p_{sh})}\left\{(s-r)\cdot\int_0^{(1+\varsigma)q}[(1+\varsigma)q-z]\cdot\psi(z|i)\mathrm{d}z\right.$$
$$\left.+(r-p_c)\cdot(1+\varsigma)q+p_cq\right\}\mathrm{d}\Lambda(i) \qquad (6.12)$$

Write the above expression as $F(q,\varsigma)$. Then with some calculus computations, it yields that

$$\begin{aligned}
&\frac{\partial F(q,\varsigma)}{\partial q}\\
&=-p+\beta\int_{-\infty}^{\bar{i}(q,p_{sl})}[(s-r)\cdot\Psi(q|i)+r]\,\mathrm{d}\Lambda(i)\\
&+\beta p_{sl}\cdot[1-\Lambda(\bar{i}(q,p_{sl}))]\\
&+(1-\beta)\int_{-\infty}^{\bar{i}(q,p_c)}[(s-r)\cdot\Psi(q|i)+r]\,\mathrm{d}\Lambda(i)\\
&+(1-\beta)p_c[\Lambda(\hat{i}(q,p_c))-\Lambda(\bar{i}(q,p_c))]\\
&+(1-\beta)\int_{\hat{i}(q,p_c)}^{\hat{i}(q,p_{sh})}\{(1+\varsigma)[(s-r)\cdot\Psi((1+\varsigma)q|i)+r]\\
&\qquad -p_c\varsigma\}\,\mathrm{d}\Lambda(i)\\
&+(1-\beta)\int_{\hat{i}(q,p_{sh})}^{+\infty}[-p_c\varsigma+(1+\varsigma)p_{sh}]\mathrm{d}\Lambda(i).
\end{aligned} \qquad (6.13)$$

Furthermore,

$$\begin{aligned}
&\frac{\partial^2 F(q,\varsigma)}{\partial q^2}\\
&=\beta\int_{-\infty}^{\bar{i}(q,p_{sl})}(s-r)\cdot\psi(q|i)\mathrm{d}\Lambda(i)\\
&+(1-\beta)(s-r)\left\{\int_{-\infty}^{\bar{i}(q,p_c)}\psi(q|i)\mathrm{d}\Lambda(i)\right.\\
&\qquad\left.+\int_{\hat{i}(q,p_c)}^{\hat{i}(q,p_{sh})}(1+\varsigma)^2\cdot\psi((1+\varsigma)q|i)\mathrm{d}\Lambda(i)\right\},
\end{aligned} \qquad (6.14)$$

and

$$\frac{\partial^2 F(q,\varsigma)}{\partial q \partial \varsigma}$$

$$= (1-\beta)\Big\{ (p_c - p_{sh}) \cdot \Big[\Lambda(\hat{i}(q,p_c)) - 1\Big]$$

$$+ \int_{\hat{i}(q,p_c)}^{\hat{i}(q,p_{sh})} \Big[(s-r)\cdot \Psi((1+\varsigma)q|i)$$

$$+(r-p_c) + (s-r)(1+\varsigma)q \cdot \psi((1+\varsigma)q|i) \Big] \mathrm{d}\Lambda(i) \Big\}. \tag{6.15}$$

By the definitions of $\hat{i}(q, p_{sh})$ and $\hat{i}(q, p_c)$, we know that for

$$i \in [\hat{i}(q,p_c),\ \hat{i}(q,p_{sh})],$$

the following inequality holds:

$$(s-r)\cdot \Psi((1+\varsigma)q|i) + (r - p_{sh}) \leq 0. \tag{6.16}$$

Hence, (i) of the theorem follows from (6.14) and (6.15).

If $q^* > 0$, then

$$\left.\frac{\partial F(q,\varsigma)}{\partial q}\right|_{q=q^*} = 0.$$

Note that q^* depends on ς. Therefore,

$$\frac{\mathrm{d}F(q^*,\varsigma)}{\mathrm{d}\varsigma} = \left(\left.\frac{\partial F(q,\varsigma)}{\partial q}\right|_{q=q^*} \right) \cdot \frac{\mathrm{d}q^*}{\mathrm{d}\varsigma} + \frac{\partial F(q^*,\varsigma)}{\partial \varsigma}$$

$$= (1-\beta)q^* \Big\{ (p_{sh} - p_c)\cdot [1 - \Lambda(\hat{i}(q^*, p_{sh}))]$$

$$+ \int_{\hat{i}(q^*,p_c)}^{\hat{i}(q^*,p_{sh})} [(s-r)\cdot \Psi((1+\varsigma)q^*|i) + r - p_c] \, \mathrm{d}\Lambda(i) \Big\}. \tag{6.17}$$

Similar to (6.16), we have that for $x \in [\hat{i}(q^*, p_c),\ \hat{i}(q^*, p_{sh})]$,

$$(s-r)\cdot \Psi((1+\varsigma)q^*|i) + r - p_c \geq 0.$$

Consequently, (ii) of the theorem follows from (6.17). The other cases can be proved in a same way; the details are omitted here. □

REMARK 6.8 From (6.13) and (6.14), we know that if $p_{sl} < p_c \le p_{sh}$, then the initial optimal order quantity q^* can be uniquely solved by (6.13). In a similar way, by Theorem 6.2, we can also prove that if $p_c \le p_{sl} \le p_{sh}$, then the initial optimal order quantity q^* can be uniquely solved by

$$\begin{aligned}
&-p + \beta \int_{-\infty}^{\bar{i}(q,p_c)} [(s-r)\cdot\Psi(q|i) + r]\,\mathrm{d}\Lambda(i) \\
&+\beta p_c[\Lambda(\hat{i}(q,p_c)) - \Lambda(\bar{i}(q,p_{sl}))] \\
&+\beta \int_{\hat{i}(q,p_c)}^{\hat{i}(q,p_{sl})} \{(1+\varsigma)[(s-r)\cdot\Psi((1+\varsigma)q|i) + r] - p_c\varsigma\}\,\mathrm{d}\Lambda(i) \\
&+\beta \int_{\hat{i}(q,p_{sl})}^{\infty} [-p_c\varsigma + (1+\varsigma)p_{sl}]\mathrm{d}\Lambda(i) \\
&+(1-\beta) \int_{-\infty}^{\bar{i}(q,p_c)} [(s-r)\cdot\Psi(q|i) + r]\,\mathrm{d}\Lambda(i) \\
&+(1-\beta)p_c[\Lambda(\hat{i}(q,p_c)) - \Lambda(\bar{i}(q,p_c))] \\
&+(1-\beta) \int_{\hat{i}(q,p_c)}^{\hat{i}(q,p_{sh})} \{(1+\varsigma)[(s-r)\cdot\Psi((1+\varsigma)q|i) + r] - p_c\varsigma\}\,\mathrm{d}\Lambda(i) \\
&+(1-\beta) \int_{\hat{i}(q,p_{sh})}^{+\infty} [-p_c\varsigma + (1+\varsigma)p_{sh}]\mathrm{d}\Lambda(i) = 0.
\end{aligned} \tag{6.18}$$

REMARK 6.9 We call $\mathrm{d}F(q^*,\varsigma)/\mathrm{d}\varsigma$ the *flexibility value rate*. Using (6.17), we have that if $p_{sl} < p_c \le p_{sh}$,

$$\begin{aligned}
&\frac{\mathrm{d}^2F(q^*,\varsigma)}{\mathrm{d}\varsigma^2} \\
&= (1-\beta)\cdot\Bigg\{(p_{sh}-p_c)[1-\Lambda(\hat{i}(q^*,p_{sh}))]\cdot\frac{\mathrm{d}q^*}{\mathrm{d}\varsigma} \\
&\quad+\frac{\mathrm{d}q^*}{\mathrm{d}\varsigma}\cdot\int_{\hat{i}(q^*,p_c)}^{\hat{i}(q^*,p_{sh})} [(s-r)\cdot\Psi((1+\varsigma)q^*|i) + r - p_c]\,\mathrm{d}\Lambda(i) \\
&\quad-(r-s)\left((q^*)^2 + (1+\varsigma)q^*\cdot\frac{\mathrm{d}q^*}{\mathrm{d}\varsigma}\right)\cdot \\
&\quad\cdot\int_{\hat{i}(q^*,p_c)}^{\hat{i}(q^*,p_{sh})} \psi((1+\varsigma)q^*|i)\mathrm{d}\Lambda(i)\Bigg\}.
\end{aligned} \tag{6.19}$$

Let $\tilde{\varsigma}$ be the solution of

$$\frac{\mathrm{d}^2F(q^*,\varsigma)}{\mathrm{d}\varsigma^2} = 0.$$

Then we know that the flexibility value rate is increasing on $[0, \tilde{\varsigma}]$ and nonincreasing on $(\tilde{\varsigma}, \infty)$. Thus, $\tilde{\varsigma}$ is the critical number that makes the flexibility value rate to be largest. Although the larger the flexibility factor ς is, the higher profit is, when the buyer considers the expense of flexibility, the buyer often chooses $\tilde{\varsigma}$ as the flexibility factor.

6.4.1 The Case of Worthless Information Revision

The case of worthless-information revision is that the information I observed between stage 1 and stage 2 cannot further reduce the demand uncertainty. Mathematically, the random variables I and D are independent. Hence, $\Psi(\cdot|i) = \Psi(\cdot)$ and $\Theta(\cdot,\cdot) = \Lambda(\cdot)\cdot\Psi(\cdot)$. From Theorem 6.1, $q_c^*(q, i, p_s)$ and $q_s^*(q, i, p_s)$ are independent of i. Therefore, in this subsection, we denote them as $q_c^*(q, p_s)$ and $q_s^*(q, p_s)$, respectively.

THEOREM 6.4 (WORTHLESS-INFORMATION REVISION). *In addition to Assumptions* 6.1 *and* 6.2, *we also assume that* $\Psi(\cdot|i) = \Psi(\cdot)$ *and* $\Theta(\cdot,\cdot) = \Lambda(\cdot)\cdot\Psi(\cdot)$.

(A) *If* $p \le \min\{p_{sl}, p_c, p_{sh}\}$, *then the optimal order quantities are given by*

$$\begin{aligned} q^* &= \Psi^{-1}\left(\frac{r-p}{r-s}\right), \\ q_c^*(q^*, p_{sl}) &= q_s^*(q^*, p_{sl}) = 0 \\ q_c^*(q^*, p_{sh}) &= q_s^*(q^*, p_{sh}) = 0; \end{aligned}$$

the optimal expected total order quantity is given by $\Psi^{-1}((r-p)/(r-s))$; *and the optimal expected profit is*

$$(r-s)\int_0^{q^*} z\cdot h(z)\mathrm{d}z.$$

(B) *If* $p > \min\{p_{sl}, p_c, p_{sh}\}$, *then we have the following three subcases.*

(B.1) *When* $[-p+\beta p_{sl}+(1-\beta)p_c] \ge 0$, *the optimal order quantities are given by*

$$\begin{aligned} q^* &= \Psi^{-1}\left(\frac{-p+\beta p_{sl}+(1-\beta)r}{(1-\beta)(r-s)}\right), \\ q_c^*(q^*, p_{sl}) &= 0, \\ q_s^*(q^*, p_{sl}) &= \Psi^{-1}\left(\frac{r-p_{sl}}{r-s}\right) - q^*, \\ q_c^*(q^*, p_{sh}) &= q_s^*(q^*, p_{sh}) = 0; \end{aligned}$$

the optimal expected total order quantity is given by

$$\beta\cdot\Psi^{-1}\left(\frac{r-p_{sl}}{r-s}\right) + (1-\beta)\cdot\Psi^{-1}\left(\frac{-p+\beta p_{sl}+(1-\beta)r}{(1-\beta)(r-s)}\right);$$

and the optimal expected profit is

$$(r-s)\left\{\beta\int_0^{q^*+q_s^*(q^*,p_{sl})} zh\psi(z)\mathrm{d}z + (1-\beta)\int_0^{q^*} z\psi(z)\mathrm{d}z\right\}.$$

(B.2) *When* $[-p+\beta p_{sl}+(1-\beta)p_c] < 0$ *and*

$$[-p+\beta p_{sl}+(1-\beta)p_c+(1+\varsigma)(1-\beta)(p_{sh}-p_c)] \geq 0,$$

then the optimal order quantities are given by

$$q^* = \frac{1}{1+\varsigma}\Psi^{-1}\left(\frac{(1-\beta)(1+\varsigma)(r-p_c)-p+\beta p_{sl}+(1-\beta)p_c}{(1-\beta)(1+\varsigma)(r-s)}\right),$$
$$q_c^*(q^*,p_{sl}) = 0,$$
$$q_s^*(q^*,p_{sl}) = \Psi^{-1}\left(\frac{r-p_{sl}}{r-s}\right) - q^*,$$
$$q_c^*(q^*,p_{sh}) = \varsigma q^*,$$
$$q_s^*(q^*,p_{sh}) = 0;$$

the optimal expected total order quantity is given by

$$(1-\beta)\cdot\Psi^{-1}\left(\frac{(1-\beta)(1+\varsigma)(r-p_c)-p+\beta p_{sl}+(1-\beta)p_c}{(1-\beta)(1+\varsigma)(r-s)}\right)$$
$$+\beta\cdot\Psi^{-1}\left(\frac{r-p_{sl}}{r-s}\right);$$

and the optimal expected profit is

$$(r-s)\left\{\beta\int_0^{q^*+q_s^*(q^*,p_{sl})} z\cdot\psi(z)\mathrm{d}z + (1-\beta)\int_0^{(1+\varsigma)q^*} z\cdot\psi(z)\mathrm{d}z\right\}.$$

(B.3) *When* $[-p+\beta p_{sl}+(1-\beta)p_c] < 0$ *and*

$$[-p+\beta p_{sl}+(1-\beta)p_c+(1+\varsigma)(1-\beta)(p_{sh}-p_c)] < 0,$$

the optimal order quantities are given by

$$\begin{aligned}
q^* &= 0,\\
q_c^*(q^*,p_{sl}) &= 0,\\
q_s^*(q^*,p_{sl}) &= \Psi^{-1}\left(\frac{r-p_{sl}}{r-s}\right),\\
q_c^*(q^*,p_{sh}) &= 0,\\
q_s^*(q^*,p_{sh}) &= \Psi^{-1}\left(\frac{r-p_{sh}}{r-s}\right);
\end{aligned}$$

the optimal expected total order quantity is given by

$$\beta \cdot \Psi^{-1}\left(\frac{r-p_{sl}}{r-s}\right) + (1-\beta)\cdot \Psi^{-1}\left(\frac{r-p_{sh}}{r-s}\right);$$

and the optimal expected profit is

$$(r-s)\left\{\beta \int_0^{q_s^*(q^*,p_{sl})} z\cdot\psi(z)\mathrm{d}z + (1-\beta)\int_0^{q_s^*(q^*,p_{sh})} z\cdot\psi(z)\mathrm{d}z\right\}.$$

REMARK 6.10 When $\varsigma = 0$, the results of (B.1) and (B.2) are the same. Furthermore, if $p > \beta p_{sl} + (1-\beta)p_{sh}$, then $p > \beta p_{sl} + (1-\beta)p_c$. Therefore, when $\varsigma = 0$, from Theorem 6.4 (B.2) and (B.3) we have that if $p \le \beta p_{sl} + (1-\beta)p_{sh}$, then the optimal order quantities are given by

$$\begin{aligned} q^* &= \frac{(1-\beta)r + \beta p_{sl} - p}{(1-\beta)(r-s)}, \\ q_s^*(q^*, p_{sl}) &= \Psi^{-1}\left(\frac{r-p_{sl}}{r-s}\right) - q^*; \end{aligned}$$

the optimal expected total order quantity is given by

$$(1-\beta)\cdot\Psi^{-1}\left(\frac{(1-\beta)r+\beta p_{sl}-p}{(1-\beta)(r-s)}\right) + \beta\cdot\Psi^{-1}\left(\frac{r-p_{sl}}{r-s}\right); \tag{6.20}$$

and the optimal expected profit is

$$(r-s)\left\{\beta \int_0^{q^*+q_s^*(q^*,p_{sl})} z\cdot\psi(z)\mathrm{d}z + (1-\beta)\int_0^{q^*} z\cdot\psi(z)\mathrm{d}z\right\}. \tag{6.21}$$

If $p > \beta p_{sl} + (1-\beta)p_{sh}$, then the optimal order quantities are given by

$$\begin{aligned} &q^* = 0, \quad q_s^*(q^*, p_{sl}) = \Psi^{-1}\left(\frac{r-p_{sl}}{r-s}\right), \\ &q_s^*(q^*, p_{sh}) = \Psi^{-1}\left(\frac{r-p_{sh}}{r-s}\right); \end{aligned}$$

the optimal expected total order quantity is given by

$$\beta\cdot\Psi^{-1}\left(\frac{r-p_{sl}}{r-s}\right) + (1-\beta)\cdot\Psi^{-1}\left(\frac{r-p_{sh}}{r-s}\right); \tag{6.22}$$

and the optimal expected profit is

$$(r-s)\left\{\beta\int_0^{q_s^*(q^*,p_{sl})} z\cdot\psi(z)\mathrm{d}z+(1-\beta)\int_0^{q_s^*(q^*,p_{sh})} z\cdot\psi(z)\mathrm{d}z\right\}. \tag{6.23}$$

These results are also obtained by Gurnani and Tang [12] when $\Psi(\cdot)$ is a normal distribution.

REMARK 6.11 When $\beta = 0$, the spot-market price is definitely higher than the unit price at stage 1. From Theorem 6.4 (B.1), we get that the optimal order quantity is

$$q^* = \Psi^{-1}\left(\frac{r-p}{r-s}\right);$$

the optimal expected total order quantity is

$$\Psi^{-1}\left(\frac{r-p}{r-s}\right); \tag{6.24}$$

and the optimal expected profit is

$$(r-s)\int_0^{q^*} z\psi(z)\mathrm{d}z. \tag{6.25}$$

This is the same as Theorem 4 (b) of Brown and Lee [5] with $\Psi(\cdot)$ being a normal distribution.

Proof of Theorem 6.4 Here we give only a proof of (B.1) and (B.2), since the other results in the theorem can be established similarly. Since $p > p_{sl}$ in Case B, then in view of $p < p_c$ and (6.6), we have

$$p_{sl} < p < p_c \leq p_{sh}. \tag{6.26}$$

Thus,

$$\begin{aligned}\Psi^{-1}\left(\frac{r-p_{sh}}{r-s}\right) \leq \Psi^{-1}\left(\frac{r-p_c}{r-s}\right) &\leq \Psi^{-1}\left(\frac{r-p}{r-s}\right)\\ &\leq \Psi^{-1}\left(\frac{r-p_{sl}}{r-s}\right).\end{aligned} \tag{6.27}$$

It suffices to show that when $\beta p_{sl} + (1-\beta)p_c \geq p$, q^* given in (B.1) is a maximizer of the function

$$\Pi_1(q) = -pq + \mathsf{E}\left[\Pi_2(q, q_s^*(q, P_s), q_c^*(q, P_s), I, P_s)\right];$$

and when $\beta p_{sl} + (1-\beta)p_c < p$ and $[-p + \beta p_{sl} + (1-\beta)p_c + (1+\varsigma)(1-\beta)(p_{sh} - p_c)] \geq 0$, q^* given in (B.2) is a maximizer of $\Pi_1(q)$.

First we look at the proof of (B.1). The proof is divided into three subcases.

Case B.1.1: $[q \geq \Psi^{-1}((r-p_{sl})/(r-s))]$

By Theorem 6.1,

$$q_c^*(q, p_{sl}) = q_s^*(q, p_{sl}) = q_c^*(q, p_{sh}) = q_s^*(q, p_{sh}) = 0.$$

Then

$$\Pi_1(q) = -pq + s\int_0^q (q-z)\cdot\psi(z)\mathrm{d}z + r\int_0^q z\cdot\psi(z)\mathrm{d}z + rq[1-\Psi(q)].$$

This implies that

$$\begin{aligned}\frac{\mathrm{d}\Pi_1(q)}{\mathrm{d}q} &= -p + s\cdot\Psi(q) + r[1-\Psi(q)] \\ &= r - p - (r-s)\cdot\Psi(q) \\ &\leq r - p - (r - p_{sl}) \\ &< 0.\end{aligned}$$

Hence, $\Pi_1(q)$ is decreasing in $[\Psi^{-1}((r-p_{sl})/(r-s)),\ \infty)$.

Case B.1.2: $[\Psi^{-1}((r-p_c)/(r-s)) \leq q < \Psi^{-1}((r-p_{sl})/(r-s))]$

It follows from Theorem 6.1 that

$$\begin{aligned}\Pi_1(q) &= -pq + V_l(r, s, p_{sl}) + \beta p_{sl} q \\ &\quad +(1-\beta)\Big[s\int_0^q (q-z)\cdot\psi(z)\mathrm{d}z \\ &\qquad +r\int_0^q z\cdot\psi(z)\mathrm{d}z + rq\cdot(1-\Psi(q))\Big],\end{aligned}$$

where

$$\begin{aligned}&V_l(r, s, p_{sl}) \\ &= \beta\Big\{-(r-s)\int_0^{\Psi^{-1}((r-p_{sl})/(r-s))}\left[\Psi^{-1}\left(\frac{r-p_{sl}}{r-s}\right) - z\right]\psi(z)\mathrm{d}z \\ &\quad +(r-p_{sl})\cdot\Psi^{-1}\left(\frac{r-p_{sl}}{r-s}\right)\Big\}.\end{aligned}$$

Therefore,

$$\begin{aligned}\frac{\mathrm{d}\Pi_1(q)}{\mathrm{d}q} &= -p + \beta p_{sl} + (1-\beta)[r - r\cdot\Psi(q) + s\cdot\Psi(q)] \\ &\leq -p + \beta p_{sl} + (1-\beta)p_c.\end{aligned}$$

This implies that $\Pi_1(q)$ is increasing on the interval $[\Psi^{-1}((r-p_c)/(r-s)),\ q^*]$ and decreasing on the interval $[q^*,\ \Psi^{-1}((r-p_{sl})/(r-s))]$.

Case B.1.3: $[q < \Psi^{-1}((r-p_c)/(r-s))]$

Proceeding as in Case B.1.2, we can show that $\Pi_1(q)$ is increasing on the interval $[0,\ \Psi^{-1}((r-p_c)/(r-s))]$.

Combining Cases 1–3 completes the proof for (B.1).

Finally, we look at (B.2). Similarly, the proof is also divided into several cases.

Case B.2.1: $[q < \Psi^{-1}((r-p_{sh})/(r-s))$ and $(1+\varsigma)q < \Psi^{-1}((r-p_{sh})/(r-s))]$

$\Pi_1(q)$ can be written as

$$\begin{aligned}\Pi_1(q) &= -pq + V_l(r,s,p_{sl}) + \beta p_{sl}q \\ &\quad +(1-\beta)\Bigg\{-(r-s)\cdot \\ &\quad \cdot\int_0^{\Psi^{-1}((r-p_{sh})/(r-s))}\left[\Psi^{-1}\left(\frac{r-p_{sh}}{r-s}\right)-z\right]\psi(z)\mathrm{d}z \\ &\quad +r\cdot\Psi^{-1}\left(\frac{r-p_{sh}}{r-s}\right)-p_c\varsigma q \\ &\quad -p_{sh}\left[\Psi^{-1}\left(\frac{r-p_{sh}}{r-s}\right)-(1+\varsigma)q\right]\Bigg\}.\end{aligned}$$

Consequently, by $[-p+\beta p_{sl}+(1-\beta)p_c+(1+\varsigma)(1-\beta)(p_{sh}-p_c)] \geq 0$,

$$\frac{\mathrm{d}\Pi_1(q)}{\mathrm{d}q} = -p+\beta p_{sl}+(1-\beta)\left[-p_c\varsigma+p_{sh}(1+\varsigma)\right] \geq 0.$$

So $\Pi_1(q)$ is increasing for q satisfying $q < \Psi^{-1}((r-p_{sh})/(r-s))$ and $(1+\varsigma)q < \Psi^{-1}((r-p_{sh})/(r-s))$.

Case B.2.2: $[q < \Psi^{-1}((r-p_{sh})/(r-s)), (1+\varsigma)q \geq \Psi^{-1}((r-p_{sh})/(r-s))$ and $(1+\varsigma)q < \Psi^{-1}((r-p_c)/(r-s))]$

Under this case, $\Pi_1(q)$ can be written as

$$\begin{aligned}\Pi_1(q) &= -pq + V_l(r,s,p_{sl}) + \beta p_{sl}q \\ &\quad +(1-\beta)\Bigg\{-(r-s)\int_0^{(1+\varsigma)q}\left[(1+\varsigma)q-z\right]\cdot\psi(z)\mathrm{d}z \\ &\quad +r\cdot(1+\varsigma)q-p_c\varsigma q\Bigg\}.\end{aligned}$$

Consequently,

$$\begin{aligned}\frac{\mathrm{d}\Pi_1(q)}{\mathrm{d}q} &= -p+\beta p_{sl} \\ &\quad +(1-\beta)\left[-(r-s)(1+\varsigma)\cdot\Psi((1+\varsigma)q)+r\cdot(1+\varsigma)-p_c\varsigma\right].\end{aligned}$$

In the following, if $(1+\varsigma)\Psi^{-1}((r-p_{sh})/(r-s)) \geq \Psi^{-1}((r-p_c)/(r-s))$, we go to Cases B.2.3 and B.2.5–B.2.7. If $(1+\varsigma)\Psi^{-1}((r-p_{sh})/(r-s)) < \Psi^{-1}((r-p_c)/(r-s))$, we go to Cases B.2.4–B.2.7.

Case B.2.3: $[q < \Psi^{-1}((r-p_{sh})/(r-s))$ and $(1+\varsigma)q \geq \Psi^{-1}((r-p_c)/(r-s))]$

We have

$$\begin{aligned}\Pi_1(q) &= -pq + V_l(r,s,p_{sl}) + \beta p_{sl} q \\ &\quad +(1-\beta)\Big\{-(r-s)\cdot \\ &\qquad \cdot \int_0^{\Psi^{-1}((r-p_c)/(r-s))} \left[\Psi^{-1}\left(\frac{r-p_c}{r-s}\right) - z\right]\cdot\psi(z)\mathrm{d}z \\ &\qquad +r\cdot\Psi^{-1}\left(\frac{r-p_c}{r-s}\right) - p_c\left[\Psi^{-1}\left(\frac{r-p_c}{r-s}\right)-q\right]\Big\}.\end{aligned}$$

Then

$$\frac{\mathrm{d}\Pi_1(q)}{\mathrm{d}q} = -p + \beta p_{sl} + (1-\beta)p_c < 0.$$

Case B.2.4: $[\Psi^{-1}((r-p_{sh})/(r-s)) \leq q \leq \Psi^{-1}((r-p_c)/(r-s))$, and $(1+\varsigma)q < \Psi^{-1}((r-p_c)/(r-s))]$

We have

$$\begin{aligned}&\Pi_1(q) \\ &= -pq + V_l(r,s,p_{sl}) + \beta p_{sl} q \\ &\quad +(1-\beta)\Big\{-(r-s)\int_0^{(1+\varsigma)q} [(1+\varsigma)q - z]\cdot\psi(z)\mathrm{d}z \\ &\quad +r(1+\varsigma)q - p_c\varsigma q\Big\}.\end{aligned}$$

Consequently,

$$\begin{aligned}&\frac{\mathrm{d}\Pi_1(q)}{\mathrm{d}q} \\ &= -p + \beta p_{sl} \\ &\quad +(1-\beta)\left[-(r-s)(1+\varsigma)\cdot\Psi((1+\varsigma)q) + r\cdot(1+\varsigma) - p_c\varsigma\right].\end{aligned}$$

Case B.2.5: $[\Psi^{-1}((r-p_{sh})/(r-s)) \leq q \leq \Psi^{-1}((r-p_c)/(r-s))$, and $(1+\varsigma)q \geq \Psi^{-1}((r-p_c)/(r-s))]$

We have

$$\begin{aligned}
&\Pi_1(q) \\
&= -pq + V_l(r,s,p_{sl}) + \beta p_{sl} q \\
&\quad +(1-\beta)\Big\{ -(r-s)\cdot \\
&\quad \cdot \int_0^{\Psi^{-1}((r-p_c)/(r-s))} \left[\Psi^{-1}\left(\frac{r-p_c}{r-s}\right) - z\right]\psi(z)\mathrm{d}z \\
&\quad + r\cdot\Psi^{-1}\left(\frac{r-p_c}{r-s}\right) - p_c\cdot\left[\Psi^{-1}\left(\frac{r-p_c}{r-s}\right) - q\right]\Big\}.
\end{aligned}$$

Then

$$\frac{\mathrm{d}\Pi_1(q)}{\mathrm{d}q} = -p + \beta p_{sl} + (1-\beta)p_c < 0.$$

Case B.2.6: $[\Psi^{-1}((r-p_c)/(r-s)) < q \le \Psi^{-1}((r-p_{sl})/(r-s))]$
We have

$$\begin{aligned}
\Pi_1(q) &= -pq + V_l(r,s,p_{sl}) + \beta p_{sl} q \\
&\quad +(1-\beta)\Big\{ -(r-s)\int_0^q [q-z]\cdot\psi(z)\mathrm{d}z + rq\Big\}.
\end{aligned}$$

Consequently,

$$\begin{aligned}
\frac{\mathrm{d}\Pi_1(q)}{\mathrm{d}q} &= -p + \beta p_{sl} + (1-\beta)[-(r-s)\cdot\Psi(q) + r] \\
&\le -p + \beta p_{sl} + (1-\beta)p_c \\
&< 0.
\end{aligned}$$

Case B.2.7: $[q > \Psi^{-1}((r-p_{sl})/(r-s))]$
We have

$$\begin{aligned}
\Pi_1(q) &= -pq + \beta\Big\{ -(r-s)\int_0^q [q-z]\cdot\psi(z)\mathrm{d}z + rq\Big\} \\
&\quad +(1-\beta)\Big\{ -(r-s)\int_0^q [q-z]\cdot\psi(z)\mathrm{d}z + rq\Big\}.
\end{aligned}$$

Consequently,

$$\begin{aligned}
\frac{\mathrm{d}\Pi_1(q)}{\mathrm{d}q} &= -p + \beta[-(r-s)\cdot\Psi(q) + r] + (1-\beta)[-(r-s)\cdot\Psi(q) + r] \\
&\le -p + \beta p_{sl} + (1-\beta)p_c \\
&< 0.
\end{aligned}$$

According to $(1+\varsigma)\Psi^{-1}((r-p_{sh})/(r-s)) \geq \Psi^{-1}((r-p_c)/(r-s))$ or $(1+\varsigma)\Psi^{-1}((r-p_{sh})/(r-s)) < \Psi^{-1}((r-p_c)/(r-s))$, (B.2) follows from Cases B.2.1–B.2.3 and B.2.5–B.2.7 or Cases B.2.1–B.2.2 and B.2.4–B.2.7, respectively. □

We now provide intuitive insights into the various results obtained in Theorem 6.4. Case A addresses the situation when the initial unit-order cost is less than the lowest possible market price. In this case, if the observed information is useless, then the buyer gains nothing by delaying his purchase to stage 2. Thus, the entire purchase is made at stage 1, and nothing is purchased at stage 2. Indeed, in this case, the contract is of no value.

In Case B, we have (6.26). Clearly, $q_c^*(q^*, p_{sl}) = 0$ in this case.

In (B.1), the expected relevant price at stage 2 is clearly $\beta p_{sl} + (1-\beta)p_c$, and it is higher than the initial price p. Therefore, the buyer will buy a sufficiently large quantity q^* at the initial price p so that he would not need to buy any quantity at all when the market price is high. Moreover, q^* will not be too large to prohibit the buyer from taking advantage of buying in the market when the spot price is low.

We now consider (B.2) and (B.3). Note that since $\varsigma > 0$, the condition

$$p > \beta p_{sl} + (1-\beta)p_{sh} + (1-\beta)\varsigma \cdot (p_{sh} - p_c) \tag{6.28}$$

in (B.3) implies $p > \beta p_{sl} + (1-\beta)p_c$. Thus, in both cases (B.2) and (B.3), $\beta p_{sl} + (1-\beta)p_c$ is lower than the initial price p. In contrast to (B.1), it seems reasonable, therefore, to reduce or completely postpone the purchase to stage 2 in (B.2) and (B.3). The (B.3) condition (6.28), however, also implies $p > \beta p_{sl} + (1-\beta)p_{sh}$. This says that the expected market price at stage 2 is lower than the initial price p, which argues for a complete postponement of the purchase. Consequently, the initial purchase quantity is zero, and the entire respective newsvendor quantity is bought from the market depending on the prevailing market price at stage 2.

This leaves us with (B.2), where we still have $p > \beta p_{sl} + (1-\beta)p_c$, but we do not have (6.28). In other words, the high market price p_{sh} is not low enough for (6.28) to hold and thus argues perhaps for a reduction in the initial purchase amount rather than a complete postponement. Let us therefore consider an initial purchase of one unit at stage 1 and ς unit at stage 2. Clearly, the purchase of ς unit at stage 2 will take place at p_{sl} when the market price is low and at p_c when the market price is high. Thus, the per unit expected cost of a reduced purchase at stage 1 followed by an additional purchase up to the contracted amount is

$$\frac{p + \beta\varsigma p_{sl} + (1-\beta)\varsigma p_c}{1+\varsigma}.$$

On the other hand, a complete postponement of the purchase of a unit to stage 2 has the expected cost

$$\beta p_{sl} + (1-\beta)p_{sh}.$$

Thus, if

$$\frac{p + \beta\varsigma p_{sl} + (1-\beta)\varsigma p_c}{1+\varsigma} < \beta p_{sl} + (1-\beta)p_{sh}$$

—that is, if

$$\begin{aligned} p &< \beta p_{sl} + (1-\beta)p_{sh} + (1-\beta)\varsigma \cdot (p_{sh} - p_c) \\ &= \beta p_{sl} + (1-\beta)p_c + (1-\beta)(1+\varsigma)(p_{sh} - p_c), \end{aligned} \tag{6.29}$$

then it is better to reduce the initial purchase than to postpone it completely. This is precisely the result obtained in (B.2).

By comparing our result with (6.21), it is possible to demonstrate that the difference between the contract and no contract is

$$(1-\beta)(r-s)\int_{q^*}^{(1+\varsigma)q^*} z \cdot \psi(z)\mathrm{d}z, \tag{6.30}$$

if $[-p + \beta p_{sl} + (1-\beta)p_c] < 0$ and

$$[-p + \beta p_{sl} + (1-\beta)p_c + (1+\varsigma)(1-\beta)(p_{sh} - p_c)] \geq 0.$$

We denote this gap as the *value of flexibility*. Equation (6.30) indicates that the value of flexibility is always positive. As long as the prices of different sources satisfy the following condition, $[-p + \beta p_{sl} + (1-\beta)p_c] < 0$ and

$$[-p + \beta p_{sl} + (1-\beta)p_c + (1+\varsigma)(1-\beta)(p_{sh} - p_c)] \geq 0,$$

the above observation reveals the fact that it is beneficial for the buyer to seek a supply contract even demand information revision is worthless.

If $p > \min\{p_{sl}, p_c, p_{sh}\}$ and

$$[-p + \beta p_{sl} + (1-\beta)p_c] \geq 0$$

—that is, if the contract-unit price is high, it follows from Theorem 6.4 that the profits are the same for both contract and no-contract case. As a result, the value of quantity flexibility is zero. Similarly, if $[-p + \beta p_{sl} + (1-\beta)p_c] < 0$ and

$$[-p + \beta p_{sl} + (1-\beta)p_c + (1+\varsigma)(1-\beta)(p_{sh} - p_c)] < 0,$$

the value of quantity flexibility is also zero.

6.4.2 The Case of Perfect Information Revision

In this subsection, we study the second extreme case where the information revision is perfect. The perfect-information revision represents a scenario such that the demand D can be completely determined by the information I observed between stage 1 and stage 2. In other words, it is possible to characterize the demand D by information I, e.g., $D = \tau(I)$. Let $\Theta(\cdot)$ be the distribution function of D. Parallel to Theorem 6.4, for the case of perfect-information revision, we present the following theorem.

THEOREM 6.5 (PERFECT-INFORMATION REVISION) *In addition to Assumptions* 6.1 *and* 6.2, *we also assume that* $D = \tau(I)$.

(A) *If* $p_{sl} \le p_c$, *then the optimal order quantity* q^* *at stage 1 is the solution of the following equation*

$$\begin{aligned}
&-p+\beta p_{sl}+(1-\beta)p_c+(1-\beta)(1+\varsigma)(p_{sh}-p_c)\\
&+[s-\beta p_{sl}-(1-\beta)p_c]\cdot\Theta(q)\\
&+(1-\beta)(1+\varsigma)(p_c-p_{sh})\cdot\Theta((1+\varsigma)q)=0
\end{aligned} \tag{6.31}$$

with the convenience $q^* = 0$ *if the solution of* (6.31) *does not exist. The optimal order quantities at stage* 2 *are*

$$\begin{aligned}
q_c^*(q^*, i, p_{sl}) &= 0,\\
q_s^*(q^*, i, p_{sl}) &= [\tau(i)-q^*]^+,\\
q_c^*(q^*, i, p_{sh}) &= [\tau(i)-q^*]^+ \wedge (\varsigma q^*),\\
q_s^*(q^*, i, p_{sh}) &= [\tau(i)-(1+\varsigma)q^*]^+,
\end{aligned}$$

the optimal expected total order quantity is given by

$$q^* + \int_{q^*}^{\infty} [z-q^*]\mathrm{d}\Theta(z),$$

and the optimal expected profit is

$$\begin{aligned}
&r\cdot\mathsf{E}[D]-s\int_0^{q^*} z\mathrm{d}\Theta(z)-\beta p_{sl}\int_{q^*}^{\infty} z\mathrm{d}\Theta(z)\\
&-(1-\beta)\left\{p_c\int_{q^*}^{(1+\varsigma)q^*} z\mathrm{d}\Theta(z)+p_{sh}\int_{(1+\varsigma)q^*}^{\infty} z\mathrm{d}\Theta(z)\right\}.
\end{aligned} \tag{6.32}$$

(B) *If* $p_{sl} > p_c$, *then the optimal order quantity* q^* *at stage* 1 *is the solution of the following equation*

$$
\begin{aligned}
&-p - \varsigma p_c + (1+\varsigma)[\beta p_{sl} + (1-\beta)p_{sh}] + (s - p_c)\cdot\Theta(q) \\
&+(1+\varsigma)[p_c - \beta p_{sl} - (1-\beta)p_{sh}]\cdot\Theta((1+\varsigma)q) = 0
\end{aligned}
\tag{6.33}
$$

with the same convenience stated in (A). *The optimal order quantities at stage 2 are*

$$
\begin{aligned}
q_c^*(q^*, i, p_{sl}) &= [\tau(i) - q^*]^+ \wedge (\varsigma q^*), \\
q_s^*(q^*, i, p_{sl}) &= [\tau(i) - (1+\varsigma)q^*]^+, \\
q_c^*(q^*, i, p_{sh}) &= [\tau(i) - q^*]^+ \wedge (\varsigma q^*), \\
q_s^*(q^*, i, p_{sh}) &= [\tau(i) - (1+\varsigma)q^*]^+,
\end{aligned}
$$

the optimal expected total order quantity is

$$
q^* + \int_{q^*}^{\infty} [z - q^*]\mathsf{d}\Theta(z),
$$

and the optimal expected profit is

$$
\begin{aligned}
&r\cdot\mathsf{E}[D] - s\int_0^{q^*} z\mathsf{d}\Theta(z) - p_c\int_{q^*}^{(1+\varsigma)q^*} z\mathsf{d}\Theta(z) \\
&-[\beta p_{sl} + (1-\beta)p_{sh}]\int_{(1+\varsigma)q^*}^{\infty} z\mathsf{d}\Theta(z).
\end{aligned}
\tag{6.34}
$$

Proof The proof of the theorem is the same as the proof of Theorem 6.4. □

REMARK 6.12 When $\varsigma = 0$, from Theorem 6.5 (A) we get that if $p \geq \beta p_{sl} + (1-\beta)p_{sh}$, then the optimal order quantities

$$
\begin{aligned}
q^* &= 0, \\
q_s^*(q^*, i, p_{sl}) &= \tau(i), \\
q_s^*(q^*, i, p_{sh}) &= \tau(i),
\end{aligned}
$$

the optimal expected total order quantity is given by $\mathsf{E}[D]$, and the optimal expected profit is

$$
r\cdot\mathsf{E}[D] - [\beta p_{sl} + (1-\beta)p_{sh}]\cdot\mathsf{E}[D];
\tag{6.35}
$$

and if $p < \beta p_{sl} + (1-\beta)p_{sh}$, then the optimal order quantities

$$\begin{aligned} q^* &= \Theta^{-1}\left(\frac{\beta p_{sl} + (1-\beta)p_{sh} - p}{\beta p_{sl} + (1-\beta)p_{sh} - s}\right), \\ q_s^*(q^*, i, p_{sl}) &= (\tau(i) - q^*)^+, \\ q_s^*(q^*, i, p_{sh}) &= (\tau(i) - q^*)^+; \end{aligned}$$

the optimal expected total order quantity is given by

$$q^* + \int_{q^*}^{\infty} [z - q^*] \mathsf{d}\Theta(z);$$

and the optimal expected profit is

$$r \cdot \mathsf{E}[D] - s \int_0^{q^*} z \mathsf{d}\Theta(z) - \beta p_{sl} \int_{q^*}^{\infty} z \mathsf{d}\Theta(z) \\ -(1-\beta)p_{sh} \int_{q^*}^{\infty} z \mathsf{d}\Theta(z). \quad (6.36)$$

Gurnani and Tang [12] also get these results when $\Theta(\cdot)$ is normal distribution.

REMARK 6.13 If $\beta = 0$—that is, the spot-market price is p_{sh} with probability one—this equals to that the buyer completely knows the spot-market price at stage 1. If $p_c < p_{sh}$, from Theorem 6.5 we get that the optimal order quantity q^* at stage 1 is the solution of the following equation

$$\begin{aligned} 0 &= -p + p_c + (1+\varsigma)(p_{sh} - p_c) + [s - p_c] \cdot \Theta(q) \\ &\quad +(1+\varsigma)(p_c - p_{sh}) \cdot \Theta((1+\varsigma)q) \end{aligned} \quad (6.37)$$

with the convenience $q^* = 0$ if the solution of (6.37) does not exist. The optimal order quantities at stage 2 are

$$\begin{aligned} q_c^*(q^*, i, p_{sh}) &= [\tau(i) - q^*]^+ \wedge (\varsigma q^*), \\ q_s^*(q^*, i, p_{sh}) &= [\tau(i) - (1+\varsigma)q^*]^+, \end{aligned}$$

the optimal expected total order quantity is given by

$$q^* + \int_{q^*}^{\infty} [z - q^*] \mathsf{d}\Theta(z),$$

and the optimal expected profit is

$$r \cdot \mathsf{E}[D] - s \int_0^{q^*} z \mathsf{d}\Theta(z) - p_c \int_{q^*}^{(1+\varsigma)q^*} z \mathsf{d}\Theta(z) \\ +p_{sh} \int_{(1+\varsigma)q^*}^{\infty} z \mathsf{d}\Theta(z). \quad (6.38)$$

Suppose that $\Theta(\cdot)$ is normal distribution. Compared with Theorem 4 (a) of Brown and Lee [5], because the buyer loses flexibility in the contract purchase at stage 2 if it does not purchase anything at stage 1, to hedge this flexibility it has to purchase some quantity at stage 1. Thus the results obtained here are different from Theorem 4 (a) of Brown and Lee [5].

LEMMA 6.1 *With Assumptions* 6.1 *and* 6.2, *and the condition* $p_{sl} \leq p_c$, *equation* (6.31) *has a solution* $q^* > 0$ *if and only if* (6.29) *holds.*

Proof Setting $q = 0$ in (6.31) and using (6.29) and the fact that $\Theta(0) = 0$, we obtain

$$\begin{aligned}
&-p + \beta p_{sl} + (1-\beta)p_c + (1-\beta)(1+\varsigma)(p_{sh} - p_c) \\
&+[s - \beta p_{sl} - (1-\beta)p_c] \cdot \Theta(0) - (1-\beta)(1+\varsigma)(p_{sh} - p_c) \cdot \Theta(0) \\
&\quad = -p + \beta p_{sl} + (1-\beta)p_c + (1-\beta)(1+\varsigma)(p_{sh} - p_c) \\
&\quad > 0.
\end{aligned} \tag{6.39}$$

In view of $\lim_{q\to\infty} \Theta(q) = 1$ and Assumption (6.1), we have

$$\begin{aligned}
&-p + \beta p_{sl} + (1-\beta)p_c + (1-\beta)(1+\varsigma)(p_{sh} - p_c) \\
&+[s - \beta p_{sl} - (1-\beta)p_c] \cdot \lim_{q\to\infty} \Theta(q) \\
&-(1-\beta)(1+\varsigma)(p_{sh} - p_c) \cdot \lim_{q\to\infty} \Theta((1+\varsigma)q) \\
&\quad = s - p \\
&\quad < 0.
\end{aligned} \tag{6.40}$$

Taking the derivative of the left-hand side of (6.31) with respect to q, we obtain

$$\begin{aligned}
&\mathrm{d}\Big\{[s - \beta p_{sl} - (1-\beta)p_c] \cdot \Theta(q) \\
&\quad -(1-\beta)(1+\varsigma)(p_{sh} - p_c) \cdot \Theta((1+\varsigma)q)\Big\} \Big/ \mathrm{d}q \\
&\quad = [s - \beta p_{sl} - (1-\beta)p_c] \cdot \frac{\mathrm{d}\Theta(q)}{\mathrm{d}q} \\
&\qquad -(1-\beta)(1+\varsigma)^2(p_{sh} - p_c) \cdot \left.\frac{\mathrm{d}\Theta(x)}{\mathrm{d}x}\right|_{x=(1+\varsigma)q}.
\end{aligned} \tag{6.41}$$

Since $p_{sl} \leq p_c$ and $s < p_{sl}$ as assumed in (6.1), we have

$$\begin{aligned}
s - \beta p_{sl} - (1-\beta)p_c &\leq s - \beta p_{sl} - (1-\beta)p_{sl} \\
&= s - p_{sl} \\
&< 0.
\end{aligned}$$

Thus, the derivative in (6.41) is strictly negative. The lemma follows from (6.39)–(6.41). □

LEMMA 6.2 *With Assumptions* 6.1 *and* 6.2, *and the condition* $p_{sl} > p_c$, *equation* (6.33) *has a solution* $q^* > 0$ *if and only if*

$$-p - \varsigma p_c + (1+\varsigma)[\beta p_{sl} + (1-\beta)p_{sh}] > 0. \tag{6.42}$$

Proof Setting $q = 0$ in (6.33) and using $\Theta(0) = 0$ and (6.42), we have

$$\begin{aligned}
&-p - \varsigma p_c + (1+\varsigma)[\beta p_{sl} + (1-\beta)p_{sh}] \\
&+(s-p_c)\cdot\Theta(0) + (1+\varsigma)[p_c - \beta p_{sl} - (1-\beta)p_{sh}]\cdot\Theta(0) \\
&\quad = -p - \varsigma p_c + (1+\varsigma)[\beta p_{sl} + (1-\beta)p_{sh}] \\
&\quad > 0,
\end{aligned} \tag{6.43}$$

and

$$\begin{aligned}
&-p - \varsigma p_c + (1+\varsigma)[\beta p_{sl} + (1-\beta)p_{sh}] + (s-p_c)\cdot\lim_{q\to\infty}\Theta(q) \\
&+(1+\varsigma)[p_c - \beta p_{sl} - (1-\beta)p_{sh}]\cdot\lim_{q\to\infty}\Theta((1+\varsigma)q) \\
&= -p + s \\
&< 0.
\end{aligned} \tag{6.44}$$

Furthermore, taking the derivative of the left-hand side of (6.33) with respect to q and using the facts $s < p < p_c, p_{sh} \geq p_{sl}$, and the condition $p_{sl} > p_c$, we obtain

$$\begin{aligned}
&\mathrm{d}\Big\{ -p - \varsigma p_c + (1+\varsigma)[\beta p_{sl} + (1-\beta)p_{sh}] + (s-p_c)\cdot\Theta(q) \\
&+(1+\varsigma)[p_c - \beta p_{sl} - (1-\beta)p_{sh}]\cdot\Theta((1+\varsigma)q)\Big\}\Big/\mathrm{d}q \\
&= (1+\varsigma)^2[p_c - \beta p_{sl} - (1-\beta)p_{sh}]\cdot\left.\frac{\mathrm{d}\Theta(x)}{\mathrm{d}x}\right|_{x=(1+\varsigma)q} \\
&\quad +(s-p_c)\cdot\frac{\mathrm{d}\Theta(q)}{\mathrm{d}q} \\
&< 0.
\end{aligned} \tag{6.45}$$

The lemma follows from (6.43)–(6.45). □

THEOREM 6.6 *Under Assumptions* 6.1 *and* 6.2, *the flexibility value is either zero or a decreasing function of* β *in both the worthless and the perfect information cases.*

Proof First, consider the case of worthless information. Using Theorem 6.4, we know that the flexibility value is zero if any one of the following conditions holds.

(i) $p \leq p_{sl}$;
(ii) $p > p_{sl}$ and $\beta p_{sl} + (1-\beta)p_c \geq p$;
(iii) $p > p_{sl}$ and $p \geq \beta p_{sl} + (1-\beta)p_{sh} + (1-\beta)\varsigma(p_{sh} - p_c)$.

Thus, we need only to prove the theorem in case (B.2) of Theorem 6.4—that is, when $p > p_{sl}$ and

$$\beta p_{sl} + (1-\beta)p_c < p \leq \beta p_{sl} + (1-\beta)p_{sh} + (1-\beta)\varsigma(p_{sh} - p_c).$$

In this case, the flexibility value is obtained in (6.30), which is clearly decreasing in β.

Now consider the case of perfect information. We must consider the following four cases:

(A.1) $p_{sl} \leq p_c$, $p \leq \beta p_{sl} + (1-\beta)p_{sh}$;
(A.2) $p_{sl} \leq p_c$, $\beta p_{sl} + (1-\beta)p_{sh} < p < \beta p_{sl} + (1-\beta)p_{sh} + (1-\beta)\varsigma(p_{sh} - p_c)$;
(A.3) $p_{sl} \leq p_c$, $\beta p_{sl} + (1-\beta)p_{sh} + (1-\beta)\varsigma(p_{sh} - p_c) \leq p$;
(B) $p_{sl} > p_c$.

The optimal solutions in the first three cases (A.1), (A.2), and (A.3) are given in Theorem 6.5 (A), and the optimal solution in case B is given in Theorem 6.5 (B). In (A.3), we know from Lemma 6.1 that $q^* = 0$, which implies that the flexibility value is zero. Below we provide the details of the proof only in case (A.1), since the proofs in cases (A.2) and (B) follow in the same way.

In case (A.1), if there is no contract, then we would have $\varsigma = 0$. Then the condition of the case implies that the inequality (6.29) is satisfied with $\varsigma = 0$. By Lemma 6.1, therefore, the optimal q^* in Theorem 6.5 (A) would be given by solving (6.31) with $\varsigma = 0$, which we write as

$$q_0^* = \Theta^{-1}\left(\frac{-p + \beta p_{sl} + (1-\beta)p_{sh}}{-s + \beta p_{sl} + (1-\beta)p_{sh}}\right). \tag{6.46}$$

Moreover from Theorem 6.5 (A), the optimal order quantity at stage 2 regardless of the market price could be

$$[\tau(i) - q_0^*]^+, \quad i = 1, 2.$$

By (6.46), we have

$$\begin{aligned}
&-p+\beta p_{sl}+(1-\beta)p_c+(1-\beta)(1+\varsigma)(p_{sh}-p_c)\\
&+[s-\beta p_{sl}-(1-\beta)p_c]\cdot\Theta(q_0^*)\\
&+(1-\beta)(1+\varsigma)(p_c-p_{sh})\cdot\Theta((1+\varsigma)q_0^*)\\
&\quad=(p_{sh}-p_c)(1-\beta)[\Theta(q_0^*)-(1+\varsigma)\cdot\Theta((1+\varsigma)q_0^*)]\\
&\qquad+(1-\beta)(p_{sh}-p_c)\delta\\
&\quad<0.
\end{aligned}\tag{6.47}$$

Thus, the solution q^* given by (6.31) with $\varsigma > 0$ is smaller than q_0^* ordered in the absence of a contract. In other words, the buyer purchases less at stage 1 when he has a contract ($\varsigma > 0$). Then from Theorem 6.5 (A) and equation (6.32), the difference of the expected profits with and without the contract is

$$\begin{aligned}
&(1-\beta)\Big\{-p_c\int_{q^*}^{(1+\varsigma)q^*} z\mathrm{d}\Theta(z)\\
&\qquad+p_{sh}\cdot\mathrm{sign}((1+\varsigma)q^*-q_0^*)\cdot\int_{((1+\varsigma)q^*)\wedge q_0^*}^{((1+\varsigma)q^*)\vee q_0^*} z\mathrm{d}\Theta(z)\Big\}\\
&+(s-\beta p_{sl})\int_{q^*}^{q_0^*} z\mathrm{d}\Theta(z),
\end{aligned}\tag{6.48}$$

where

$$\mathrm{sign}(x)=\begin{cases}1, & \text{if } x>0,\\ -1, & \text{if } x<0,\\ 0, & \text{if } x=0.\end{cases}$$

From (6.48), we know that under (A.1), the contract improves the buyer's expected profit. Furthermore, the smaller the value of β is, the larger the value of flexibility is. This completes the proof. □

6.5. Impact of Forecast Accuracy

In this section, we investigate the impact of forecast accuracy. We start with an alternative definition of the accuracy for forecast.

DEFINITION 6.1 *Consider two random variables X and Y. We say that X is of a* higher increasing convex order *than Y, denoted by $X \geq_{\mathrm{ic}} Y$, if*

$$\mathsf{E}[H(X)] \geq \mathsf{E}[H(Y)] \tag{6.49}$$

for all nondecreasing convex function $H(\cdot)$.

Clearly, if $\mathsf{E}[X] = \mathsf{E}[Y]$ and $X \geq_{\text{ic}} Y$, then

$$\mathsf{Var}(X) \geq \mathsf{Var}(Y). \tag{6.50}$$

Furthermore, $X \geq_{\text{ic}} Y$ if and only if there exists a random variable ε, with $\mathsf{E}[\varepsilon|Y] \geq 0$ almost surely, such that

$$X = Y + \varepsilon.$$

That is, X has more noise than Y (see Brumelle and Vickson [6]).

These two facts may give us an intuitive explanation of why X is said to be of a higher increasing convex order than Y. For more discussion on increasing convex order, the readers are referred to Song [23] and Shaked and Shanthikumar [22].

Consider two systems 1 and 2, which face demands D^1 and D^2, respectively. We assume all other parameters to be the same for both systems. For simplicity, we also assume that both systems observe the same signal I in updating their respective demands. To be specific, demands D^1 and D^2, following Chapter 3, can be written as

$$D^1 = \varphi^1(I, R^1) \text{ and } D^2 = \varphi^2(I, R^2),$$

where R^1 and R^2 are independent random variables. Then $\varphi^k(i, R^k)$ represents the updated demand based on the observed information i of I for system k, $k = 1, 2$. Furthermore, we say that the demand forecast for system 2 is *more accurate* under the increasing convex order than the demand forecast for system 1, if $\varphi^1(i, R^1) \geq_{\text{ic}} \varphi^2(i, R^2)$ for each observed value i. It follows, therefore, that if $\mathsf{E}[\varphi^1(i, R^1)] = \mathsf{E}[\varphi^2(i, R^2)]$ and $\varphi^1(i, R^1) \geq_{\text{ic}} \varphi^2(i, R^2)$ for each i, then the variance of the updated demand of system 1 is larger than that of system 2 for each i. In this case, we can now prove the intuitive result that the expected profit of a system with more accurate forecasts than another's is higher.

THEOREM 6.7 *If for each observed value i of I, $\mathsf{E}[\varphi^1(i, R^1)] = \mathsf{E}[\varphi^2(i, R^2)]$ and $\varphi^1(i, R^1) \geq_{\text{ic}} \varphi^2(i, R^2)$, then the expected profit for system 1 is lower than that for system 2, ceteris paribus.*

REMARK 6.14 This theorem claims an intuitive fact—that is, the more accurate the information that the buyer will obtain is, the more profit that the buyer will earn eventually is.

Proof of Theorem 6.7 Let $\Pi_2^k(q, q_c, q_s, I, P_s)$ be the conditional expected profit, as defined in (6.3), of system k at stage 2 given I and P_s. If we could show that for any given $q \geq 0$ and any observed value (i, p_s) of (I, P_s),

$$\max_{\substack{0 \leq q_s < \infty \\ 0 \leq q_c \leq q}} \Pi_2^1(q, q_c, q_s, i, p_s) \leq \max_{\substack{0 \leq q_s < \infty \\ 0 \leq q_c \leq q}} \Pi_2^2(q, q_c, q_s, i, p_s), \tag{6.51}$$

then

$$\mathsf{E}\left(\max_{\substack{0\le q_s<\infty\\ 0\le q_c\le \varsigma q}} \Pi_2^1(q,q_c,q_s,I,P_s)\right)$$
$$\le \mathsf{E}\left(\max_{\substack{0\le q_s<\infty\\ 0\le q_c\le \varsigma q}} \Pi_2^2(q,q_c,q_s,I,P_s)\right), \tag{6.52}$$

and in turn,

$$\max_{q\ge 0}\left\{-pq+\mathsf{E}\left(\max_{\substack{0\le q_s<\infty\\ 0\le q_c\le \varsigma q}} \Pi_2^1(q,q_c,q_s,I,P_s)\right)\right\}$$
$$\le \max_{q\ge 0}\left\{-pq+\mathsf{E}\left(\max_{\substack{0\le q_s<\infty\\ 0\le q_c\le \varsigma q}} \Pi_2^2(q,q_c,q_s,I,P_s)\right)\right\}.$$

Thus, we have the theorem if we prove (6.51). To this end, it is sufficient to show that for any given $q \ge 0$, $q_s \ge 0$, and q_c, $0 \le q_c \le \varsigma q$,

$$\Pi_2^1(q,q_c,q_s,i,p_s) \le \Pi_2^2(q,q_c,q_s,i,p_s). \tag{6.53}$$

To prove (6.53), let $\Psi^k(z|i)$ and $\psi^k(z|i)$ be the conditional distribution and the conditional density of D^k for system k given i, respectively. That is, $\Psi^k(z|i)$ and $\psi^k(z|i)$ are distribution and density of $\varphi^k(i,R^k)$, respectively. Note that by (6.5),

$$\begin{aligned}\Pi_2^k(q,q_c,q_s,i,p_s) &= \int_0^{q+q_c+q_s} -(r-s)\left[q+q_c+q_s-z\right]\cdot\psi^k(z|i)\mathrm{d}z\\ &\quad +r\cdot(q+q_c+q_s)-p_sq_s-p_cq_c\\ &= -(r-s)\int_{q+q_c+q_s}^{\infty}\left[z-(q+q_c+q_s)\right]\cdot\psi^k(z|i)\mathrm{d}z\\ &\quad +(r-s)\cdot\mathsf{E}[\varphi^k(i,R^k)]+s\cdot(q+q_c+q_s)\\ &\quad -p_sq_s-p_cq_c.\end{aligned} \tag{6.54}$$

Note that $[z-(q+q_c+q_s)]^+$ is a nondecreasing convex function of z. Hence, in view of our assumptions $\mathsf{E}[\varphi^1(i,R^1)] = \mathsf{E}[\varphi^2(i,R^2)]$ and $\varphi^1(i,R^1) \ge_{\text{ic}} \varphi^2(i,R^2)$, we have

$$\begin{aligned}\Pi_2^1(q,q_c,q_s,i,p_s) &= -(r-s)\int_{q+q_c+q_s}^{\infty}\left[z-(q+q_c+q_s)\right]\cdot\psi^1(z|i)\mathrm{d}z\\ &\quad +(r-s)\cdot\mathsf{E}[\varphi^1(i,R^1)]+s\cdot(q+q_c+q_s)\\ &\quad -p_sq_s-p_cq_c\end{aligned}$$

$$
\begin{aligned}
&\le -(r-s)\int_{q+q_c+q_s}^{\infty}[z-(q+q_c+q_s)]\cdot\psi^2(z|i)\mathrm{d}z \\
&\quad +(r-s)\cdot\mathsf{E}[\varphi^2(i,R^2)]+s\cdot(q+q_c+q_s) \\
&\quad -p_s q_s - p_c q_c \\
&= \Pi_2^2(q,q_c,q_s,i,p_s).
\end{aligned}
\tag{6.55}
$$

This proves (6.53) as required. □

To investigate the impact of the forecast accuracy on the optimal expected total-order quantity, we introduce another definition to describe forecast accuracy.

DEFINITION 6.2 *Consider two nonnegative random variables X and Y satisfying $\mathsf{E}[X]=\mathsf{E}[Y]$ that have distributions F_X and F_Y with densities f_X and f_Y. Suppose that X and Y are either both continuous or both discrete. We say that X is* more variable *than Y, denoted by $X \ge_{\text{var}} Y$, if*

$$\mathcal{S}(f_X - f_Y) = 2 \text{ with sign sequence } +,-,+ \tag{6.56}$$

—that is, there exist $0<\alpha_1<\alpha_2<\infty$ such that $f_X(t)-f_Y(t)>0$ when $t\in(0,\alpha_1)$, $f_X(t)-f_Y(t)<0$ when $t\in(\alpha_1,\alpha_2)$, and $f_X(t)-f_Y(t)>0$ when $t\in(\alpha_2,\infty)$. Here the notation $\mathcal{S}(f(t))$ means the number of sign changes of a function $f(\cdot)$ as t increases from 0 to ∞.

For further discussion on the property of more variability, see Song [23] and Whitt [26]. Note that (6.56) implies

$$\mathcal{S}(F_X - F_Y) = 1 \text{ with sign sequence } +,-. \tag{6.57}$$

Furthermore, from $\mathsf{E}[X]=\mathsf{E}[Y]$ and (6.56), it is possible to show that

$$\mathsf{E}(X-\mathsf{E}[X])^2 > \mathsf{E}(Y-\mathsf{E}[Y])^2. \tag{6.58}$$

See also Song [23] and Ross [17]. As the variance measures the deviation of a random variable from its mean, so (6.58) motivates why X is known to be more variable than Y if X and Y satisfy (6.56).

Let T^k be the total quantity ordered by system k, $k=1,2$. Note that T^1 and T^2 are random variables. We have the following theorem.

THEOREM 6.8 *Under Assumptions* 6.1 *and* 6.2, *if $\varphi^1(i,R^1)\ge_{\text{var}}\varphi^2(i,R^2)$ and $\varsigma=0$, then there is a positive θ such that*
(i) *when $(r-p_{sl})/(r-s)\le\theta$, we have $\mathsf{E}[T^1]\le\mathsf{E}[T^2]$;*
(ii) *when $(r-p_{sh})/(r-s)\ge\theta$, we have $\mathsf{E}[T^1]\ge\mathsf{E}[T^2]$.*

Proof Let $q_s^{*k}(q,i,p_s)$ be the optimal order quantity by system k at stage 2, when the observed value of (I,P_s) is (i,p_s), $k=1,2$. It follows from Proposition 4.11 of Song [23] that for fixed q and I, there exists a $\theta(I)$ such that when

$$\frac{r-p_{sl}}{r-s} \leq \theta, \tag{6.59}$$

then

$$q_s^{*1}(q,i,p_{sl}) \leq q_s^{*2}(q,i,p_{sl}), \quad q_s^{*1}(q,i,p_{sh}) \leq q_s^{*2}(q,i,p_{sh}), \tag{6.60}$$

and when

$$\frac{r-p_{sh}}{r-s} \geq \theta, \tag{6.61}$$

then

$$q_s^{*1}(q,i,p_{sl}) \geq q_s^{*2}(q,i,p_{sl}), \quad q_s^{*1}(q,i,p_{sh}) \geq q_s^{*2}(q,i,p_{sh}). \tag{6.62}$$

Let q^{*k} be the optimal order quantity by system k at stage 1, $k=1,2$. If $q^{*k}>0$, by (6.13), then $q^{*k}>0$ must be the solution of the following equation with respect to q:

$$\begin{aligned}
&-p+\beta p_{sl}+(1-\beta)p_{sh} \\
&+\beta\int_{-\infty}^{\bar{i}^k(q,p_{sl})}\Big[(s-r)\cdot\Psi^k(q|i)+(r-p_{sl})\Big]\cdot \mathrm{d}\Lambda(i) \\
&+(1-\beta)\int_{-\infty}^{\bar{i}^k(q,p_{sh})}\Big[(s-r)\cdot\Psi^k(q|i)+(r-p_{sh})\Big]\cdot \mathrm{d}\Lambda(i)=0,
\end{aligned} \tag{6.63}$$

where $\bar{i}^k(q,p_{sl})$ and $\bar{i}^k(q,p_{sh})$ are defined by

$$\Psi^k\left(q|\bar{i}^k(q,p_{sl})\right)=\frac{r-p_{sl}}{r-s} \quad \text{and} \quad \Psi^k\left(q|\bar{i}^k(q,p_{sh})\right)=\frac{r-p_{sh}}{r-s}.$$

Let $\bar{q}_{sl}^1$ satisfy

$$\Psi^1\left(\bar{q}_{sl}^1|\bar{i}^2(q,p_{sl})\right)=\frac{r-p_{sl}}{r-s}.$$

Then using (6.60),

$$\bar{q}_{sl}^1 \leq q,$$

which, in view of the monotonicity of $\Psi^k(q|i)$, implies that if (6.59) holds,

$$\bar{i}^1(q,p_{sl}) \geq \bar{i}^2(q,p_{sl}). \tag{6.64}$$

Similarly, if (6.59) holds,

$$\bar{i}^1(q, p_{sh}) \geq \bar{i}^2(q, p_{sh}). \tag{6.65}$$

Going along the same line of the proof of (6.64), we can prove that if (6.61) holds, then

$$\bar{i}^1(q, p_{sl}) \leq \bar{i}^2(q, p_{sl}), \;\; \bar{i}^1(q, p_{sh}) \leq \bar{i}^2(q, p_{sh}). \tag{6.66}$$

For $i \in [\bar{i}^2(q, p_{sh}), \; \bar{i}^1(q, p_{sh})]$, by the monotonicity of $\Psi^k(q|i)$,

$$\begin{aligned} \Psi^2(q|i) \leq \Psi^2\left(q|\bar{i}^2(q, p_{sh})\right) &= \frac{r - p_{sh}}{r - s} \\ &= \Psi^1\left(q|\bar{i}^1(q, p_{sh})\right) \\ &\leq \Psi^1(q|i). \end{aligned}$$

Thus, from $\varphi^1(i, R^1) \geq_{\text{var}} \varphi^2(i, R^2)$, for any $i \leq \bar{i}^1(q, p_{sh})$, if (6.59) holds, then

$$\Psi^1(q|i) \geq \Psi^2(q|i). \tag{6.67}$$

Therefore,

$$\begin{aligned} &\int_{-\infty}^{\bar{i}^1(q,p_{sl})} \left[(s-r)\cdot\Psi^1(q|i) + (r - p_{sl})\right] \mathrm{d}\Lambda(i) \\ &\leq \int_{-\infty}^{\bar{i}^2(q,p_{sl})} \left[(s-r)\cdot\Psi^2(q|i) + (r - p_{sl})\right] \mathrm{d}\Lambda(i), \end{aligned} \tag{6.68}$$

and

$$\begin{aligned} &\int_{-\infty}^{\bar{i}^1(q,p_{sh})} \left[(s-r)\cdot\Psi^1(q|i) + (r - p_{sh})\right] \mathrm{d}\Lambda(i) \\ &\leq \int_{-\infty}^{\bar{i}^2(q,p_{sh})} \left[(s-r)\cdot\Psi^2(q|i) + (r - p_{sh})\right] \mathrm{d}\Lambda(i). \end{aligned} \tag{6.69}$$

Thus, the result $q^{*1} \leq q^{*2}$ follows directly from

$$\begin{aligned} 0 \;=\; & -p + \beta p_{sl} + (1-\beta)p_{sh} \\ & +\beta \int_{-\infty}^{\bar{i}^2(q^{*2},p_{sl})} \left[(s-r)\cdot\Psi^2(q^{*2}|i) + (r - p_{sl})\right] \mathrm{d}\Lambda(i) \\ & +(1-\beta) \int_{-\infty}^{\bar{i}^2(q^{*2},p_{sh})} \left[(s-r)\cdot\Psi^2(q^{*2}|i) + (r - p_{sh})\right] \mathrm{d}\Lambda(i) \end{aligned}$$

$$
\begin{aligned}
> \; & -p + \beta p_{sl} + (1-\beta)p_{sh} \\
& + \beta \int_{-\infty}^{\bar{i}^1(q^{*1}, p_{sl})} \left[(s-r)\cdot \Psi^1(q^{*2}|i) + (r - p_{sl}) \right] \mathrm{d}\Lambda(i) \\
& + (1-\beta) \int_{-\infty}^{\bar{i}^1(q^{*1}, p_{sh})} \left[(s-r)\cdot \Psi^1(q^{*2}|i) + (r - p_{sh}) \right] \mathrm{d}\Lambda(i).
\end{aligned}
$$

The first part of the theorem is proved. The second part can be proved in a similar way. □

REMARK 6.15 There are many commonly used demand distributions having the relationship $\varphi^1(i, R^1) \geq_{\text{var}} \varphi^2(i, R^2)$. For example, $\varphi^1(i, R^1)$ is uniform $(a_1 + i,\ b_1 + i)$ and $\varphi^2(i, R^2)$ is uniform $(a_2 + i,\ b_2 + i)$ with $a_1 < a_2, b_1 > b_2$ and $a_1 + b_1 = a_2 + b_2$.

6.6. Multiperiod Problems

In this section, we generalize the problem investigated above to the multiple-period case. Formally, at the beginning of period m (stage $m.1$, $1 \leq m \leq n$), there are n periods, with the knowledge of unit price p^m, the contract-unit price p_c^m of the future optional purchase, the distributions of the spot-market price, and the customer demand. The buyer makes a decision of initial purchase q^m. The buyer is also aware of that the information of the customer demand and the spot-market price will be updated at stage $m.2$ between the beginnings of periods m and $(m+1)$. At stage $m.2$, the uncertainty of the customer demand is reduced.

At stage $m.2$, it is possible for the buyer to make an adjustment in responding to the new information obtained between stage $m.1$ and stage $m.2$. The buyer can purchase additional product q_c^m with $q_c^m \leq \varsigma^m q^m$ at the contract price p_c^m. Moreover, the buyer can purchase the same product from a spot market at the market price. We further assume that the market price can be modeled as a random variable with geometric distribution—that is, the market price P_s^m has two possible cases—lower market price p_{sl}^m with probability β^m and higher market price p_{sh}^m with probability $1 - \beta^m$. Thus, the decision at stage $m.2$ is to choose the purchase quantity q_s^m from the spot market and q_c^m on-contract when the market price is observed. Of course, $q_c^m \leq \varsigma^m q^m$. Note that the degree of quantity flexibility is determined by the flexibility bound ς^m and the initial purchase quantity q^m jointly.

Finally, at the end of the period, the customer demand realizes. The buyer is assumed to lose revenue r^m for each of unsatisfied demand, and the excess inventory is assumed to be carried over but with the unit cost $\hat{p}^m$. At the end of the last period (period n), the excess inventory is assumed to have a salvage value of s^n. To avoid trivial cases, similar to Section 6.2, we assume

$$r^m > \max\{p_{sl}^m, p_c^m, p_{sh}^m, p^m, \hat{p}^m\}, \quad 1 \leq m \leq n;$$

and

$$s^n < \min\{p_{sl}^n, p_c^n, p_{sh}^n, p^n\}.$$

We use D^m to denote the initial demand forecast at stage $m.1$ and I^m to represent the information observed between stage $m.1$ and stage $m.2$. A time line of the system dynamics and the ordering decisions is illustrated in Figure 6.2. Let

$$\begin{aligned}
\Theta^m(\cdot,\cdot) &= \text{the joint distribution function of } D^m \text{ and } I^m; \\
\theta^m(\cdot,\cdot) &= \text{the joint density function of } D^m \text{ and } I^m; \\
\Lambda^m(\cdot) &= \text{the marginal distribution function of } I^m; \\
\lambda^m(\cdot) &= \text{the marginal density function of } I^m; \\
\Psi^m(\cdot|i^m) &= \text{the conditional distribution function of } D^m \text{ given } I^m = i^m; \\
\psi^m(\cdot|i^m) &= \text{the conditional density function of } D^m \text{ given } I^m = i^m.
\end{aligned}$$

Let x^{m-1} be the initial inventory level of period m. With the notation given above, then the profit obtained at period m $(1 \leq m \leq n-1)$ is

$$-p^m q^m + \mathsf{E}\left[\Pi_2^m(x^{m-1}, q^m, q_c^m, q_s^m, I^m, P_s^m)\right],$$

where

$$\begin{aligned}
&\Pi_2^m(x^{m-1}, q^m, q_c^m, q_s^m, I^m, P_s^m) \\
&= \mathsf{E}\Bigg\{\Bigg[r^m \cdot \left(D^m \wedge (x^{m-1} + q^m + q_c^m + q_s^m)\right) \\
&\qquad -p^m \cdot (x^{m-1} + q^m + q_c^m + q_s^m - D^m)^+ \\
&\qquad -p_c^m q_c^m - p_s^m q_s^n\Bigg]\Bigg|(I^m, P_s^m)\Bigg\}.
\end{aligned} \tag{6.70}$$

The profit obtained at the last period is

$$-p^n q^n + \mathsf{E}\left[\Pi_2^n(x^{n-1}, q^n, q_c^n, q_s^n, I^n, P_s^n)\right],$$

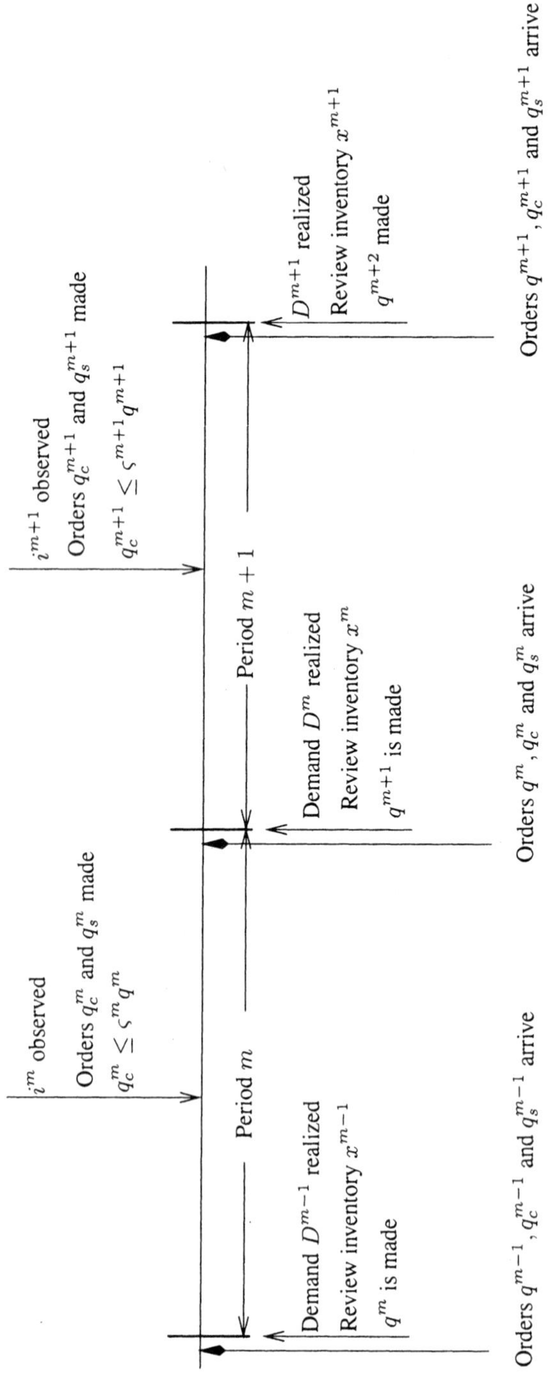

Figure 6.2. A time line for the execution of a quantity-flexibility contract

where

$$\begin{aligned}
&\Pi_2^n(x^{n-1}, q^n, q_c^n, q_s^n, I^n, P_s^n) \\
&= \mathsf{E}\Big\{\Big[r^n \cdot (D^n \wedge (x^{n-1} + q^n + q_c^n + q_s^n)) \\
&\quad + s^n \cdot (x^{n-1} + q^n + q_c^n + q_s^n - D^n)^+ \\
&\quad - p_c^n q_c^n - p_s^n q_s^n\Big] \Big| (I^n, P_s^n)\Big\}.
\end{aligned} \tag{6.71}$$

Let $\Pi^n(x^{n-1})$ be the maximum profit at period n with the initial inventory level x^{n-1}—that is,

$$\begin{aligned}
&\Pi_1^n(x^{n-1}) \\
&= \max_{q^n \geq 0}\left\{-p^n q^n + \mathsf{E}\left[\max_{\substack{0 \leq q_s^n < \infty \\ 0 \leq q_c^n \leq \varsigma^n q^n}} \Pi_2^n(x^{n-1}, q^n, q_c^n, q_s^n, I^n, P_s^n)\right]\right\}.
\end{aligned} \tag{6.72}$$

Similarly, let $\Pi_1^m(x^{m-1})$ be the maximum profit from period m to the last period with the initial inventory level x^{m-1}. Then

$$\begin{aligned}
&\Pi_1^m(x^{m-1}) \\
&= \max_{q^m \geq 0}\Big\{ - p^m q^m + \mathsf{E}\Big(\max_{\substack{0 \leq q_s^m < \infty \\ 0 \leq q_c^m \leq \varsigma^m q^m}} \Pi_2^m(x^{m-1}, q^m, q_c^m, q_s^m, I^m, P_s^m) \\
&\quad + \mathsf{E}\Big[\Pi_1^{m+1}(x^{m-1} + q^m + q_c^m + q_s^m - D^m)\Big|(I^m, P_s^m)\Big]\Big)\Big\}.
\end{aligned} \tag{6.73}$$

It is direct to verify that $\Pi_1^m(x^{m-1})$ is concave. Based on the concavity of $\Pi_1^m(x^{m-1})$, similar to Theorem 6.1, we also have the following result.

THEOREM 6.9 *There are Q^{m*}, $Q_c^{m*}(i^m, p_{sl}^m)$, $Q_s^{m*}(i^m, p_{sl}^m)$, $Q_c^{m*}(i^m, p_{sh}^m)$, and $Q_s^{m*}(i^m, p_{sh}^m)$, which are independent of x^{m-1}, such that*

(i) *if $p_{sl}^m \leq p_c^m \leq p_{sh}^m$, then the optimal reaction at stage m.1 is*

$$q^{m*} = (Q^{m*} - x^{m-1})^+.$$

The optimal reaction at stage m.2 is to order all additional required product from the spot market if the market price turns out to be low. If the market price is high, the optimal reaction is to order additional product on contract and to *order from the spot market only when the required product exceeds the quantity flexibility bound—that is,*

$$\begin{aligned} q_c^{m*}(q^{m*}, i^m, p_{sl}^m) &= 0, \\ q_s^{m*}(q^{m*}, i^m, p_{sl}^m) &= \left[Q_c^{m*}(i^m, p_{sl}^m) - q^{m*} - x^{m-1}\right]^+, \\ q_c^{m*}(q^{m*}, i^m, p_{sh}^m) &= (\varsigma^m q^{m*}) \wedge \left[Q_c^{m*}(i^m, p_{sh}^m) - q^{m*} - x^{m-1}\right]^+, \\ q_s^{m*}(q^{m*}, i^m, p_{sh}^m) &= \left[Q_s^{m*}(i^m, p_{sh}^m) - (1+\varsigma^m) q^{m*} - x^{m-1}\right]^+; \end{aligned}$$

(ii) *if* $p_c^m \le p_{sl}^m \le p_{sh}^m$, *then the optimal reaction at stage* $m.1$ *is*

$$q^{m*} = (Q^{m*} - x^{m-1})^+,$$

and the optimal reaction at stage $m.2$ *is to order additional product on contract and to order the product from the spot market only when required product exceeds the quantity flexibility bound—that is,*

$$\begin{aligned} q_c^{m*}(q^{m*}, i^m, p_{sl}^m) &= (\varsigma^m q^{m*}) \wedge \left[Q_c^{m*}(i^m, p_{sl}^m) - q^{m*} - x^{m-1}\right]^+, \\ q_s^{m*}(q^{m*}, i^m, p_{sl}^m) &= \left[Q_s^{m*}(i^m, p_{sl}^m) - (1+\varsigma^m) q^{m*} - x^{m-1}\right]^+, \\ q_c^{m*}(q^{m*}, i^m, p_{sh}^m) &= (\varsigma^m q^{m*}) \wedge \left[Q_c^{m*}(i^m, p_{sh}^m) - q^{m*} - x^{m-1}\right]^+, \\ q_s^{m*}(q^{m*}, i^m, p_{sh}^m) &= \left[Q_s^{m*}(i^m, p_{sh}^m) - (1+\varsigma^m) q^{m*} - x^{m-1}\right]^+. \end{aligned}$$

Proof The proof is similar to Theorem 6.4, and is therefore omitted. □

REMARK 6.16 Suppose that for each period, market prices are i.i.d., demands are i.i.d. and demand forecasts are i.i.d.— that is, for all m,

$$\beta^m = \beta, \ \ \Lambda^m(\cdot) = \Lambda(\cdot) \ \text{ and } \ \Psi^m(\cdot|\cdot) = \Psi(\cdot|\cdot).$$

Then the optimal purchase quantities are of myopic. Formally, there exists a pair

$$(Q^*, Q_c^*(i, p_{sl}), Q_s^*(i, p_{sl}), Q_c^*(i, p_{sh}), Q_s^*(i, p_{sh}))$$

such that for all m,

$$\begin{aligned} &Q^{m*} = Q^*, \ \ Q_c^{m*}(i, p_{sl}) = Q_c^*(i, p_{sl}), \ \ Q_s^{m*}(i, p_{sl}) = Q_s^*(i, p_{sl}), \\ &Q_c^{m*}(i, p_{sh}) = Q_c^*(i, p_{sh}), \ \ Q_s^{m*}(i, p_{sh}) = Q_s^*(i, p_{sh}). \end{aligned}$$

6.7. Numerical Example

In this section, let us consider that the joint distribution of information I and demand D, $\Theta(\cdot,\cdot)$ is a bivariate normal distribution with means μ and η, standard deviations τ and σ, and the correlation coefficient ρ. Then, the resulting marginal density $\Lambda(\cdot)$ is normal with mean μ and standard variation

τ. The conditional density $\psi(\cdot|i)$ is normal with mean $\eta + \rho\sigma(i - \mu)/\tau$ and standard deviation $\sigma\sqrt{1-\rho^2}$. Formally,

$$\begin{aligned}\psi(x|i) &= \frac{\mathrm{dP}(D \leq x|I = i)}{\mathrm{d}x} \\ &= \frac{1}{\sqrt{2\pi}\sigma\sqrt{1-\rho^2}} \exp\left\{-\frac{\left[x - \eta - \rho\sigma\frac{i-\mu}{\tau}\right]^2}{2\sigma^2(1-\rho^2)}\right\},\end{aligned} \tag{6.74}$$

(see pages 22-28 of Bickel and Doksum [4]). Noting that $\sigma\sqrt{1-\rho^2} \leq \sigma$, similar to Fisher and Raman [10], the bivariate normal distribution indicates how information I enables the retailer to obtain more accurate demand forecast with the variance measure. Now if $0 \leq \rho < 1$, then D is conditionally stochastically increasing with respect to I, and if $-1 \leq \rho < 0$, then D is conditionally stochastically decreasing with respect to I. Let D_i denote the random variable with the density (6.74). When we need to stress the dependence on ρ, D_i and $\psi(x|i)$ are written as $D_i(\rho)$ and $\psi(\rho, x|i)$, respectively.

In this section, we assume that

$$0 \leq \rho < 1.$$

From the above discussion, we know that the quality of information I can be represented by a convenient single-parameter ρ, the correlation coefficient. If $\rho = 0$, then I and D are completely uncorrelated (independent). Thus, the realization of I provides no information about the final demand D. If $\rho = 1$, however, I and D are completely correlated, and the realization of I gives perfect information about the value of the final demand D. For values of ρ between 0 and 1, although the relationship between magnitude of ρ and magnitude of information quality is not clear; however, we have that the larger ρ is, the smaller the variance of $D_i(\rho)$ is—that is, we can reduce the uncertainty of D in terms of ρ.

Given ρ_1 and ρ_2 with $0 < \rho_1 < \rho_2 < 1$, we can directly verify that for any observation of $I = i$,

$$\lim_{|x| \to +\infty} \frac{\psi(\rho_1, x|i)}{\psi(\rho_2, x|i)} = +\infty.$$

Hence, by

$$\frac{1}{\sqrt{2\pi}\sqrt{1-\rho_1^2}} \leq \frac{1}{\sqrt{2\pi}\sqrt{1-\rho_2^2}},$$

we have

$$D_i(\rho_1) \geq_{\text{var}} D_i(\rho_2). \tag{6.75}$$

If we take $F_1(x) = x^2$, then

$$\mathsf{E}\left[F_1(D_i(\rho_1))\right] > \mathsf{E}\left[F_1(D_i(\rho_2))\right]. \tag{6.76}$$

On the other hand, let i satisfy $i > \mu$, and we choose $\varepsilon > 0$ such that

$$\begin{aligned} &\sigma\sqrt{1-\rho_1^2} + (\varepsilon - 1)\left[\eta + \rho_1\sigma\frac{i-\mu}{\tau}\right] \\ &< \sigma\sqrt{1-\rho_2^2} + (\varepsilon - 1)\left[\eta + \rho_2\sigma\frac{i-\mu}{\tau}\right]. \end{aligned}$$

Then taking $F_2(x) = x^2 + \varepsilon x$, we have

$$\mathsf{E}\left[F_2(D_i(\rho_1))\right] < \mathsf{E}\left[F_2(D_i(\rho_2)\right]. \tag{6.77}$$

Combining (6.76) and (6.77), we know that $D_i(\rho_1) \geq_{\text{ic}} D_i(\rho_2)$ is not true.

However, we can directly verify the following monotonicity properties with respect to ρ.

PROPOSITION 6.1 *Assume that Assumption* 6.1 *holds and that* $0 \leq \rho < 1$. *Then we have*

(i) *the initial optimal order quantity* q^* *is nonincreasing in* ρ, *and*

(ii) *the optimal expected profit* π^* *is increasing in* ρ.

Proof Let

$$\bar{i}(q, p_{sl}) = \mu + \frac{\tau}{\rho\sigma}\left[q - \eta - \sigma\sqrt{1-\rho^2}\Phi^{-1}\left(\frac{r - p_{sl}}{r - s}\right)\right],$$

$$\hat{i}(q, p_{sl}) = \mu + \frac{\tau}{\rho\sigma}\left[(1+\varsigma)q - \eta - \sigma\sqrt{1-\rho^2}\Phi^{-1}\left(\frac{r - p_{sl}}{r - s}\right)\right],$$

$$\bar{i}(q, p_c) = \mu + \frac{\tau}{\rho\sigma}\left[q - \eta - \sigma\sqrt{1-\rho^2}\Phi^{-1}\left(\frac{r - p_c}{r - s}\right)\right],$$

$$\hat{i}(q, p_c) = \mu + \frac{\tau}{\rho\sigma}\left[(1+\varsigma)q - \eta - \sigma\sqrt{1-\rho^2}\Phi^{-1}\left(\frac{r - p_c}{r - s}\right)\right],$$

$$\bar{i}(q, p_{sh}) = \mu + \frac{\tau}{\rho\sigma}\left[q - \eta - \sigma\sqrt{1-\rho^2}\Phi^{-1}\left(\frac{r - p_{sh}}{r - s}\right)\right],$$

$$\hat{i}(q, p_{sh}) = \mu + \frac{\tau}{\rho\sigma}\left[(1+\varsigma)q - \eta - \sigma\sqrt{1-\rho^2}\Phi^{-1}\left(\frac{r - p_{sh}}{r - s}\right)\right].$$

Consider the case $p_{sl} \leq p_c \leq p_{sh}$. Let $\hat{q}$ be the solution of

$$\begin{aligned} 0 \;=\; & -p + \beta\Big[p_{sl} + (r - p_{sl}) \cdot \Lambda(\bar{i}(q, p_{sl})) \\ & \qquad -(r - s)\int_{-\infty}^{\bar{i}(q,p_{sl})} \Psi(q|i)\mathrm{d}\Lambda(i)\Big] \end{aligned}$$

$$+(1-\beta)\Big[-p_c\varsigma + p_{sh}(1+\varsigma) + (r-p_c)\cdot\Lambda(\bar{i}(q,p_c))$$
$$+(1+\varsigma)(r-p_{sh})\cdot\Lambda(\hat{i}(q,p_{sh}))$$
$$-(r-p_c)(1+\varsigma)\cdot\Lambda(\hat{i}(q,p_c))$$
$$-(r-s)\int_{-\infty}^{\bar{i}(q,p_c)}\Psi(q|i)\mathrm{d}\Lambda(i)$$
$$-(1+\varsigma)(r-s)\int_{\hat{i}(q,p_c)}^{\hat{i}(q,p_{sh})}\Psi((1+\varsigma)q|i)\mathrm{d}\Lambda(i)\Big] \tag{6.78}$$

with the convenience $\hat{q}=0$ if the solution of (6.78) does not exist. Define, for each i,

$$\hat{t}(i,p_{sl}) = \eta + \rho\sigma\frac{i-\mu}{\tau} + \sigma\sqrt{1-\rho^2}\Phi^{-1}\left(\frac{r-p_{sl}}{r-s}\right),$$
$$\hat{t}(i,p_{c}) = \eta + \rho\sigma\frac{i-\mu}{\tau} + \sigma\sqrt{1-\rho^2}\Phi^{-1}\left(\frac{r-p_{c}}{r-s}\right),$$
$$\hat{t}(i,p_{sh}) = \eta + \rho\sigma\frac{i-\mu}{\tau} + \sigma\sqrt{1-\rho^2}\Phi^{-1}\left(\frac{r-p_{sh}}{r-s}\right).$$

If

$$p_{sl}\le p_c\le p_{sh}, \tag{6.79}$$

then by (6.13) and Theorem 6.2, the optimal order policy is given by

$$q^* = \hat{q},$$

and

$$q_c^*(q^*,i,p_{sl}) = 0,$$

$$q_s^*(q^*,i,p_{sl}) = \begin{cases} 0, & \text{if } i\le \bar{i}(q^*,p_{sl}),\\ \hat{t}(i,p_{sl})-q^*, & \text{if } i> \bar{i}(q^*,p_{sl}),\end{cases}$$

$$q_c^*(q^*,i,p_{sh}) = \begin{cases} 0, & \text{if } i\le \bar{i}(q^*,p_{c}),\\ \hat{t}(i,p_{c})-q^*, & \text{if } \bar{i}(q^*,p_c)< i\le \hat{i}(q^*,p_{c}),\\ \varsigma q^*, & \text{if } i> \hat{i}(q^*,p_{c}),\end{cases}$$

$$q_s^*(q^*,i,p_{sh}) = \begin{cases} 0, & \text{if } i\le \hat{i}(q^*,p_{sh}),\\ \hat{t}(i,p_{sh})-(1+\varsigma)q^*, & \text{if } i> \hat{i}(q^*,p_{sh}).\end{cases}$$

Furthermore, by (6.12), the optimal expected profit is given by

$$\beta(r-s)\left\{\int_{-\infty}^{\bar{i}(q^*,p_{sl})}\left[\int_{-\infty}^{q^*} t\cdot\psi(z|i)\mathrm{d}z\right]\mathrm{d}\Lambda(i)\right.$$
$$\left.+\int_{\bar{t}(q^*,p_{sl})}^{\infty}\left[\int_{-\infty}^{\hat{i}(i,p_{sl})} z\cdot\psi(z|i)\mathrm{d}z\right]\mathrm{d}\Lambda(i)\right\}$$
$$+(1-\beta)(r-s)\left\{\int_{-\infty}^{\bar{i}(q^*,p_c)}\left[\int_{-\infty}^{q^*} z\cdot\psi(z|i)\mathrm{d}z\right]\mathrm{d}\Lambda(i)\right.$$
$$+\int_{\bar{i}(q^*,p_c)}^{\hat{i}(q^*,p_c)}\left[\int_{-\infty}^{\hat{t}(i,p_c)} z\cdot\psi(z|i)\mathrm{d}z\right]\mathrm{d}\Lambda(i)$$
$$+\int_{\hat{i}(q^*,p_c)}^{\hat{i}(q^*,p_{sh})}\left[\int_{-\infty}^{(1+\varsigma)q^*} z\cdot\psi(z|i)\mathrm{d}z\right]\mathrm{d}\Lambda(i)$$
$$\left.+\int_{\hat{i}(q^*,p_{sh})}^{\infty}\left[\int_{-\infty}^{\hat{t}(q^*,p_{sh})} z\cdot\psi(z|i)\mathrm{d}z\right]\mathrm{d}\Lambda(i)\right\}. \tag{6.80}$$

To get the case $p_c \leq p_{sl} \leq p_{sh}$, in view of (6.18), let $\bar{q}$ be the solution of

$$-p+\beta\Bigg[-p_c\varsigma+p_{sl}\cdot(1+\varsigma)+(r-p_c)\cdot\Lambda(\bar{i}(q,p_c))$$
$$+(1+\varsigma)(r-p_{sl})\cdot\Lambda(\hat{i}(q,p_{sl}))$$
$$-(r-p_c)(1+\varsigma)\cdot\Lambda(\hat{i}(q,p_c))-(r-s)\int_{-\infty}^{\bar{i}(q,p_c)}\Psi(q|i)\mathrm{d}\Lambda(i)$$
$$-(1+\varsigma)(r-s)\int_{\hat{i}(q,p_c)}^{\hat{i}(q,p_{sh})}\Psi((1+\varsigma)q|i)\mathrm{d}\Lambda(i)\Bigg]$$
$$+(1-\beta)\Bigg[-p_c\varsigma+p_{sh}\cdot(1+\varsigma)+(r-p_c)\cdot\Lambda(\bar{i}(q,p_c))$$
$$+(1+\varsigma)(r-p_{sh})\cdot\Lambda(\hat{i}(q,p_{sh}))$$
$$-(r-p_c)(1+\varsigma)\cdot\Lambda(\hat{i}(q,p_c))-(r-s)\int_{-\infty}^{\bar{i}(q,p_c)}\Psi(q|i)\mathrm{d}\Lambda(i)$$
$$-(1+\varsigma)(r-s)\int_{\hat{i}(q,p_c)}^{\hat{i}(q,p_{sh})}\Psi((1+\varsigma)q|i)\mathrm{d}\Lambda(i)\Bigg]=0 \tag{6.81}$$

with the convenience $\bar{q} = 0$ if the solution of (6.81) does not exist. If

$$p_c \le p_{sl} \le p_{sh}, \tag{6.82}$$

then it follows from (6.18) and Theorem 6.2 that the optimal order policy is given by

$$q^* = \bar{q},$$

and

$$q_c^*(q^*, i, p_{sl}) = \begin{cases} 0, & \text{if } i \le \bar{i}(q^*, p_c), \\ \hat{t}(i, p_c) - q^*, & \text{if } \bar{i}(q^*, p_c) < i \le \hat{i}(q^*, p_c), \\ \varsigma q^*, & \text{if } i > \hat{i}(q^*, p_c), \end{cases}$$

$$q_s^*(q^*, i, p_{sl}) = \begin{cases} 0, & \text{if } i \le \hat{i}(q^*, p_{sl}), \\ \hat{t}(i, p_{sl}) - (1+\varsigma)q^*, & \text{if } i > \hat{i}(q^*, p_{sl}), \end{cases}$$

$$q_c^*(q^*, i, p_{sh}) = \begin{cases} 0, & \text{if } i \le \bar{i}(q^*, p_c), \\ \hat{t}(i, p_c) - q^*, & \text{if } \bar{i}(q^*, p_c) < i \le \hat{i}(q^*, p_c), \\ \varsigma q^*, & \text{if } i > \hat{i}(q^*, p_c), \end{cases}$$

$$q_s^*(q^*, i, p_{sh}) = \begin{cases} 0, & \text{if } i \le \hat{i}(q^*, p_{sh}), \\ \hat{t}(i, p_{sh}) - (1+\varsigma)q^*, & \text{if } i > \hat{i}(q^*, p_{sh}). \end{cases}$$

Furthermore, the optimal expected profit is given by

$$\begin{aligned} \beta(r-s)\Bigg\{ & \int_{-\infty}^{\bar{i}(q^*,p_c)} \left[\int_{-\infty}^{q^*} z \cdot \psi(z|i) \mathrm{d}z \right] \mathrm{d}\Lambda(i) \\ & + \int_{\bar{i}(q^*,p_c)}^{\hat{i}(q^*,p_c)} \left[\int_{-\infty}^{\hat{t}(i,p_c)} z \cdot \psi(z|i) \mathrm{d}z \right] \mathrm{d}\Lambda(i) \\ & + \int_{\hat{i}(q^*,p_c)}^{\hat{i}(q^*,p_{sl})} \left[\int_{-\infty}^{(1+\varsigma)q^*} z \cdot \psi(z|i) \mathrm{d}z \right] \mathrm{d}\Lambda(i) \\ & + \int_{\hat{i}(q^*,p_{sl})}^{\infty} \left[\int_{-\infty}^{\hat{t}(q^*,p_{sl})} z \cdot \psi(z|i) \mathrm{d}z \right] \mathrm{d}\Lambda(i) \Bigg\} \end{aligned}$$

$$
\begin{aligned}
+(1-\beta)(r-s)\Bigg\{ & \int_{-\infty}^{\bar{i}(q^*,p_c)} \left[\int_{-\infty}^{q^*} z\cdot\psi(z|i)\mathrm{d}z\right]\mathrm{d}\Lambda(i) \\
& + \int_{\bar{i}(q^*,p_c)}^{\hat{i}(q^*,p_c)} \left[\int_{-\infty}^{\hat{t}(i,p_c)} z\cdot\psi(z|i)\mathrm{d}z\right]\mathrm{d}\Lambda(i) \\
& + \int_{\hat{i}(q^*,p_c)}^{\hat{i}(q^*,p_{sh})} \left[\int_{-\infty}^{(1+\varsigma)q^*} z\cdot\psi(z|i)\mathrm{d}z\right]\mathrm{d}\Lambda(i) \\
& + \int_{\hat{i}(q^*,p_{sh})}^{\infty} \left[\int_{-\infty}^{\hat{t}(q^*,p_{sh})} z\cdot\psi(z|i)\mathrm{d}z\right]\mathrm{d}\Lambda(i)\Bigg\}.
\end{aligned}
\tag{6.83}
$$

Consider the proof of (i) for the case $p_{sl} \le p_c \le p_{sh}$. Write the right-side of (6.78) as $H_1(q,\rho)$. Note that

$$
\begin{aligned}
\lim_{q\to+\infty} H_1(q,\rho) &= -p+\beta\Big[p_{sl}+(r-p_{sl})-(r-s)\Big] \\
&\quad +(1-\beta)\Big[-p_c\varsigma+p_{sh}(1+\varsigma)+(r-p_c) \\
&\quad -(1+\varsigma)(r-p_c)-(r-s)+(1+\varsigma)(r-p_{sh})\Big] \\
&= s-p<0.
\end{aligned}
\tag{6.84}
$$

On the other hand, for all $q \ge 0$,

$$
\begin{aligned}
&\frac{\partial H_1(q,\rho)}{\partial q} \\
&= -\beta(r-s)\int_{-\infty}^{\bar{i}(q,p_{sl})} \psi(q|i)\mathrm{d}\Lambda(i) \\
&\quad -(1-\beta)(r-s)\Bigg\{\int_{-\infty}^{\bar{i}(q,p_c)} \psi(q|i)\mathrm{d}\Lambda(i) \\
&\qquad + \int_{\hat{i}(q,p_c)}^{\hat{i}(q,p_{sh})} \psi((1+\varsigma)q|i)\mathrm{d}\Lambda(i)\Bigg\} \\
&< 0,
\end{aligned}
\tag{6.85}
$$

where we use

$$\begin{aligned}
\Psi\left(q|\bar{i}(q,p_c)\right) &= \frac{r-p_c}{r-s}, \\
\Psi\left(q|\hat{i}(q,p_{sl})\right) &= \frac{r-p_{sl}}{r-s}, \\
\Psi\left((1+\varsigma)q|\hat{i}(q,p_c)\right) &= \frac{r-p_c}{r-s}, \\
\Psi\left((1+\varsigma)q|\hat{i}(q,p_{sh})\right) &= \frac{r-p_{sh}}{r-s}.
\end{aligned}$$

Therefore, by (6.84), we get that

$$q^* > 0 \quad \text{if and only if} \quad H_1(0,\rho) > 0. \tag{6.86}$$

Let

$$m_1(\rho) = \frac{-\eta - \rho\sigma(u-\mu)/\tau}{\sigma\sqrt{1-\rho^2}}.$$

Furthermore,

$$\begin{aligned}
&\frac{\partial H_1(0,\rho)}{\partial \rho} \\
&= -\beta(r-s)\int_{-\infty}^{\bar{i}(0,p_{sl})} \psi(0|i)\cdot\sigma\sqrt{1-\rho^2}\frac{\mathrm{d}m_1(\rho)}{\mathrm{d}\rho}\mathrm{d}\Lambda(i) \\
&\quad -(1-\beta)(r-s)\left\{\int_{-\infty}^{\bar{i}(0,p_c)} \psi(0|i)\cdot\sigma\sqrt{1-\rho^2}\frac{\mathrm{d}m_1(\rho)}{\mathrm{d}\rho}\mathrm{d}\Lambda(i)\right. \\
&\quad \left. +(1+\varsigma)\int_{\hat{i}(0,p_c)}^{\hat{i}(0,p_{sh})} \psi(0|i)\cdot\sigma\sqrt{1-\rho^2}\frac{\mathrm{d}m_1(\rho)}{\mathrm{d}\rho}\mathrm{d}\Lambda(i)\right\}.
\end{aligned} \tag{6.87}$$

By some simple integral calculation, we have

$$\begin{aligned}
&-\beta(r-s)\int_{-\infty}^{\bar{i}(0,p_{sl})} \psi(0|i)\cdot\sigma\sqrt{1-\rho^2}\frac{\mathrm{d}m_1(\rho)}{\mathrm{d}\rho}\mathrm{d}\Lambda(i) \\
&= -\beta(r-s)\int_{-\infty}^{\bar{i}(0,p_{sl})} \frac{1}{2\pi\tau}\exp\left\{-\frac{\eta^2}{2\sigma^2}\right\}\cdot\exp\left\{-\frac{\left[\sigma\frac{i-\mu}{\tau}+\rho\eta\right]^2}{2\sigma^2(1-\rho^2)}\right\} \\
&\qquad \cdot\left(-\frac{\sigma\frac{i-\mu}{\tau}+\rho\eta}{\sigma(1-\rho^2)^{3/2}}\right)\mathrm{d}i \\
&< 0,
\end{aligned} \tag{6.88}$$

and similarly,

$$-(1-\beta)(r-s)\left\{\int_{-\infty}^{\bar{i}(0,p_c)} \psi(0|i)\cdot\frac{\mathrm{d}m_1(\rho)}{\mathrm{d}\rho}\mathrm{d}\Lambda(i) + (1+\varsigma)\int_{\hat{i}(0,p_c)}^{\hat{i}(0,p_{sh})} \psi(0|i)\cdot\frac{\mathrm{d}m_1(\rho)}{\mathrm{d}\rho}\mathrm{d}\Lambda(i)\right\} < 0. \tag{6.89}$$

Combining (6.87)–(6.89), we obtain that

$$\frac{\partial H_1(0,\rho)}{\partial\rho} < 0. \tag{6.90}$$

Therefore, there exists a ρ_0 $(0 \le \rho_0 \le 1)$ such that for $\rho \in (0,\ \rho_0)$,

$$H_1(0,\rho) > 0, \tag{6.91}$$

and for $\rho \in (\rho_0,\ 1)$,

$$H_1(0,\rho) \le 0. \tag{6.92}$$

It follows from (6.86) that

$$q^* \neq 0 \quad \text{for } \rho \in (0,\ \rho_0), \tag{6.93}$$

and

$$q^* = 0 \ \text{ for } \rho \in (\rho_0,\ 1). \tag{6.94}$$

Consequently, q^* is the unique solution of equation

$$H_1(q,\rho) = 0 \ \text{ for } \rho \in (0,\ \rho_0).$$

To prove (i) of the proposition, from (6.93)–(6.94), it suffices to prove that for $\rho \in (0,\ \rho_0)$

$$\frac{\mathrm{d}q^*}{\mathrm{d}\rho} < 0. \tag{6.95}$$

We will use the implicit derivative formula to prove (6.95). Let

$$\begin{aligned} m_2(\rho) &= \frac{q-\eta-\rho\sigma(u-\mu)/\tau}{\sigma\sqrt{1-\rho^2}}, \\ m_3(\rho) &= \frac{(1+\varsigma)q-\eta-\rho\sigma(u-\mu)/\tau}{\sigma\sqrt{1-\rho^2}}. \end{aligned}$$

By the right side of (6.78), similar to (6.87),

$$\begin{aligned}
&\frac{\partial H_1(q,\rho)}{\partial \rho} \\
&= -\beta(r-s)\int_{-\infty}^{\bar{i}(q,p_{sl})} \psi(q|i)\cdot\sigma\sqrt{1-\rho^2}\frac{\mathrm{d}m_2(\rho)}{\mathrm{d}\rho}\mathrm{d}\Lambda(i) \\
&\quad -(1-\beta)(r-s)\left\{\int_{-\infty}^{\bar{i}(q,p_c)} \psi(q|i)\cdot\sigma\sqrt{1-\rho^2}\frac{\mathrm{d}m_2(\rho)}{\mathrm{d}\rho}\mathrm{d}\Lambda(i)\right. \\
&\quad \left. + \int_{\hat{i}(q,p_c)}^{\hat{i}(q,p_{sh})} \psi((1+\varsigma)q|i)\cdot\sigma\sqrt{1-\rho^2}\frac{\mathrm{d}m_3(\rho)}{\mathrm{d}\rho}\mathrm{d}\Lambda(i)\right\}.
\end{aligned} \tag{6.96}$$

Similar to (6.90), we can show by (6.96) that

$$\frac{\partial H_1(q,\rho)}{\partial \rho} < 0. \tag{6.97}$$

Consequently, by the implicit derivative formula, (6.85) and (6.97) yield that

$$\frac{\mathrm{d}q^*}{\mathrm{d}\rho} = -\frac{\partial H_1(q,\rho)}{\partial q}\bigg/\frac{\partial H_1(q,\rho)}{\partial \rho} > 0, \tag{6.98}$$

which proves (6.95). Thus we prove (i) for the case $p_{sl} \le p_c \le p_{sh}$. Going along the same line, by using (6.81), (i) for the case $p_c \le p_{sl} \le p_{sh}$ can be proved.

Similar to the above proof of (i), we can prove (ii) by using (6.80) and (6.83). The detailed proof is omitted here. □

Next, we carry out a series numerical experiment. The objective is twofold: first, to implement the algorithm for the general case where the close-form solution does not exist; second, investigate how factors such as the quality of information and the distribution of the spot-market price affect the optimal solutions.

For a given set of contract parameters (such as contract price and flexibility factor ς) and spot-market parameters, the quality of information revision is the key factor for the optimal decision. The effect of the quality of information revision is illustrated in sequel. We define the degree of forecast improvement $f(\rho)$ as $f(\rho) = 1 - \sqrt{1-\rho^2}$. The parameters for the numerical example are listed in Table 6.1

Figure 6.3 depicts the order-quantity decision with respect to the quality-of-information revision. In the numerical experiment, we choose a different

p_{sh}	p_c	p_{sl}	p	r	s	μ	τ	η	σ
5000	4000	3000	3000	10000	800	10000	7500	10000	7500

Table 6.1. Parameters used in the numerical experiment

contract flexibility factor ς and β, which is the probability of taking the low price at the spot market. It is intuitive to observe that the order quantity reduces as the quality of information update improves. In addition, when the probability of taking the low price at the spot market is high, the order quantity is low as well. The reaction to the flexible factor ς is interesting. Take the case of $\beta = 0.8$ and the flexibility factor $\varsigma = 0.2$ as an example. Compared with the case of the larger flexibility factor ($\varsigma = 0.6$), when the information quality is low, the optimal order strategy is to order less; when the information quality is high, the optimal order strategy is to order more. This is because the initial order is not adequate when the information quality is poor. The contract flexibility may not provide a sufficient cushion. On the other hand, when the quality of information is good, it is beneficial to make a larger initial order, which will provide a larger buffer for potential changes.

6.8. Concluding Remarks

In this chapter, we have studied single and multiperiod quantity-flexible contracts that allow an initial order at the beginning of a period, a forecast revision in the middle of the period, and further purchases on contract and in the spot market before the demand is realized at the end of the period. The additional purchase quantity on the contract at a contractual price is limited by the specified flexibility limit. Any amount, however, can be purchased on the spot market at the prevailing market price. The initial purchase quantity at a given price is based on the demand distribution, the market-price distribution, the contractual prices, the flexibility level, and the possibility of a forecast revision before additional purchase. We provide optimal initial orders and the optimal feedback quantities to be purchased following the demand-forecast revisions. We examine the impact of the information quality and the flexibility on the optimal decisions. We measure the value of flexibility and provide conditions when this value is positive.

We would like to point out that this is an initial study of optimal management and design for the flexibility contracts. We have made a number of simplifications in this study. In our model, the inventory carry-over induces a temporal relationship between the earlier and later periods. On the contrary, by allowing lost sales, the demand temporal relationship is not modeled. We also ignore the

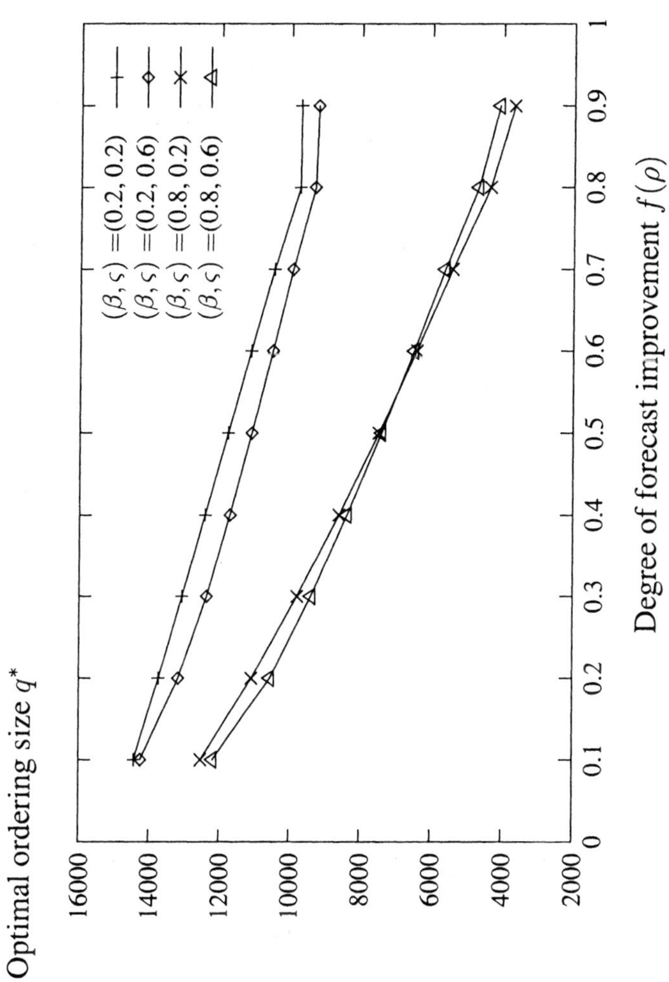

Figure 6.3. The optimal order quantity as a function of the quality of information

fixed order cost in our model. Although we believe that it is still trackable, by taking these factors into consideration, the k-convex function and the dynamics of fully backlogged demand would make the problem more challenging. Another stimulating model could include an exercise price for the contract. In addition, we allow only two possible market prices, high and low, which are geometrically distributed. This could be extended to allow for a range of prices having a general probability distribution as well.

6.9. Notes

The chapter is based on Sethi, Yan, and Zhang [21].

Eppen and Iyer [9] study a special form of the quantity-flexibility contract that allows the retailer to return a portion of its purchase to the supplier. Bassok and Anupindi [3] analyze a single-product periodic-review inventory system with a minimum-quantity contract, which stipulates that the cumulative purchase over the life of the contract must exceed a specified minimum quantity to qualify for a price discount. They demonstrate that the optimal inventory policy for the buyer is an order-up-to type and that the order-up-to level can be determined by a newsvendor model. Anupindi and Bossok [1] extend this work to the case of multiple products. In this case, the supply contract requires that the total purchase amount in dollars exceed a specified minimum to obtain the price discount. Tsay [24] studies incentives, causes of inefficiency, and possible ways of performance improvement in a quantity-flexibility contract. In particular, Tsay [24] investigates order revisions in response to new demand information, where the information is a location parameter of the demand distribution. Tsay and Lovejoy [25] investigate the quantity-flexibility contracts in more complex settings of multiple players, multiple demand periods, and demand-forecast updates. They study issues relating to desired levels of flexibility and local and systemwide performances.

Similar to the structure of quantity-flexibility contracts, a form of take-or-pay provision has been used in many long-term natural resources and energy-supply contracts (Tsay [24]). A take-or-pay contract is an agreement between a buyer and a supplier, which often specifies a minimum quantity that the buyer must purchase (take) and the maximum quantity that the buyer can obtain (pay) over the contract period. Brown and Lee [5] note that capacity-reservation agreements, common in the semiconductor industry, have a similar structure. Brown and Lee examine how much capacity should be reserved as take and how much capacity should be reserved for future (pay).

Related research has been carried out in the area of inventory management with demand-forecast updates. It is possible to classify this research into three categories. The first category uses Bayesian analysis. Bayesian models were first introduced in the inventory literature by Dvoretzky, Kiefer, and Wolfowitz [8]. Eppen and Iyer [9] analyze a quick-response program in a fashion-buying problem by using the Bayesian rule to update demand distributions. The use of time-series models in demand-forecast updating characterizes the second category, which includes the papers by Johnson and Thompson [15] and Lovejoy [16]. The third category is concerned with forecast revisions. This approach is developed and used by Hausmann [13], Sethi and Sorger [18], Heath and Jackson [14], Donohue [7], Yan, Liu, and Hsu [27], Gurnani and Tang [12], Barnes-Schuster, Bassok, and Anupindi [2], Gallego and Özer [11], Sethi, Yan, and Zhang [19, 20], and others. We refer the readers to a more detailed review in Chapter 1.

References

[1] R. Anupindi and Y. Bassok. Approximations for multiproduct contracts with stochastic demands and business volume discounts: Single supplier case. *IIE Transactions*, 30:723–734, 1998.

[2] D. Barnes-Schuster, Y. Bassok, and R. Anupindi. Coordination and flexibility in supply contracts with options. *Manufacturing and Service Operations Management*, 4:171–207, 2002.

[3] Y. Bassok and R. Anupindi. Analysis of supply contracts with total minimum commitment. *IIE Transactions*, 29:373–382, 1997.

[4] P. Bickel and K. Doksum. *Mathematical Statisitics:Basic Ideas and Selected Topics*. Holden Day Publishers, San Francisco, CA, 1977.

[5] A.O. Brown and H.L. Lee. Optimal pay-to-delay capacity reservation with application to the semiconductor industry. Working Paper, Stanford University, Stanford, CA, 1997.

[6] L. Brumelle and R. Vickson. A unified approach to stochastic dominance. *Stochastic Optimization Models in Finance*, W. Ziemba and R. Vickson (editors). Academic Press, New York, 1975.

[7] K.L. Donohue. Efficient supply contracts for fashion goods with forecast updating and two production modes. *Management Science*, 46:1397–1411, 2000.

[8] A. Dvoretzky, J. Kiefer, and J. Wolfowitz. The inventory problem: ii. Case of unknown distributions of demand. *Econometrica*, 20:450–466, 1952.

[9] G.D. Eppen and A.V. Iyer. Improved fashion buying with Bayesian updates. *Operations Research*, 45:805–819, 1997.

[10] M. Fisher and A. Raman. Reducing the cost of demand uncertainty through accurate response to early sales. *Operations Research*, 44:87–99, 1996.

[11] G. Gallego and Ö. Özer. Integrating replenishment decisions with advance demand information. *Management Science*, 47:1344–1360, 2001.

[12] H. Gurnani and C.S. Tang. Note: optimal ordering decisions with uncertain cost and demand forecast updating. *Management Science*, 45:1456–1462, 1999.

[13] W.H. Hausman. Sequential decision problems: A model to exploit existing forecasters. *Management Science*, 16:B93-B111, 1969.

[14] D. Heath and P. Jackson. Modeling the evolution of demand forecast with application to safety stock analysis in production/distribution systems. *IIE Transactions*, 26:17–30, 1994.

[15] G.D. Johnson and H. Thompson. Optimality of myopic inventory policies for certain dependent demand processes. *Management Science*, 21:1303–1307, 1975.

[16] W.S. Lovejoy. Myopic policies for some inventory models with uncertain demand distributions. *Management Science*, 36:724–738, 1990.

[17] S. Ross. *Stochastic Processes.* John Wiley, New York, 1983.

[18] S.P. Sethi and G. Sorger. A theory of rolling horizon decision making. *Annals of Operations Research*, 29:387–416, 1991.

[19] S.P. Sethi, H. Yan, and H. Zhang. Peeling layers of an onion: An inventory model with multiple delivery modes and forecast updates. *Journal of Optimization Theory and Applications*, 108: 253–281, 2001.

[20] S.P. Sethi, H. Yan, and H. Zhang. Inventory models with fixed costs,forecast updates and two delivery modes. *Operations Research*, 51:321–328, 2003.

[21] S.P. Sethi, H. Yan, and H. Zhang. Quantity-flexibility contracts: Optimal decisions with information updates. *Decision Sciences*, 35:691–712, 2004.

[22] M. Shaked and J.G. Shanthikumar. *Stochastic Orders and Their Applications.* Academic Press, New York, 1994.

[23] J. Song. The effect of lead time uncertainty in a simple stochastic inventory model. *Management Science*, 40:603–613, 1994.

[24] A. Tsay. The quantity flexibility contract and supplier-customer incentives. *Management Science*, 45:1339–1358, 1999.

[25] A. Tsay and W.S. Lovejoy. Quantity-flexibility contracts and supply chain performance. *Manufacturing and Service Operations Management*, 1:89–111, 1999.

[26] W. Whitt. Uniform conditional variability ordering of probability distributions. *Journal of Applied Probability*, 22:619–633, 1985.

[27] H. Yan, K. Liu, and A. Hsu. Optimal ordering in a dual-supplier system with demand forecast updates. *Production and Operations Management*, 12:30–45, 2003.

Chapter 7

PURCHASE CONTRACT MANAGEMENT: FIXED EXERCISE COST

7.1. Introduction

In this chapter, we consider a single-period two-stage model in which any purchase on contract at stage 2 incurs a fixed cost of exercising the contract. Once the contract is exercised, the buyer may increase the initial order quantity by any amount at a higher unit cost or cancel some or all of the ordered items for a lower-than-cost refund. In contrast, the previous chapter considers a multiperiod model with two decision-making stages in each period and the extend of flexibility as a decision variable at stage 1, but with no contract exercise cost. The model under consideration in this chapter involves newsvendor formulas and an (s, S) type policy. These features allow us to obtain an explicit solution that provides a number of valuable insights into a better purchase-contract management. When there are other means of hedging demand uncertainties such as product substitution, we establish the value of the purchase contract. We prove that the optimal cost function is monotone with respect to the contract-exercise cost. In addition, we demonstrate the asymptotic property of the cost function—namely, that the cost converges to a fixed value when the contract-exercise cost is sufficiently large. These findings provide benchmarks in determining strategies for hedging demand uncertainties.

The chapter is organized as follows. Section 7.2 introduces the notation and formulates the problem. After a discussion of factors that are involved in decision making, we formulate the problem of deriving the optimal initial order at stage 1 and the optimal adjustment policy at stage 2 as a two-stage dynamic programming problem. Section 7.3 proves that the optimal policy at stage 2 is a generalized (s, S) policy. While the problem at stage 1 for any given demand distribution can be solved numerically, we provide closed-form solutions for a popular class of demand distributions in Section 7.4 and explicit solutions for uniform distributions in Section 7.5. The explicit nature of these optimal solutions facilitates the application of sensitivity analysis in deciding about the investment the buyer can make in improving the forecast update and contract parameters. Moreover, it is also possible to determine a critical contract-exercise cost above which the contract is not as desirable as an available hedging alternative. Section 7.6 summarizes the chapter and points to future research directions. The chapter is concluded with notes in Section 7.7.

7.2. Problem Formulation

A purchase contract is an agreement between a supplier and a buyer that specifies terms of purchase and delivery. In the agreement, the buyer indicates an intended order quantity with an understanding that changing this initial order quantity at a later stage is subject to a fixed contract-exercise cost and a higher variable reorder cost or a lower-than-cost refund. Therefore, we divide the buyer's decision process into two stages. In the first stage, the buyer places an initial order. In the second stage, based on the improved demand forecast and the decision made in the first stage, the buyer may adjust the initial order upward at a cost no less than the initial cost or downward with a refund value that is lower than the initial cost. In addition, a fixed exercise cost is also incurred if any adjustment is made. The items with the confirmed quantity at stage 2 are delivered at the end of the second stage.

Specifically, the buyer faces the following cost parameters: a cost of $c_1 > 0$ per unit for items ordered at stage 1 and a cost of $c_{21} > 0$ per unit for items ordered at stage 2. On the other hand, if a unit of the stage 1 order is cancelled at stage 2, it is modeled as a negative order at stage 2. In this case, the buyer has either a refund value or must pay a cancellation cost. We model this situation by letting c_{22} denote the refund value or the cancellation cost per unit at stage 2, where c_{22} is the unit refund when $c_{22} \geq 0$ and $-c_{22}$ is the unit cancellation cost when $c_{22} < 0$. This phenomenon of cancellation cost is common when the merchandise is perishable or environment-polluting. It is reasonable to assume that $c_{21} \geq c_1 \geq c_{22}$. In addition, there is a fixed contract-exercise cost K at stage 2 for adjustment to the initial order quantity. Furthermore, as is standard, we assume a unit shortage cost of $p > c_{21}$ for any unsatisfied demand, since otherwise it would be trivially optimal for the buyer not to order additional items at stage 2. A unit holding cost of $h > \min\{0, -c_{22}\}$ is charged for excess inventory. Note that when $c_{22} < 0$, the unit holding cost must be more expensive than the unit cancellation cost $-c_{22}$, otherwise it would be optimal not to cancel any part of the initial order. We leave out two trivial cases, $p = c_{21}$ and $h = -c_{22} > 0$, for which the optimal solution is straightforward.

Let $D \geq 0$ represent the random demand. Let $I \geq 0$ represent an information observed in stage 1 with cumulative distribution $\Lambda(\cdot)$ and density $\lambda(\cdot)$. Let $\Psi(\cdot|i)$ denote the cumulative distribution function of D given $I = i$ with $\psi(\cdot|i)$ as the corresponding density. The information I represents an improved forecast (in terms of the conditional distribution) of the demand D.

Denote q_k as the order quantity in stage k, $k = 1, 2$. We can write the conditional expected cost at stage 2 as

$$\Pi_2(q_1, q_2, I) = \begin{cases} K + c_2(q_2) \cdot q_2 + \mathsf{E}\left[h \cdot (q_1 + q_2 - D)^+ \right. \\ \qquad\qquad \left. + p \cdot (D - q_1 - q_2)^+ \middle| \, I\right], & \text{if } q_2 \neq 0, \\ \mathsf{E}\left[h \cdot (q_1 - D)^+ + p \cdot (D - q_1)^+ \middle| \, I\right], & \text{if } q_2 = 0, \end{cases}$$

where

$$c_2(q_2) = \begin{cases} c_{21}, & \text{if } \; q_2 \geq 0, \\ c_{22}, & \text{if } \; q_2 < 0. \end{cases}$$

Note that q_2 is a history-dependent decision variable and that $\Pi_2(q_1, q_2, I)$ is a random variable. For $I = i$, we write the conditional expected cost, which we note is not a random variable, as

$$\Pi_2(q_1, q_2, i) = \Pi_2(q_1, q_2, I)\Big|_{I=i}. \tag{7.1}$$

The total expected cost is

$$c_1 q_1 + \mathsf{E}\left[\Pi_2(q_1, q_2, I)\right]. \tag{7.2}$$

Our purpose is to minimize the total expected cost—that is, the value function π_1^* is defined as follows:

$$\pi_1^* = \min_{\substack{q_1 \geq 0 \\ q_2 \geq -q_1}} \left\{ c_1 q_1 + \mathsf{E}\left[\min_{q_2 \geq -q_1} \left\{ \Pi_2(q_1, q_2, i) \right\} \right] \right\}.$$

The dynamic programming equations for this problem are

$$\pi_2^*(q_1, i) = \min_{q_2 \geq -q_1} \left\{ \Pi_2(q_1, q_2, i) \right\}, \tag{7.3}$$

$$\pi_1^* = \min_{q_1 \geq 0} \left\{ \Pi_1(q_1) \right\}, \tag{7.4}$$

where

$$\Pi_1(q_1) = c_1 q_1 + \mathsf{E}\left[\pi_2^*(q_1, I)\right]. \tag{7.5}$$

7.3. Optimal Solution for Stage 2

In this section, we explore structural properties of the second-stage cost function in our contract model. First, we consider the case when there is no fixed cost of exercising the contract. Thus, with $K = 0$, the cost function (7.1) reduces to

$$G_2(q_1, q_2, i) = \begin{cases} G_2^+(q_1, q_2, i), & q_2 \geq 0, \\ G_2^-(q_1, q_2, i), & q_2 < 0, \end{cases} \tag{7.6}$$

where

$$\begin{aligned} G_2^+(q_1, q_2, i) &= c_{21}q_2 + \mathsf{E}\left[h \cdot (q_1 + q_2 - D)^+ \right. \\ &\qquad \left. +p \cdot (D - q_1 - q_2)^+ \middle| I = i\right], \\ G_2^-(q_1, q_2, i) &= c_{22}q_2 + \mathsf{E}\left[h \cdot (q_1 + q_2 - D)^+ \right. \\ &\qquad \left. +p \cdot (D - q_1 - q_2)^+ \middle| I = i\right]. \end{aligned}$$

Note that

$$\lim_{q_2 \downarrow 0} G_2^+(q_1, q_2, i) = \lim_{q_2 \uparrow 0} G_2^-(q_1, q_2, i) = G_2(q_1, 0, i)$$

and therefore that $G_2(q_1, q_2, i)$ is continuous with respect to q_2. We present the optimal solution in the following lemma.

LEMMA 7.1 *The cost function $G_2(q_1, q_2, i)$ is convex in q_2 and differentiable except at $q_2 = 0$. The optimal adjustment at stage 2 without the fixed exercise cost is*

$$q_2^*(q_1, i) = \begin{cases} \Sigma_1(i) - q_1, & \textit{if } q_1 < \Sigma_1(i), \\ 0, & \textit{if } \Sigma_1(i) \leq q_1 \leq \Sigma_2(i), \\ \Sigma_2(i) - q_1, & \textit{if } q_1 > \Sigma_2(i), \end{cases} \tag{7.7}$$

where

$$0 \leq \Sigma_1(i) = \Psi^{-1}(\beta_1|i) \leq \Sigma_2(i) = \Psi^{-1}(\beta_2|i), \tag{7.8}$$

with

$$\beta_k = \frac{p - c_{2k}}{p + h}, \quad k = 1, 2. \tag{7.9}$$

Proof Let

$$\hat{G}(q_1, q_2, i) = \mathsf{E}\left[h \cdot (q_1 + q_2 - D)^+ + p \cdot (D - q_1 - q_2)^+ \middle| I = i\right],$$

and

$$\bar{G}(q_1, q_2, i) = \begin{cases} c_{21} \cdot q_2, & \text{if} \quad q_2 \geq 0, \\ c_{22} \cdot q_2, & \text{if} \quad q_2 < 0. \end{cases}$$

We know that $\hat{G}(q_1, q_2, i)$ is convex and differentiable and that $\bar{G}(q_1, q_2, i)$ is convex and differentiable except at $q_2 = 0$. Therefore, the convexity of $G_2(q_1, q_2, i)$ in q_2 and its nondifferentiability at $q_2 = 0$ follow from

$$G_2(q_1, q_2, i) = \hat{G}(q_1, q_2, i) + \bar{G}(q_1, q_2, i).$$

For $q_2 > 0$, setting

$$\frac{\partial G_2^+(q_1, q_2, i)}{\partial q_2} = (c_{21} - p) + (p + h)\Psi(q_1 + q_2|i) = 0 \tag{7.10}$$

provides the minimizer $\Sigma_1(i)$ as specified in (7.8) and (7.9). Likewise, for $q_2 < 0$, using $G_2^-(q_1, q_2, i)$ instead of $G_2^+(q_1, q_2, i)$ in (7.10) gives us $\Sigma_2(i) \geq \Sigma_1(i)$ as specified in (7.8) and (7.9). It is obvious from (7.9) and (7.10) that $q_2 \geq -q_1$. Clearly, then, the policies delivered for $q_1 < \Sigma_1(i)$ and $q_1 > \Sigma_2(i)$ in (7.7) are optimal. Furthermore, when $\Sigma_1(i) \leq q_1 \leq \Sigma_2(i)$, the optimal solution can only be $q_2 = 0$. Instead of (7.10), the optimality condition in this case, because of the nondifferentiability at $q_2 = 0$, is

$$0 \in \Big[(c_{22} - p) + (p + h)\Psi(q_1 + q_2|i),\ (c_{21} - p) + (p + h)\Psi(q_1 + q_2|i)\Big],$$

which clearly holds when $q_1 \in [\Sigma_1(i),\ \Sigma_2(i)]$. □

REMARK 7.1 Note that q_1 can be considered to be the inventory level at the beginning of stage 2. Then the stage 2 problem when $q_2^*(q_1, i) \geq 0$ is easily seen to be the standard newsvendor problem. The result in Lemma 7.1 could therefore be considered as an extension of the newsvendor problem when returns are allowed.

REMARK 7.2 When $p = c_{21}$ and $h = -c_{22}$, there is a unique optimal order quantity $q_2^*(q_1, i) = 0$.

In Figure 7.1, we depict the cost $G_2(q_1, q_2, i)$ as a function of q_2 with five different values of q_1. From (7.6), we see that each cost curve consists of two pieces: $G_2^-(q_1, q_2, i)$ for $q_2 < 0$ and $G_2^+(q_1, q_2, i)$ for $q_2 \geq 0$. The cost curve

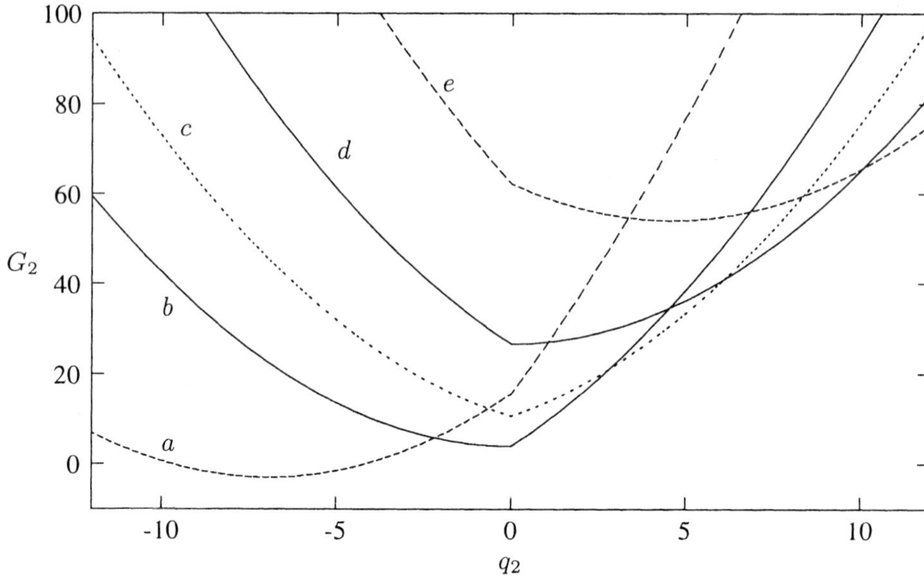

Figure 7.1. Curves of $G_2(q_1, q_2, i)$ as a function of q_2 when q_1 takes different values

a has an interior minimum of $G_2^-(q_1, q_2, i)$ at $q_2 < 0$. In this case, $q_1 > \Sigma_2(i)$ and $q_2^* = \Sigma_2(i) - q_1 < 0$. When q_1 decreases to $\Sigma_2(i)$, we get the cost curve *b* where the interior minimum of $G_2^-(q_1, q_2, i)$ is obtained at $q_2^* = 0$. Going to the other side, when $q_1 < \Sigma_1(i)$, we have cost curve *e*. Here $G_2^+(q_1, q_2, i)$ takes an interior minimum at $q_2^* = \Sigma_1(i) - q_1 > 0$. When q_1 increases to $\Sigma_1(i)$, we get the cost curve *d* where $G_2^+(q_1, q_2, i)$ has an interior minimum at $q_2^* = 0$. The remaining cost curve *c* represents the case of a q_1 such that $\Sigma_1(i) \leq q_1 \leq \Sigma_2(i)$. Here, the minimum is at $q_2^* = 0$. This minimum is a boundary minimum of both $G_2^+(q_1, q_2, i)$ and $G_2^-(q_1, q_2, i)$. In fact, the interior minimum of $G_2^-(q_1, q_2, i)$ would be at some $q_2 > 0$ if it was applicable. Then likewise, the interior minimum of $G_2^+(q_1, q_2, i)$ would be at $q_2 < 0$ if it was applicable.

We now return to the case when $K > 0$. The cost function $\Pi_2(q_1, q_2, i)$ is discontinuous at $q_2 = 0$. Consider the difference between the cost of no ordering and that of bringing the inventory level up to $\Sigma_1(i)$ or down to $\Sigma_2(i)$ depending on whether $q_1 < \Sigma_1(i)$ or $q_1 > \Sigma_2(i)$, respectively.

LEMMA 7.2 *With $\Sigma_1(i)$ and $\Sigma_2(i)$ given by* (7.8), *we have*

(i) *$G_2^+(q_1, 0, i) - K - G_2^+(q_1, \Sigma_1(i) - q_1, i)$ is strictly convex and decreasing in q_1 for all $q_1 < \Sigma_1(i)$, and $G_2^-(q_1, 0, i) - K - G_2^-(q_1, \Sigma_2(i) - q_1, i)$ is strictly convex and increasing in q_1 for all $q_1 > \Sigma_2(i)$;*

(ii) *there exist a unique* $\sigma_1(i) < \Sigma_1(i)$ *such that*

$$G_2^+(\sigma_1(i), 0, i) = K + G_2^+(\sigma_1(i), \Sigma_1(i) - \sigma_1(i), i)$$

and a unique $\sigma_2(i) > \Sigma_2(i)$ *such that*

$$G_2^-(\sigma_2(i), 0, i) = K + G_2^-(\sigma_2(i), \Sigma_2(i) - \sigma_2(i), i).$$

Proof Define

$$\Delta(q_1, i) = \begin{cases} G_2^+(q_1, 0, i) - K - G_2^+(q_1, \Sigma_1(i) - q_1, i), & q_1 < \Sigma_1(i), \\ G_2^-(q_1, 0, i) - K - G_2^-(q_1, \Sigma_2(i) - q_1, i), & q_1 > \Sigma_2(i), \end{cases} \tag{7.11}$$

to be the difference between the cost of no ordering and that of bringing the inventory level up to $\Sigma_1(i)$ or down to $\Sigma_2(i)$ depending on whether $q_1 < \Sigma_1(i)$ or $q_1 > \Sigma_2(i)$, respectively.

(i) Since $G_2^+(q_1, 0, i)$ is strictly convex in q_1 and $G_2^+(q_1, \Sigma_1(i) - q_1, i)$ is linear in q_1, $\Delta(q_1, i)$ is strictly convex in q_1 for $q_1 < \Sigma_1(i)$. Similarly, $\Delta(q_1, i)$ is strictly convex in q_1 for $q_1 > \Sigma_2(i)$. Differentiating $\Delta(q_1, i)$ with respect to q_1 gives

$$\frac{\mathrm{d}\Delta(q_1, i)}{\mathrm{d}q_1} = \begin{cases} (p+h)\Psi(q_1|i) - (p - c_{21}), & \text{if } q_1 < \Sigma_1(i), \\ (p+h)\Psi(q_1|i) - (p - c_{22}), & \text{if } q_1 > \Sigma_2(i). \end{cases}$$

For all $q_1 < \Sigma_1(i)$, $\Psi(q_1|i) \le \Psi(\Sigma_1(i)|i) = (p - c_{21})/(p+h)$, and for all $q_1 > \Sigma_2(i)$, $\Psi(q_1|i) \ge \Psi(\Sigma_2(i)|i) = (p - c_{22})/(p+h)$. Thus, $\Delta(q_1, i)$ is decreasing in q_1 when $q_1 < \Sigma_1(i)$ and increasing in q_1 when $q_1 > \Sigma_2(i)$.

(ii) From (7.11),

$$\lim_{q_1 \uparrow \Sigma_1(i)} \Delta(q_1, i) = -K < 0.$$

Also, it is easy to see that since $p > c_{21}$, $\lim_{q_1 \downarrow -\infty} \Delta(q_1, i) = +\infty$. In view of the fact that $\Delta(q_1, i)$ is decreasing in q_1 for $q_1 < \Sigma_1(i)$, there exists a $\sigma_1(i)$ as stipulated in the statement (ii) of the lemma.

The proof of (ii) for the existence of the required $\sigma_2(i)$ is similar. Here our assumption of $h > -c_{22}$ implies that $\Delta(q_1, i) \to +\infty$ as $q_1 \to +\infty$. □

Based on above preliminaries, we present the main result of this section as follows.

THEOREM 7.1 *The optimal policy at Stage 2 is*

$$q_2^*(q_1, i) = \begin{cases} \Sigma_1(i) - q_1, & \textit{if } q_1 < \sigma_1(i), \\ 0, & \textit{if } \sigma_1(i) \le q_1 \le \sigma_2(i), \\ \Sigma_2(i) - q_1, & \textit{if } q_1 > \sigma_2(i), \end{cases} \tag{7.12}$$

where $\Sigma_1(i)$ and $\Sigma_2(i)$ are defined by (7.8)*, and $\sigma_1(i)$ and $\sigma_2(i)$ are given by Lemma* 7.2.

Proof The proof requires a number of cases to deal with. Recall that

$$G_2(q_1, 0, i) = G_2^+(q_1, 0, i) = G_2^-(q_1, 0, i).$$

Case 1: $[q_1 < \Sigma_1(i)]$
By Lemma 7.1,

$$\begin{aligned} &\min\left\{K + \inf_{q_2<0}\left\{G_2^-(q_1, q_2, i)\right\}, \quad G_2(q_1, 0, i), \right. \\ &\qquad \left. K + \inf_{q_2>0}\left\{G_2^+(q_1, q_2, i)\right\}\right\} \\ &= \min\left\{K + G_2^-(q_1, 0, i), \quad G_2(q_1, 0, i), \right. \\ &\qquad \left. K + G_2^+(q_1, \Sigma_1(i) - q_1, i)\right\} \\ &= \min\left\{G_2^+(q_1, 0, i), \quad K + G_2^+(q_1, \Sigma_1(i) - q_1, i)\right\}. \end{aligned}$$

By (ii) of Lemma 7.2, when $q_1 < \sigma_1(i)$, we have

$$G_2^+(q_1, 0, i) > K + G_2^+(q_1, \Sigma_1(i) - q_1, i).$$

Thus, $q_2^*(q_1, i) = \Sigma_1(i) - q_1$. When $q_1 \ge \sigma_1(i)$, we have

$$G_2^+(q_1, 0, i) \le K + G_2^+(q_1, \Sigma_1(i) - q_1, i),$$

and therefore $q_2^*(q_1, i) = 0$.

Case 2: $[q_1 > \Sigma_2(i)]$
By Lemma 7.1,

$$\begin{aligned} &\min\left\{K + \inf_{q_2<0}\left\{G_2^-(q_1, q_2, i)\right\}, \quad G_2(q_1, 0, i), \right. \\ &\qquad \left. K + \inf_{q_2>0}\left\{G_2^+(q_1, q_2, i)\right\}\right\} \end{aligned}$$

$$= \min\Big\{K + G_2^-(q_1, \Sigma_2(i) - q_1, i), \quad G_2(q_1, 0, i),$$
$$K + G_2^+(q_1, 0, i)\Big\}$$
$$= \min\Big\{K + G_2^-(q_1, \Sigma_2(i) - q_1, i), \quad G_2^-(q_1, 0, i)\Big\}.$$

By (ii) of Lemma 7.2, we have $q_2^*(q_1, i) = \Sigma_2(i) - q_1$ when $q_1 > \sigma_2(i)$, and $q_2^*(q_1, i) = 0$ when $\Sigma_2(i) < q_1 \le \sigma_2(i)$.

Case 3: $[\Sigma_1(i) \le q_1 \le \Sigma_2(i)]$

By Lemma 7.1,

$$\min\Big\{K + \inf_{q_2<0}\Big\{G_2^-(q_1, q_2, i)\Big\}, \quad G_2(q_1, 0, i),$$
$$K + \inf_{q_2>0}\Big\{G_2^+(q_1, q_2, i)\Big\}\Big\}$$
$$= \min\Big\{K + G_2^-(q_1, 0, i), \quad G_2(q_1, 0, i), \quad K + G_2^+(q_1, 0, i)\Big\}$$
$$= G_2(q_1, 0, i).$$

Thus, $q_2^*(q_1, i) = 0$. □

This policy is a composition of two (s, S)-type policies—one for increasing the initial order and the other for decreasing the initial order. We term this policy as $(\sigma_1(i), \Sigma_1(i); \sigma_2(i), \Sigma_2(i))$ or more simply as $(\sigma_1, \Sigma_1; \sigma_2, \Sigma_2)(i)$ policy. The parameters $\sigma_1(i)$ and $\Sigma_1(i)$ are reorder point and order-up-to level, whereas $\sigma_2(i)$ and $\Sigma_2(i)$ are reduction point and reduce-down-to level, respectively. In other words, the buyer increases the initial order to raise it to $\Sigma_1(i)$ when the initial order is lower than $\sigma_1(i)$, the buyer decreases the initial order to reduce it down to $\Sigma_2(i)$ when the initial order is higher than $\sigma_2(i)$, and the buyer takes no action when the initial order is within the interval $[\sigma_1(i), \ \sigma_2(i)]$.

The $(\sigma_1, \Sigma_1; \sigma_2, \Sigma_2)(i)$ policy can be considered as a generalized newsvendor problem with a piecewise linear order or cancellation cost and a fixed cost. For newsvendor models with piecewise linear cost, with set-up cost, or with cancellation, respectively, see Porteus [7].

REMARK 7.3 The optimal order-up-to level $\Sigma_1(i)$ and reduce-down-to level $\Sigma_2(i)$ do not depend on the fixed contract-exercise cost K. In contrast, the reorder point $\sigma_1(i)$ and the reduction point $\sigma_2(i)$ depend on K. Intuitively, the optimal order-up-to and reduce-down-to levels strike a balance between overordering and underordering, while the reorder and reduction points measure the tradeoff between inventory or shortage cost and the fixed contract-exercise cost.

REMARK 7.4 When $p = c_{21}$ and $h = -c_{22}$, there is a unique optimal policy. If $p = c_{21}$, taking no action when $q_1 < \Sigma_1(i)$ is clearly an optimal policy. Likewise, if $h = -c_{22}$, taking no action is optimal when $q_1 > \Sigma_2(i)$.

7.4. Optimal Solution for a Class of Demand Distributions

In this section, we explore the optimal policy under the following assumptions about the information and the conditional demand distribution given the information.

DEFINITION 7.1 *For any given two random variables X and Y, X is called a* location parameter *of the conditional distribution Y given $X = x$ if for any $x_1 \geq x_2$,*

$$\mathsf{P}\left(Y \leq y \mid X = x_1\right) = \mathsf{P}\left(Y \leq y - x_1 + x_2 \mid X = x_2\right).$$

ASSUMPTION 7.1 *The information I is a location parameter of the conditional distribution D.*

A location parameter specifies an abscissa location point of a distribution. Usually, a location parameter is the midpoint or lower endpoint. Many distributions including exponential and uniform can be characterized by a location parameter.

If i is the location parameter of the conditional distribution D, then it is clear that for $i_2 \geq i_1$, we have

$$\psi(\eta|i_2) = \psi(\eta - i_2 + i_1|i_1), \tag{7.13}$$

$$\Psi(\eta|i_2) = \Psi(\eta - i_2 + i_1|i_1). \tag{7.14}$$

In writing (7.13) and (7.14), we understand that $\Psi(x|i) = \psi(x|i) = 0$ when $x \leq 0$, in view of the fact that $D \geq 0$. We can now prove the following result.

THEOREM 7.2 *Under Assumption 7.1, for any integrable function $H(\cdot)$ and $i_2 \geq i_1$,*

$$\mathsf{E}\left[H(x - D)|i_2\right] = \mathsf{E}\left[H(x - D - (i_2 - i_1))|i_1\right]. \tag{7.15}$$

Furthermore, define $w(x, i) = \mathsf{E}[H(x - D)|i]$. Then for $i_2 \geq i_1$,

$$\begin{aligned} w(x, i_2) = \mathsf{E}\left[H(x - D)|i_2\right] &= \mathsf{E}\left[H(x - D - (i_2 - i_1))|i_1\right] \\ &= w(x - i_2 + i_1, i_1). \end{aligned} \tag{7.16}$$

Proof Using (7.13), we have

$$\begin{aligned}
\mathsf{E}\left[H(x-D)|i_2\right] &= \int_{-\infty}^{+\infty} H(x-\eta)\cdot\psi(\eta|i_2)\mathrm{d}\eta \\
&= \int_{-\infty}^{+\infty} H(x-\eta)\cdot\psi(\eta-i_2+i_1|i_1)\mathrm{d}\eta \\
&= \int_{-\infty}^{+\infty} H(x-\eta-(i_2-i_1))\cdot\psi(\eta|i_1)\mathrm{d}\eta \\
&= \mathsf{E}\left[H(x-D-(i_2-i_1))|i_1\right].
\end{aligned}$$

The proof of the second part follows trivially from the first. □

REMARK 7.5 The analysis in this chapter goes through for the case when the conditional distribution of demand D given the information is approximated by a normal distribution. In this case, we would choose the mean as the location parameter.

Relation (7.16) says that the value of a function $w(\cdot,i_2)$ at a point x given i_2 can be obtained by evaluating the function $w(\cdot,i_1)$ at the point $x-i_2+i_1$. Geometrically speaking, $w(\cdot,i_2)$ is nothing but the function $w(\cdot,i_1)$ shifted to the right by an amount i_2-i_1. This immediately gives us the following corollary.

COROLLARY 7.1 *Under Assumption* 7.1, *the optimal order-up-to level* $\Sigma_1(\cdot)$ *and reduce-down-to level* $\Sigma_2(\cdot)$ *satisfy*

$$\Sigma_k(i_2)=\Sigma_k(i_1)+i_2-i_1 \textit{ for any } i_2\geq i_1,\ k=1,2.$$

For the optimal reorder and reduction points, similarly, we have the following results.

COROLLARY 7.2 *Under Assumption* 7.1, *the optimal reorder and reduction points satisfy*

$$\sigma_k(i_2)=\sigma_k(i_1)+i_2-i_1, \quad \textit{for } i_2\geq i_1, \quad k=1,2. \tag{7.17}$$

Proof We prove only the case when $k=1$. The proof for the case $k=2$ is similar.

$$\begin{aligned}
G_2^+(q_1,q_2|\psi_2) &= \mathsf{E}\left[c_{21}q_2+h\cdot(q_1+q_2-D)^+\right. \\
&\qquad \left.+p\cdot(D-q_1-q_2)^+|i_2\right] \\
&= \mathsf{E}\left[c_{21}q_2+h\cdot(q_1+q_2-D-i_2+i_1)^+\right. \\
&\qquad \left.+p\cdot(D+i_2-i_1-q_1-q_2)^+\big|\, i_1\right] \\
&= G_2^+(q_1-i_2+i_1,q_2|i_1).
\end{aligned} \tag{7.18}$$

By Lemma 7.2, $\sigma_1(\cdot)$ satisfies

$$G_2^+\left(\sigma_1(i_1), 0\middle|i_1\right) = K + G_2^+\left(\sigma_1(i_1), \Sigma_1(i_1) - \sigma_1(i_1)\middle|i_1\right), \qquad (7.19)$$

$$G_2^+\left(\sigma_1(i_2), 0\middle|i_2\right) = K + G_2^+\left(\sigma_1(i_2), \Sigma_1(i_2) - \sigma_1(i_2)\middle|i_2\right). \qquad (7.20)$$

Subtracting (7.19) from (7.20) gives

$$\begin{aligned} &G_2^+\left(\sigma_1(i_2), 0\middle|i_2\right) - G_2^+\left(\sigma_1(i_1), 0\middle|i_1\right) \\ &= G_2^+\left(\sigma_1(i_2), \Sigma_1(i_2) - \sigma_1(i_2)\middle|i_2\right) \\ &-G_2^+\left(\sigma_1(i_1), \Sigma_1(i_1) - \sigma_1(i_1)\middle|i_1\right). \end{aligned} \qquad (7.21)$$

Using (7.18), definition of G_2^+, and Corollary 7.1 on the right-hand side of (7.21) yields,

$$\begin{aligned} &G_2^+\left(\sigma_1(i_2) - i_2 + i_1, \Sigma_1(i_2) - \sigma_1(i_2)\middle|i_1\right) \\ &-G_2^+\left(\sigma_1(i_1), \Sigma_1(i_1) - \sigma_1(i_1)\middle|i_1\right) \\ &= c_{21}\left[\sigma_1(i_2) - \sigma_1(i_1) - i_2 + i_1\right]. \end{aligned} \qquad (7.22)$$

Using (7.18) on the left-hand side of (7.21) as well as combining (7.22) yields

$$\begin{aligned} &G_2^+\left(\sigma_1(i_2) - i_2 + i_1, 0\middle|i_1\right) - G_2^+\left(\sigma_1(i_1), 0\middle|i_1\right) \\ &= c_{21}\left[\sigma_1(i_2) - \sigma_1(i_1) - i_2 + i_1\right]. \end{aligned} \qquad (7.23)$$

It is clear that $\sigma_1(i_2) = \sigma_1(i_1) + (i_2 - i_1)$ is a solution of (7.23). □

Corollaries 7.1 and 7.2 imply that both the reorder and reduction points can be expressed as the summation of the location parameter i and a constant term that is independent of i. That is, if $i \in [0, +\infty)$,

$$\sigma_k(i) = i + u_k, \quad k = 1, 2, \qquad (7.24)$$

where $u_k = \sigma_k(0)$.

Based on the optimal policy (7.12) at stage 2 and using (7.24), we write the optimal cost function at stage 2 as follows:

$$\pi_2^*(q_1, i) = \begin{cases} K + G_2^+(q_1, \Sigma_1(i) - q_1, i), \\ \qquad\qquad \text{if } i > q_1 - u_1, \\ G_2(q_1, 0, i), \quad \text{if } q_1 - u_2 \le i \le q_1 - u_1, \\ K + G_2^-(q_1, \Sigma_2(i) - q_1, i), \\ \qquad\qquad \text{if } i < q_1 - u_2. \end{cases} \qquad (7.25)$$

From the definition of $\sigma_k(i)$ in Lemma 7.2 and the fact noted in connection with (7.6), we can see that $\pi_2^*(q_1, i)$ is a continuous function of i. Using (7.25) in (7.5), we obtain

$$\begin{aligned}\Pi_1(q_1) &= c_1 q_1 + \int_0^{q_1-u_2} \Big[K + G_2^-\left(q_1, \Sigma_2(i) - q_1, i\right)\Big] \mathrm{d}\Lambda(i) \\ &\quad + \int_{q_1-u_2}^{q_1-u_1} G_2(q_1, 0, i) \mathrm{d}\Lambda(i) \\ &\quad + \int_{q_1-u_1}^{\infty} \Big[K + G_2^+\left(q_1, \Sigma_1(i) - q_1, i\right)\Big] \mathrm{d}\Lambda(i).\end{aligned}$$

We take its derivative with respect to q_1,

$$\begin{aligned}&\frac{\mathrm{d}\Pi_1(q_1)}{\mathrm{d}q_1} \\ &= c_1 + \Big[K - \int_0^{q_1-u_2} c_{22} \mathrm{d}\Lambda(i) \\ &\quad + G_2^-\left(q_1, \Sigma_2(i) - q_1, i\right)\Big] \cdot \lambda(i)\Big|_{i=q_1-u_2} \\ &\quad + G_2(q_1, 0, q_1 - u_1) \cdot \lambda(q_1 - u_1) \\ &\quad - G_2(q_1, 0, q_1 - u_2) \cdot \lambda(q_1 - u_2) \\ &\quad + \int_{q_1-u_2}^{q_1-u_1} \Big[(p+h)\Psi(q_1|i) - p\Big] \mathrm{d}\Lambda(i) - \int_{q_1-u_1}^{\infty} c_{21} \mathrm{d}\Lambda(i) \\ &\quad - \Big[K + G_2^+\left(q_1, \Sigma_1(i) - q_1, i\right)\Big] \cdot \lambda(i)\Big|_{i=q_1-u_1} \\ &= c_1 - \int_0^{q_1-u_2} c_{22} \mathrm{d}\Lambda(i) - \int_{q_1-u_2}^{\infty} c_{21} \mathrm{d}\Lambda(i) \\ &\quad + \int_{q_1-u_2}^{q_1-u_1} \Big[(p+h)\Psi(q_1|i) - (p - c_{21})\Big] \mathrm{d}\Lambda(i) \\ &\quad + \Delta(q_1, i)\lambda(i)\Big|_{i=q_1-u_1} - \Delta(q_1, i)\lambda(i)\Big|_{i=q_1-u_2} \\ &= c_1 - \int_0^{q_1-u_2} c_{22} \mathrm{d}\Lambda(i) - \int_{q_1-u_2}^{\infty} c_{21} \mathrm{d}\Lambda(i) \\ &\quad + \int_{q_1-u_2}^{q_1-u_1} \Big[(p+h)\Psi(q_1|i) - (p - c_{21})\Big] \mathrm{d}\Lambda(i),\end{aligned}$$

(7.26)

where the second equality uses the definitions of $c_2(\cdot)$ and $\Delta(q_1, i)$, and the last equality is obtained by noting that $\Delta(q_1, i)|_{i=q_1-u_k} = \Delta\big(\sigma_k(i), i\big) = 0$ for $k = 1, 2$.

We can now state the main result of this section.

THEOREM 7.3 *With Assumption* 7.1, *there exists an initial-order quantity* q_1^* *that minimizes* $\Pi_1(q_1)$. *Moreover, the optimal initial-order quantity* q_1^* *satisfies*

$$\begin{aligned} 0 &= c_1 - \int_0^{q_1^*-u_2} c_{22} \mathrm{d}\Lambda(i) - \int_{q_1^*-u_2}^{\infty} c_{21} \mathrm{d}\Lambda(i) \\ &+ \int_{q_1^*-u_2}^{q_1^*-u_1} \Big[(p+h)\Psi(q_1^*|i) - (p - c_{21})\Big] \mathrm{d}\Lambda(i). \qquad (7.27) \end{aligned}$$

Proof The existence of q_1^* follows from the facts that

$$\begin{aligned} \lim_{q_1\to 0} \frac{\mathrm{d}\Pi_1(q_1)}{\mathrm{d}q_1} &= \lim_{q_1\to 0} c_1 - \int_0^{q_1-u_2} c_{22}\mathrm{d}\Lambda(i) - \int_{q_1-u_2}^{\infty} c_{21}\mathrm{d}\Lambda(i) \\ &= c_1 - c_{21} \\ &< 0 \end{aligned}$$

and

$$\begin{aligned} \lim_{q_1\to \infty} \frac{\mathrm{d}\Pi_1(q_1)}{\mathrm{d}q_1} &= c_1 - \int_0^{\infty} c_{22}\mathrm{d}\Lambda(i) \\ &= c_1 - c_{22} \\ &> 0. \end{aligned}$$

□

REMARK 7.6 If the density function $\lambda(\cdot)$ is a *Pólya frequency function* of order 2 (PF_2), then the total expected cost $\Pi_1(q_1)$ is a unimodal function of q_1, and the optimal initial order quantity q_1^* is uniquely determined by (7.27).

The definition and properties of PF_2 functions can be found in Karlin [6] and Porteus [7] . We should point out that unimodality of $\Pi_1(q_1)$ holds also for some densities that are not PF_2. A particular case is that of uniform distribution, treated below.

7.5. Analysis for Uniformly Distributed Demand

In this section, we assume a uniformly distributed demand. We obtain a closed-form solution for both stages 1 and 2. With the closed-form solution, we further investigate the values of the forecast update and the purchase contract.

7.5.1 Optimal Solution

Here we study the purchase-contract problem defined in Section 7.2 under the assumptions that I is uniformly distributed over the interval $\left[\gamma - \frac{a}{2},\ \gamma + \frac{a}{2}\right]$ and, given $I = i$, that D follows the uniform distribution over the interval $\left[i - \frac{\varepsilon a}{2},\ i + \frac{\varepsilon a}{2}\right]$. Thus,

$$\begin{aligned}
\lambda(i) &= \frac{1}{a}, & i &\in \left[\gamma - \frac{a}{2},\ \gamma + \frac{a}{2}\right],\\
\Lambda(i) &= \frac{1}{a}\left(i - \gamma + \frac{a}{2}\right), & i &\in \left[\gamma - \frac{a}{2},\ \gamma + \frac{a}{2}\right],\\
\psi(\eta|i) &= \frac{1}{\varepsilon a}, & \eta &\in \left[i - \frac{\varepsilon a}{2},\ i + \frac{\varepsilon a}{2}\right],\\
\Psi(\eta|i) &= \frac{1}{\varepsilon a}\left(\eta - i + \frac{\varepsilon a}{2}\right), & \eta &\in \left[i - \frac{\varepsilon a}{2},\ i + \frac{\varepsilon a}{2}\right],
\end{aligned}$$

where $0 \le \varepsilon \le 1$ represents the fact of reduction in the forecast errors at the updating stage. The value of ε can be obtained from either the buyer's experience or regression methods. Note that all of these functions are assumed to be zero outside their respective ranges defined above.

REMARK 7.7 The above specifications do not imply that the unconditional distribution of D is uniform. Indeed, the unconditional density function of D is given by

$$\begin{cases}
\frac{1}{\varepsilon a^2}\left[\eta - \left(\gamma - \frac{a}{2} - \frac{\varepsilon a}{2}\right)\right], & \text{if } \eta \in \left[\gamma - \frac{a}{2} - \frac{\varepsilon a}{2},\ \gamma - \frac{a}{2} + \frac{\varepsilon a}{2}\right],\\
\frac{1}{a}, & \text{if } \eta \in \left(\gamma - \frac{a}{2} + \frac{\varepsilon a}{2},\ \gamma + \frac{a}{2} - \frac{\varepsilon a}{2}\right),\\
-\frac{1}{\varepsilon a^2}\left[\eta - \left(\gamma + \frac{a}{2} + \frac{\varepsilon a}{2}\right)\right], & \text{if } \eta \in \left[\gamma + \frac{a}{2} - \frac{\varepsilon a}{2},\ \gamma + \frac{a}{2} + \frac{\varepsilon a}{2}\right].
\end{cases}$$

Depending on the amount of total order $q_1 + q_2$, there exist three possibilities for the expression of the cost function $G_2(q_1, q_2, i)$ defined in (7.6). If the total order is below the lower bound of the conditional demand given the information, then the inventory cannot meet all of the demand, and a penalty is incurred. We denote the cost as $G_2^p(q_1, q_2, i)$ in this case. If the total order exceeds the upper bound of the conditional demand given the information, then there will be some leftover inventory after the demand is met, and a holding cost applies. In this case, we denote the cost as $G_2^h(q_1, q_2, i)$. Finally, if the total order is in the range of the conditional demand given the information, then both penalty and holding costs are incurred. In this case, we denote the cost as $G_2^b(q_1, q_2, i)$. It is easy to see that

$$G_2(q_1,q_2,i) = \begin{cases} G_2^p(q_1,q_2,i), & \text{if } q_1+q_2 \le i-\frac{\varepsilon a}{2}, \\ G_2^b(q_1,q_2,i), & \text{if } i-\frac{\varepsilon a}{2} < q_1+q_2 < i+\frac{\varepsilon a}{2}, \\ G_2^h(q_1,q_2,i), & \text{if } q_1+q_2 \ge i+\frac{\varepsilon a}{2}, \end{cases} \tag{7.28}$$

where

$$G_2^p(q_1,q_2,i) = c_2(q_2)\cdot q_2 + p\int_{i-\frac{\varepsilon a}{2}}^{i+\frac{\varepsilon a}{2}} \frac{\eta-q_1-q_2}{\varepsilon a}\mathrm{d}\eta,$$

$$G_2^b(q_1,q_2,i) = c_2(q_2)\cdot q_2 + h\int_{i-\frac{\varepsilon a}{2}}^{q_1+q_2} \frac{q_1+q_2-\eta}{\varepsilon a}\mathrm{d}\eta + p\int_{q_1+q_2}^{i+\frac{\varepsilon a}{2}} \frac{\eta-q_1-q_2}{\varepsilon a}\mathrm{d}\eta,$$

$$G_2^h(q_1,q_2,i) = c_2(q_2)\cdot q_2 + h\int_{i-\frac{\varepsilon a}{2}}^{i+\frac{\varepsilon a}{2}} \frac{q_1+q_2-\eta}{\varepsilon a}\mathrm{d}\eta.$$

As a special case of Theorem 7.1, we have the following corollary.

COROLLARY 7.3 *The optimal policy at stage 2 is $(\sigma_1,\Sigma_1;\sigma_2,\Sigma_2)(i)$, where the order-up-to level $\Sigma_1(i)$ and the reduce-down-to level $\Sigma_2(i)$ are given by*

$$\Sigma_k(i) = i+\varepsilon a\Big(\beta_k-\frac{1}{2}\Big), \quad \beta_k = \frac{p-c_{2k}}{p+h}, \quad k=1,2, \tag{7.29}$$

and the reorder point $\sigma_1(i)$ and the reduction point $\sigma_2(i)$ are given by

$$\sigma_1(i) = \begin{cases} \Sigma_1(i)-\mu(K), & \text{if } \mu(K) < \varepsilon a\beta_1, \\ \Sigma_1(i)-\Big(\frac{K}{p-c_{21}}+\frac{\varepsilon a\beta_1}{2}\Big), & \text{if } \mu(K) \ge \varepsilon a\beta_1, \end{cases} \tag{7.30}$$

$$\sigma_2(i) = \begin{cases} \Sigma_2(i)+\mu(K), \\ \qquad \text{if } \mu(K) < \varepsilon a(1-\beta_2), \\ \Sigma_2(i)+\Big[\frac{K}{h+c_{22}}+\frac{\varepsilon a(1-\beta_2)}{2}\Big], \\ \qquad \text{if } \mu(K) \ge \varepsilon a(1-\beta_2), \end{cases} \tag{7.31}$$

where

$$\mu(K) = \sqrt{\frac{2\varepsilon aK}{p+h}}. \tag{7.32}$$

Proof We first prove (7.29). By Theorem 7.1, the two levels $\Sigma_1(i)$ and $\Sigma_2(i)$ are the minimum of G_2. To prove (7.29), equating to zero the derivative at $q_2 \neq 0$, we obtain

$$\frac{\partial G_2^p(q_1, q_2, i)}{\partial q_2} = c_2(q_2) - p, \tag{7.33}$$

$$\frac{\partial G_2^h(q_1, q_2, i)}{\partial q_2} = h + c_2(q_2), \tag{7.34}$$

$$\begin{aligned}\frac{\partial G_2^b(q_1, q_2, i)}{\partial q_2} &= [c_2(q_2) - p] + (p+h) \cdot \Psi(q_1 + q_2 | i) \\ &= [c_2(q_2) - p] + (p+h)\left[q_1 + q_2 - i + \frac{\varepsilon a}{2}\right] \cdot \frac{1}{\varepsilon a}.\end{aligned} \tag{7.35}$$

Under the assumptions that $p > c_{21}$ and $h > -c_{22}$, $G_2^p(q_1, q_2, i)$ is strictly decreasing in q_2, and $G_2^h(q_1, q_2, i)$ is strictly increasing in q_2. Therefore, it is impossible for $G_2^p(q_1, q_2, i)$ and $G_2^h(q_1, q_2, i)$ to get a first-order condition for a minimum, which leads to

$$\arg\min_{q_2 \geq 0} \left\{ G_2(q_1, q_2, i) \right\} = \arg\min_{q_2 \geq 0} \left\{ G_2^b(q_1, q_2, i) \right\}. \tag{7.36}$$

This, in view of (7.38), proves (7.29).

Find the reorder/reduction points $\sigma_1(i)$ and $\sigma_2(i)$. This will be divided into several steps.

Step 1 When $q_1 < i - \frac{\varepsilon a}{2}$, let

$$G_2^p(q_1, 0, i) = K + G_2^b(q_1, \Sigma_1(i) - q_1, i).$$

Some simple calculations give $q_1 = \Sigma_1(i) - \left(\frac{K}{p - c_{21}} + \frac{\varepsilon a \beta_1}{2}\right)$. Since

$$\frac{\partial}{\partial q_1}\left\{ G_2^p(q_1, 0, i) - \left[K + G_2^b(q_1, \Sigma_1(i) - q_1, i)\right]\right\} = c_{21} - p < 0$$

—that is, $G_2^p(q_1, 0, i) - \left[K + G_2^b(q_1, \Sigma_1(i) - q_1, i)\right]$ is strictly decreasing in q_1, we have that $\Sigma_1(i) - \left(\frac{K}{p - c_{21}} + \frac{\varepsilon a \beta_1}{2}\right)$ is the unique solution of

$$G_2^p(q_1, 0, i) = K + G_2^b(q_1, \Sigma_1(i) - q_1, i).$$

Step 2 When $i - \frac{\varepsilon a}{2} \leq q_1 < \Sigma_1(i)$, let

$$G_2^b(q_1, 0, i) = K + G_2^b(q_1, \Sigma_1(i) - q_1, i),$$

and we obtain $q_1 = \Sigma_1(i) - \mu(K)$. Since

$$\begin{aligned} &\frac{\partial}{\partial q_1}\Big\{G_2^b(q_1, 0, i) - \Big[K + G_2^b(q_1, \Sigma_1(i) - q_1, i)\Big]\Big\} \\ &= \frac{p+h}{\varepsilon a}\Big(q_1 - i + \frac{\varepsilon a}{2}\Big) + (c_{21} - p) \\ &< \frac{p+h}{\varepsilon a}\Big(\Sigma_1(i) - i + \frac{\varepsilon a}{2}\Big) + (c_{21} - p) \\ &= 0, \end{aligned} \tag{7.37}$$

where the inequality makes use of the condition $q_1 < \Sigma_1(i)$—namely,

$$G_2^b(q_1, 0, i) - \Big[K + G_2^b(q_1, \Sigma_1(i) - q_1, i)\Big]$$

is strictly decreasing in q_1—we have that $\Sigma_1(i) - \mu(K)$ is the unique solution of

$$G_2^b(q_1, 0, i) = K + G_2^b(q_1, \Sigma_1(i) - q_1, i).$$

Step 3 When

$$\Sigma_2(i) < q_1 < i + \frac{\varepsilon a}{2},$$

similar to Step 2, then $\Sigma_2(i) + \mu(K)$ is a unique solution of

$$G_2^b(q_1, 0, i) = K + G_2^b(q_1, \Sigma_2(i) - q_1, i).$$

Step 4 When

$$i + \frac{\varepsilon a}{2} \le q_1 \le \gamma + \frac{a}{2} + \frac{\varepsilon a}{2},$$

then $\Sigma_2(i) + \left[\frac{K}{h + c_{22}} + \frac{\varepsilon a(1 - \beta_2)}{2}\right]$ is a unique solution of

$$G_2^h(q_1, 0, i) = K + G_2^b(q_1, \Sigma_2(i) - q_1, i).$$

Finally, also by using the equivalent conditions

$$\begin{aligned} &\Sigma_1(i) - \left(\frac{K}{p - c_{21}} + \frac{\varepsilon a \beta_1}{2}\right) \le i - \frac{\varepsilon a}{2} \\ \Longleftrightarrow\ & \frac{1}{2}\frac{p - c_{21}}{p+h}\varepsilon a - \frac{K}{p - c_{21}} \le 0 \\ \Longleftrightarrow\ & \left(\frac{p - c_{21}}{p+h}\varepsilon a\right)^2 \le \frac{2\varepsilon a K}{p+h} \\ \Longleftrightarrow\ & \mu(K) \ge \frac{p - c_{21}}{p+h}\varepsilon a \\ \Longleftrightarrow\ & \mu(K) \ge \varepsilon a \beta_1 \\ \Longleftrightarrow\ & \Sigma_1(i) - \mu(K) \le i - \frac{\varepsilon a}{2}, \end{aligned} \tag{7.38}$$

and

$$\begin{aligned}
&\Sigma_2(i) + \mu(K) \geq i + \frac{\varepsilon a}{2} \\
&\Longleftrightarrow \Sigma_2(i) + \left[\frac{K}{h + c_{22}} + \frac{\varepsilon a(1 - \beta_2)}{2}\right] \geq i + \frac{\varepsilon a}{2} \\
&\Longleftrightarrow \mu(K) \geq \varepsilon a(1 - \beta_2),
\end{aligned} \tag{7.39}$$

the reorder or reduction points can be expressed as a continuous function with respect to the fixed contract-exercise cost K as shown by (7.30) and (7.31).

By Theorem 7.1, the optimal policy is of $(\sigma_1, \Sigma_1; \sigma_2, \Sigma_2)(i)$ type. □

REMARK 7.8 Since for $q_1 \leq \sigma_1(i)$,

$$\begin{aligned}
\Sigma_1(i) - q_1 &\geq \Sigma_1(i) - \sigma_1(i) \\
&= \begin{cases} \mu(K), & \text{if } \mu(K) < \varepsilon a\beta_1, \\ \dfrac{K}{p - c_{21}} + \dfrac{\varepsilon a\beta_1}{2}, & \text{if } \mu(K) \geq \varepsilon a\beta_1, \end{cases}
\end{aligned}$$

$\mu(K)$ and $\frac{K}{p - c_{21}} + \frac{\varepsilon a\beta_1}{2}$ can be interpreted as minimum reorder lot sizes. Similarly, $\mu(K)$ and $\frac{K}{h + c_{22}} + \frac{\varepsilon a(1 - \beta_2)}{2}$ can be interpreted as minimum reduction lot sizes.

We present two extreme cases of $(\sigma_1, \Sigma_1; \sigma_2, \Sigma_2)(i)$ policy as follows.

Extreme Case 1 If $K = 0$, then $\sigma_1(i) = \Sigma_1(i)$ and $\sigma_2(i) = \Sigma_2(i)$. Thus, $(\sigma_1, \Sigma_1; \sigma_2, \Sigma_2)(i)$ policy reduces to a base-stock policy $(\Sigma_1(i); \Sigma_2(i))$, where $\Sigma_1(i)$ and $\Sigma_2(i)$ are the two base-stock levels.

Extreme Case 2 If $K \neq 0$ and $c_{21} = c_1 = c_{22} \geq 0$, then on the payment of the fixed contract-exercise cost K, the buyer can either reorder additional items at the initial unit purchase price or cancel some items to obtain a full refund. In such a situation, the optimal order-up-to level coincides with the reduce-down-to level, and the reorder point and the reduction point are symmetric about this level. The optimal initial order quantity, which is not unique, is presented explicitly later in Theorem 7.4 (iii).

In the $(\sigma_1, \Sigma_1; \sigma_2, \Sigma_2)(i)$ policy at stage 2, recall that both $\sigma_1(i)$ and $\sigma_2(i)$ depend on the fixed contract-exercise cost K. Let us assume that $\beta_1 + \beta_2 > 1$—that is, $(h+c_{22}) \leq (p-c_{21})$—for convenience in exposition. Note that a similar analysis can be carried out for the case of $\beta_1 + \beta_2 \leq 1$.

When $\beta_1 + \beta_2 > 1$, by (7.38) and (7.39), we have that if

$$\Sigma_1(i) - \left(\frac{K}{p - c_{21}} + \frac{\varepsilon a\beta_1}{2}\right) \leq i - \frac{\varepsilon a}{2},$$

then

$$\mu(K) \geq \frac{p - c_{21}}{p+h}\varepsilon a. \tag{7.40}$$

Consequently, $\beta_1 + \beta_2 > 1$ and (7.40) imply that

$$\mu(K) > \varepsilon a \cdot (1 - \beta_2).$$

Therefore, if $\beta_1 + \beta_2 > 1$, then

$$\Sigma_1(i) - \Big(\frac{K}{p - c_{21}} + \frac{\varepsilon a \beta_1}{2}\Big) \leq i - \frac{\varepsilon a}{2} \leq \Sigma_2(i) + \mu(K) \leq i + \frac{\varepsilon a}{2}$$

is impossible. Therefore, there exist the following three cases depending on the value of K.

Case 1: $[\mu(K) < \varepsilon a(1 - \beta_2)]$
From Corollary 7.3, we have

$$\begin{aligned} \sigma_1(i) &= \Sigma_1(i) - \mu(K), \\ \sigma_2(i) &= \Sigma_2(i) + \mu(K), \end{aligned}$$

and

$$i - \frac{\varepsilon a}{2} \leq \sigma_1(i) \leq \sigma_2(i) \leq i + \frac{\varepsilon a}{2}.$$

Case 2: $[\varepsilon a(1 - \beta_2) \leq \mu(K) < \varepsilon a \beta_1]$
Under this case,

$$\begin{aligned} \sigma_1(i) &= \Sigma_1(i) - \mu(K), \\ \sigma_2(i) &= \Sigma_2(i) + \left[\frac{K}{h + c_{22}} + \frac{\varepsilon a(1 - \beta_2)}{2}\right], \end{aligned}$$

and

$$i - \frac{\varepsilon a}{2} \leq \sigma_1(i) \leq i + \frac{\varepsilon a}{2} \leq \sigma_2(i).$$

Case 3: $[\mu(K) \geq \varepsilon a \beta_1]$
We have

$$\begin{aligned} \sigma_1(i) &= \Sigma_1(i) - \Big(\frac{K}{p - c_{21}} + \frac{\varepsilon a \beta_1}{2}\Big), \\ \sigma_2(i) &= \Sigma_2(i) + \left[\frac{K}{h + c_{22}} + \frac{\varepsilon a(1 - \beta_2)}{2}\right], \end{aligned}$$

and

$$\sigma_1(i) \leq i - \frac{\varepsilon a}{2} \leq i + \frac{\varepsilon a}{2} \leq \sigma_2(i).$$

In this chapter, we focus on Case 1 and on an extreme behavior of Case 3. Other cases can be similarly treated. To get the properties of $\Pi_1(q_1)$, we first summarize the optimal cost function at stage 2, $\pi_2^*(q_1, i)$.

When $q_1 < \sigma_1(i)$, the $(\sigma_1, \Sigma_1; \sigma_2, \Sigma_2)(i)$–policy requires the buyer to increase the initial order q_1 to bring the inventory level up to $\Sigma_1(i)$. Thus,

$$\begin{aligned} \pi_2^*(q_1, i) &= K + G_2^b(q_1, \Sigma_1(i) - q_1, i) \\ &= K - c_{21}q_1 + c_{21}i + \tfrac{(p-c_{21})(1-\beta_1)\varepsilon a}{2}. \end{aligned} \tag{7.41}$$

When $q_1 > \sigma_2(i)$, it is optimal for the buyer to reduce the initial order q_1 down to the inventory level $\Sigma_2(i)$. Thus,

$$\begin{aligned} \pi_2^*(q_1, i) &= K + G_2^b(q_1, \Sigma_2(i) - q_1, i) \\ &= K - c_{22}q_1 + c_{22}i + \tfrac{(p-c_{22})(1-\beta_2)\varepsilon a}{2}. \end{aligned} \tag{7.42}$$

Finally, when the initial order quantity is between the reorder point and the reduction point—that is, $\sigma_1(i) \leq q_1 \leq \sigma_2(i)$— then the $(\sigma_1, \Sigma_1; \sigma_2, \Sigma_2)(i)$–policy requires the buyer to take no action. Thus,

$$\begin{aligned} \pi_2^*(q_1, i) &= G_2(q_1, 0, i) \\ &= \begin{cases} G_2^p(q_1, 0, i), & \text{if } q_1 \leq i - \frac{\varepsilon a}{2}, \\ G_2^b(q_1, 0, i), & \text{if } i - \frac{\varepsilon a}{2} < q_1 < i + \frac{\varepsilon a}{2}, \\ G_2^h(q_1, 0, i), & \text{if } q_1 \geq i + \frac{\varepsilon a}{2}. \end{cases} \end{aligned}$$

In this last case, depending on the relative position of the initial order quantity with respect to the reorder and reduction points and the bounds of the conditional demand given the information, we have the following three scenarios.

When

$$i - \frac{\varepsilon a}{2} \leq \sigma_1(i) < q_1 < \sigma_2(i) \leq i + \frac{\varepsilon a}{2},$$

both holding and penalty costs are incurred, and the optimal cost function is

$$\begin{aligned} \pi_2^*(q_1, i) &= G_2^b(q_1, 0, i) \\ &= \frac{p+h}{2\varepsilon a}(q_1 - i)^2 + \frac{h-p}{2}(q_1 - i) + \frac{\varepsilon a(p+h)}{8}. \end{aligned} \tag{7.43}$$

In Case 1, by (7.41), (7.42), and (7.43), the optimal cost function at stage 2 is

$$\pi_2^*(q_1, i) = \begin{cases} K - c_{21}q_1 + c_{21}i + \frac{\varepsilon a}{2}(p - c_{21})(1 - \beta_1), \\ \qquad \text{if } \; i > q_1 - \varepsilon a(\beta_1 - \frac{1}{2}) + \mu(K), \\ K - c_{22}q_1 + c_{22}i + \frac{\varepsilon a}{2}(p - c_{22})(1 - \beta_2), \\ \qquad \text{if } \; i < q_1 - \varepsilon a(\beta_2 - \frac{1}{2}) - \mu(K), \\ \frac{p+h}{2\varepsilon a}(q_1 - i)^2 + \frac{h-p}{2}(q_1 - i) + \frac{\varepsilon a(p+h)}{8}, \\ \qquad \text{otherwise.} \end{cases} \tag{7.44}$$

Recall that

$$\Pi_1(q_1) = c_1 q_1 + \mathsf{E}\left[\pi_2^*(q_1, I)\right] \tag{7.45}$$

and that $\Pi_1(q_1)$ is a piecewise smooth function with respect to the initial order quantity q_1.

Furthermore, when

$$\sigma_1(i) \le i + \frac{\varepsilon a}{2} \le q_1 \le \sigma_2(i),$$

the demand is less than the initial-order quantity $w.p.1$. In this case, no penalty cost occurs—that is,

$$\pi_2^*(q_1, i) = G_2^h(q_1, 0, i) = hq_1 - hi. \tag{7.46}$$

Finally, when $\sigma_1(i) \le q_1 \le i - \frac{\varepsilon a}{2} \le \sigma_2(i)$, the demand is larger than the initial order quantity $w.p.1$. In this case, no holding cost occurs—that is,

$$\pi_2^*(q_1, i) = G_2^p(q_1, 0, i) = pi - pq_1. \tag{7.47}$$

Equations (7.41)–(7.47) provide a complete characterization of the optimal cost function at stage 2. With this, we are ready to tackle the first-stage problem.

To minimize $\Pi_1(q_1)$, we first develop its expression, calculate its derivative, and then discuss its simple analytical property. In the following, it will be seen that $\Pi_1(q_1)$ is a piecewise function over five intervals.

When

$$\gamma - \frac{a}{2} - \frac{\varepsilon a}{2} < q_1 \le \gamma - \frac{a}{2} + \varepsilon a(\beta_1 - \frac{1}{2}) - \mu(K),$$

additional order is needed. Therefore,

$$\Pi_1(q_1) = c_1 q_1 + \int_{\gamma - \frac{a}{2}}^{\gamma + \frac{a}{2}} \left[K - c_{21}q_1 + c_{21}i + \frac{\varepsilon a}{2}(p - c_{21})(1 - \beta_1)\right] \mathrm{d}\Lambda(i),$$

and

$$\frac{\mathrm{d}\Pi_1(q_1)}{\mathrm{d}q_1} = c_1 - c_{21}. \tag{7.48}$$

Thus, $\Pi_1(q_1)$ is nonincreasing since $c_{21} \geq c_1$.

When

$$\begin{cases} q_1 > \gamma - \frac{a}{2} + \varepsilon a(\beta_1 - \frac{1}{2}) - \mu(K), \\ q_1 \leq \gamma - \frac{a}{2} + \varepsilon a(\beta_2 - \frac{1}{2}) + \mu(K), \end{cases} \tag{7.49}$$

then

$$\begin{aligned} \Pi_1(q_1) &= c_1 q_1 + \int_{\gamma-\frac{a}{2}}^{q_1-\varepsilon a(\beta_1-\frac{1}{2})+\mu(K)} \Big[\frac{(p+h)}{2\varepsilon a}(q_1-i)^2 \\ &\qquad + \frac{h-p}{2}(q_1-i) + \frac{\varepsilon a(p+h)}{8}\Big]\mathrm{d}\Lambda(i) \\ &\quad + \int_{q_1-\varepsilon a(\beta_1-\frac{1}{2})+\mu(K)}^{\gamma+\frac{a}{2}} \Big[K - c_{21}q_1 + c_{21}i \\ &\qquad + \frac{\varepsilon a}{2}(p-c_{21})(1-\beta_1)\Big]\mathrm{d}\Lambda(i), \end{aligned}$$

and

$$\begin{aligned} \frac{\mathrm{d}\Pi_1(q_1)}{\mathrm{d}q_1} &= \frac{p+h}{2\varepsilon a^2}\Big\{q_1 - \Big[\gamma - \frac{a}{2} + \varepsilon a(\beta_1 - \frac{1}{2}) \\ &\qquad - \mu(K + c_{21}a - c_1 a)\Big]\Big\} \cdot (q_1 - z_1), \end{aligned} \tag{7.50}$$

where

$$z_1 = \gamma - \frac{a}{2} + \varepsilon a(\beta_1 - \frac{1}{2}) + \mu(K + c_{21}a - c_1 a).$$

Since $c_{21} \geq c_1 \geq c_{22}$,

$$\mu(K) \leq \mu(K + c_{21}a - c_1 a).$$

Thus,

$$\begin{aligned} &\gamma - \frac{a}{2} + \varepsilon a(\beta_1 - \frac{1}{2}) - \mu(K + c_{21}a - c_1 a) \\ &\leq \gamma - \frac{a}{2} + \varepsilon a(\beta_1 - \frac{1}{2}) - \mu(K). \end{aligned}$$

Because of (7.49), we have that the first term of equation (7.50) is positive. Hence, $\Pi_1(q_1)$ is nonincreasing if and only if $q_1 \le z_1$.

Suppose that

$$\gamma - \frac{a}{2} + \varepsilon a(\beta_2 - \frac{1}{2}) + \mu(K) < q_1 \le \gamma + \frac{a}{2} + \varepsilon a(\beta_1 - \frac{1}{2}) - \mu(K).$$

(Owing to the assumption $\beta_1 + \beta_2 > 1$, we have

$$\frac{a}{2} + \frac{\varepsilon a}{2}\beta_1 - \frac{\varepsilon a}{2}\beta_2 > \varepsilon a \cdot (1 - \beta_2);$$

this and the condition $\mu(K) < \varepsilon a(1 - \beta_2)$ imply that the inequality

$$\gamma - \frac{a}{2} + \varepsilon a(\beta_2 - \frac{1}{2}) + \mu(K) < \gamma + \frac{a}{2} + \varepsilon a(\beta_1 - \frac{1}{2}) - \mu(K) \tag{7.51}$$

is true). Then

$$\begin{aligned}
\Pi_1(q_1) &= c_1 q_1 + \int_{\gamma - \frac{a}{2}}^{q_1 - \varepsilon a(\beta_2 - \frac{1}{2}) - \mu(K)} \Big[K - c_{22} q_1 + c_{22} i \\
&\qquad + \frac{\varepsilon a}{2}(p - c_{22})(1 - \beta_2)\Big] \mathrm{d}\Lambda(i) \\
&+ \int_{q_1 - \varepsilon a(\beta_2 - \frac{1}{2}) - \mu(K)}^{q_1 - \varepsilon a(\beta_1 - \frac{1}{2}) + \mu(K)} \Big[\frac{(p+h)}{2\varepsilon a}(q_1 - i)^2 \\
&\qquad + \frac{h-p}{2}(q_1 - i) + \frac{\varepsilon a(p+h)}{8}\Big] \mathrm{d}\Lambda(i) \\
&+ \int_{q_1 - \varepsilon a(\beta_1 - \frac{1}{2}) + \mu(K)}^{\gamma + \frac{a}{2}} \Big[K - c_{21} q_1 + c_{21} i \\
&\qquad + \frac{\varepsilon a}{2}(p - c_{21})(1 - \beta_1)\Big] \mathrm{d}\Lambda(i),
\end{aligned}$$

and

$$\frac{\mathrm{d}\Pi_1(q_1)}{\mathrm{d}q_1} = \frac{1}{a}(c_{21} - c_{22})(q_1 - z_2), \tag{7.52}$$

where

$$z_2 = \gamma + \frac{a}{2} \cdot \frac{c_{21} + c_{22} - 2c_1}{c_{21} - c_{22}} + \frac{\varepsilon a}{2}(\beta_1 + \beta_2 - 1). \tag{7.53}$$

Clearly, $\Pi_1(q_1)$ is a convex function over this interval, and $\Pi_1(q_1)$ is nondecreasing if and only if $q_1 \geq z_2$.

When

$$\gamma + \frac{a}{2} + \varepsilon a \cdot (\beta_1 - \frac{1}{2}) - \mu(K) < q_1 \leq \gamma + \frac{a}{2} + \varepsilon a(\beta_2 - \frac{1}{2}) + \mu(K),$$

then

$$\begin{aligned}
\Pi_1(q_1) &= c_1 q_1 + \int_{\gamma-\frac{a}{2}}^{q_1-\varepsilon a\cdot(\beta_2-\frac{1}{2})-\mu(K)} \Big[K - c_{22}q_1 + c_{22}i \\
&\qquad + \frac{\varepsilon a}{2}(p - c_{22})(1-\beta_2)\Big] \mathrm{d}\Lambda(i) \\
&\quad + \int_{q_1-\varepsilon a(\beta_2-\frac{1}{2})-\mu(K)}^{\gamma+\frac{a}{2}} \Big[\frac{(p+h)}{2\varepsilon a}\cdot(q_1-i)^2 \\
&\qquad + \frac{h-p}{2}\cdot(q_1-i) + \frac{\varepsilon a(p+h)}{8}\Big] \mathrm{d}\Lambda(i)
\end{aligned}$$

and

$$\begin{aligned}
\frac{\mathrm{d}\Pi_1(q_1)}{\mathrm{d}q_1} &= \frac{p+h}{2\varepsilon a^2}\Big\{\Big[\gamma + \frac{a}{2} + \varepsilon a\cdot(\beta_2 - \frac{1}{2}) \\
&\qquad + \mu(K + c_1 a - c_{22}a)\Big] - q_1\Big\}\cdot(q_1 - z_3),
\end{aligned} \tag{7.54}$$

where

$$z_3 = \gamma + \frac{a}{2} + \varepsilon a \cdot \left(\beta_2 - \frac{1}{2}\right) - \mu(K + c_1 a - c_{22}a).$$

Since the first term of equation (7.54) is positive when $c_1 \geq c_{22}$, $\Pi_1(q_1)$ is nondecreasing *if and only if* $q_1 \geq z_3$.

When

$$\gamma + \frac{a}{2} + \varepsilon a(\beta_2 - \frac{1}{2}) + \mu(K) < q_1 \leq \gamma + \frac{a}{2} + \frac{\varepsilon a}{2},$$

then

$$\Pi_1(q_1) = c_1 q_1 + \int_{\gamma-\frac{a}{2}}^{\gamma+\frac{a}{2}} \Big[K - c_{22}q_1 + c_{22}i + \frac{\varepsilon a}{2}(p - c_{22})(1-\beta_2)\Big]\mathrm{d}\Lambda(i),$$

and

$$\frac{\mathrm{d}\Pi_1(q_1)}{\mathrm{d}q_1} = c_1 - c_{22}. \tag{7.55}$$

Since $c_1 \geq c_{22}$, $\Pi_1(q_1)$ is nondecreasing.

Based on the discussion of $\Pi_1(q_1)$ and its property, it is easy to check that $\Pi_1(q_1)$ is continuous and differentiable with respect to q_1 over the interval $[\gamma - \frac{a}{2} - \frac{\varepsilon a}{2},\ \gamma + \frac{a}{2} + \frac{\varepsilon a}{2}]$. We present below the optimal stage 1 policy in Case 1.

THEOREM 7.4 *Assume Case* 1*—that is, let* $\mu(K) < \varepsilon a(1 - \beta_2)$. *Then*
(i) $\Pi_1(q_1)$ *is unimodal with respect to* q_1;
(ii) *if* $c_{21} > c_1 > c_{22}$, *the optimal order quantity at stage* 1 *is*

$$q_1^* = \begin{cases} z_1, & \text{if } B = \min\{A, B\} < \mu(K), \\ z_2, & \text{if } \mu(K) \leq \min\{A, B\}, \\ z_3, & \text{if } A = \min\{A, B\} < \mu(K), \end{cases} \tag{7.56}$$

where

$$z_1 = \gamma - \frac{a}{2} + \varepsilon a(\beta_1 - \frac{1}{2}) + \mu(K + c_{21}a - c_1 a),$$

$$z_2 = \gamma + \frac{a}{2} \cdot \frac{c_{21} + c_{22} - 2c_1}{c_{21} - c_{22}} + \frac{\varepsilon a}{2}(\beta_1 + \beta_2 - 1),$$

$$z_3 = \gamma + \frac{a}{2} + \varepsilon a(\beta_2 - \frac{1}{2}) - \mu(K + c_1 a - c_{22}a),$$

and

$$A = \frac{c_1 - c_{22}}{c_{21} - c_{22}}a - (\beta_2 - \beta_1)\frac{\varepsilon a}{2},$$

and

$$B = \frac{c_{21} - c_1}{c_{21} - c_{22}}a - (\beta_2 - \beta_1)\frac{\varepsilon a}{2};$$

(iii) *if* $c_{21} = c_{22} = c_1$, *then each point in the interval*

$$\left[\gamma - \frac{a}{2} + \varepsilon a(\beta_2 - \frac{1}{2}) + \mu(K),\ \gamma + \frac{a}{2} + \varepsilon a(\beta_1 - \frac{1}{2}) - \mu(K)\right]$$

qualifies as an optimal stage 1 *order quantity.*

Proof (i)–(ii) In the case of $c_{21} > c_1 > c_{22}$, we prove only the case when $B = \min\{A, B\} < \mu(K)$. Other cases can be treated similarly.

First of all, $\mu(K) > B$ implies

$$(\mu(K + c_{21}a - c_1 a))^2 < \left(\mu(K) + \varepsilon a \frac{c_{21} - c_{22}}{p + h}\right)^2,$$

which implies that

$$z_1 < \gamma - \frac{a}{2} + \varepsilon a\left(\beta_2 - \frac{1}{2}\right) + \mu(K).$$

Also, note that

$$z_1 \geq \gamma - \frac{a}{2} + \varepsilon a\left(\beta_1 - \frac{1}{2}\right) - \mu(K).$$

Therefore, z_1 is in the second interval and is a possible minimum point.

Next, we can prove that

$$\begin{aligned} \mu(K) > B \iff & z_1 < \gamma - \frac{a}{2} + \varepsilon a\left(\beta_2 - \frac{1}{2}\right) + \mu(K) \\ \iff & z_2 < \gamma - \frac{a}{2} + \varepsilon a\left(\beta_2 - \frac{1}{2}\right) + \mu(K), \end{aligned} \tag{7.57}$$

where the last inequality means that z_2 is less than the lower bound of the third interval. Since

$$\begin{aligned} &\gamma - \frac{a}{2} + \varepsilon a\left(\beta_2 - \frac{1}{2}\right) + \mu(K) \\ &\quad < \gamma + \frac{a}{2} + \varepsilon a\left(\beta_1 - \frac{1}{2}\right) - \mu(K) \text{ (see (7.51))}, \end{aligned}$$

we obtain $z_2 < \gamma + \frac{a}{2} + \varepsilon a\left(\beta_1 - \frac{1}{2}\right) - \mu(K)$. We can also prove

$$\begin{aligned} & z_2 < \gamma + \frac{a}{2} + \varepsilon a\left(\beta_1 - \frac{1}{2}\right) - \mu(K) \\ \iff & z_3 < \gamma + \frac{a}{2} + \varepsilon a\left(\beta_1 - \frac{1}{2}\right) - \mu(K), \end{aligned} \tag{7.58}$$

where the last inequality means that z_3 is less than the lower bound of the fourth interval. Hence, $\Pi_1(q_1)$ cannot attain its minimum at z_2 or z_3 in this case.

Finally, let us check the monotone property interval by interval. If

$$\gamma - \frac{a}{2} - \frac{\varepsilon a}{2} \leq q_1 < \gamma - \frac{a}{2} + \varepsilon a\left(\beta_1 - \frac{1}{2}\right) - \mu(K),$$

by making use of equation (7.48), $\Pi_1(q_1)$ is nonincreasing. If

$$\gamma - \frac{a}{2} + \varepsilon a\left(\beta_1 - \frac{1}{2}\right) - \mu(K) \leq q_1 < \gamma - \frac{a}{2} + \varepsilon a\left(\beta_2 - \frac{1}{2}\right) + \mu(K),$$

recall that $\Pi_1(q_1)$ is nonincreasing if and only if $q_1 \leq z_1$, $\Pi_1(q_1)$ is nonincreasing over the interval

$$\left[\gamma - \frac{a}{2} + \varepsilon a\left(\beta_1 - \frac{1}{2}\right) - \mu(K),\ \ z_1\right),$$

and nondecreasing over the interval

$$\left[z_1,\ \ \gamma - \frac{a}{2} + \varepsilon a\left(\beta_2 - \frac{1}{2}\right) + \mu(K)\right).$$

If

$$\gamma - \frac{a}{2} + \varepsilon a\left(\beta_2 - \frac{1}{2}\right) + \mu(K) \leq q_1 < \gamma + \frac{a}{2} + \varepsilon a\left(\beta_1 - \frac{1}{2}\right) - \mu(K),$$

recall that $\Pi_1(q_1)$ is nondecreasing if and only if $q_1 \geq z_2$, $G_1(q_1)$ is always nondecreasing. If

$$\gamma + \frac{a}{2} + \varepsilon a\left(\beta_1 - \frac{1}{2}\right) - \mu(K) \leq q_1 < \gamma + \frac{a}{2} + \varepsilon a\left(\beta_2 - \frac{1}{2}\right) + \mu(K),$$

recall that $\Pi_1(q_1)$ is nondecreasing if and only if $q_1 \geq z_3$, $\Pi_1(q_1)$ is always nondecreasing. If

$$\gamma + \frac{a}{2} + \varepsilon a\left(\beta_2 - \frac{1}{2}\right) + \mu(K) \leq q_1 \leq \gamma + \frac{a}{2} + \frac{\varepsilon a}{2},$$

by making use of equation (7.55), $\Pi_1(q_1)$ is always nondecreasing.

In summary, $\Pi_1(q_1)$ is nonincreasing on the left-hand side of z_1, whereas it is nondecreasing on the right-hand side of z_1. By definition of a unimodal function, $\Pi_1(q_1)$ is unimodal and $q_1^* = z_1$.

We summarize in Table 7.1, the analysis along with those in the other two cases.

(iii) In the case of $c_{21} = c_1 = c_{22}$, note that

$$z_1 = \gamma - \frac{a}{2} + \varepsilon a\left(\beta_2 1 - \frac{1}{2}\right) + \mu(K),$$

$$z_3 = \gamma + \frac{a}{2} + \varepsilon a\left(\beta_2 - \frac{1}{2}\right) - \mu(K),$$

and for any $q_1 \in [z_1, z_3]$,

$$\frac{d\Pi_1(q_1)}{dq_1} = 0.$$

Recall that $\Pi_1(q_1)$ is nonincreasing if and only if $q_1 \leq z_1$ and nondecreasing if and only if $q_1 \geq z_3$. Thus, $\Pi_1(q_1)$ is unimodal. Hence, any $q_1 \in [z_1,\ z_3]$ is optimal. □

q_1	$\Pi_1(q_1)$ if $\mu(K) > B = \min\{A, B\}$	$\Pi_1(q_1)$ if $\mu(K) \leq \min\{A, B\}$	$\Pi_1(q_1)$ if $\mu(K) > A = \min\{A, B\}$
1st	$\downarrow$	$\downarrow$	$\downarrow$
2nd	$\downarrow$, if $q_1 \leq z_1$; $\uparrow$, if $q_1 > z_1$.	$\downarrow$	$\downarrow$
3rd	$\uparrow$	$\downarrow$, if $q_1 \leq z_2$; $\uparrow$, if $q_1 > z_2$.	$\downarrow$
4th	$\uparrow$	$\uparrow$	$\downarrow$, if $q_1 \leq z_3$; $\uparrow$, if $q_1 > z_3$.
5th	$\uparrow$	$\uparrow$	$\uparrow$

Table 7.1. Unimodality of $\Pi_1(q_1)$ when $\mu(K) < \varepsilon a \cdot (1 - \beta_2)$

We conclude this section by commenting on a subcase of Case 3 when K is sufficiently large, so that no ordering is optimal in stage 2. How large the fixed cost should be for this purpose and what is the optimal stage 1 order are specified in the following result.

THEOREM 7.5 *Assume Case 3—that is, $\mu(K) \geq \varepsilon a \beta_1$. In this case, when $K \geq (p - c_{21})\left[a - \frac{\varepsilon a}{2}(2 - \beta_1)\right]$, the optimal initial order quantity is given by*

$$q_1^* = \begin{cases} \gamma - \frac{a}{2} - \frac{\varepsilon a}{2} + a\sqrt{2\varepsilon\beta_0}, \\ \qquad \text{if } \beta_0 \leq \frac{\varepsilon}{2} \text{ and} \\ \qquad K \geq (p - c_{21}) \cdot \left(a + \frac{\varepsilon a}{2}\beta_1 - a\sqrt{2\varepsilon\beta_0}\right), \\ \gamma - \frac{a}{2} + \beta_0 a, \\ \qquad \text{if } \frac{\varepsilon}{2} < \beta_0 < 1 - \frac{\varepsilon}{2}, \\ \gamma + \frac{a}{2} + \frac{\varepsilon a}{2} - a\sqrt{2\varepsilon(1 - \beta_0)}, \\ \qquad \text{if } \beta_0 \geq 1 - \frac{\varepsilon}{2} \text{ and} \\ \qquad K \geq (h + c_{22}) \cdot \left[a + \frac{\varepsilon a}{2}(1 - \beta_2) - a\sqrt{2\varepsilon(1 - \beta_0)}\right], \end{cases}$$

where $\beta_0 = (p - c_1)/(p + h)$, and the optimal policy at stage 2 is $q_2^* = 0$.

It is also possible to obtain an explicit expression for the value function v_1 in different cases. For example, when $q_1^* = z_1$, the optimal cost

$$\begin{aligned}\pi_1^* &= K + c_1\gamma + \frac{1}{2}(c_{21} - c_1)a \\ &\quad -\frac{2}{3}\sqrt{\frac{2\varepsilon}{a(p+h)}}\left[K^{\frac{3}{2}} + \sqrt{(K + c_{21}a - c_1a)^3}\right] \\ &\quad +\frac{\varepsilon a}{2(p+h)}\left[(h+c_1)(p-c_1) + (c_{21}-c_1)^2\right].\end{aligned}$$

7.5.2 Further Analysis

With explicit optimal solutions obtained in Sections 7.4 and 7.5, it is easy to carry out a sensitivity analysis with respect to forecast and contract parameters. It is certainly of interest to improve the quality of the demand forecasts. Improving either the stage 1 forecast or the stage 2 forecast or both could result in cost reductions. For example, the marginal benefits of information updates with respect to ε and a—namely, $-\partial\pi_1^*/\partial\varepsilon$ and $-\partial\pi_1^*/\partial a$—provide an indication of the relative importance of stage 1 and 2 forecasts in the model, respectively. Given these and the costs of efforts in reducing a and ε, the buyer can easily figure out where he should put his next dollar in improving demand forecasts. As for the contract-exercise cost K, it is possible to identify a critical value K_0, such that if K exceeds (resp. is less than) K_0, the buyer should invest in improving stage 1 (resp. stage 2) forecast.

Besides the sensitivity analysis, it is also possible to select a suitable strategy to manage the tradeoffs between different hedging alternatives. As is shown in Section 7.2, there exist a number of alternatives in hedging demand uncertainty (for example, in addition to the purchase contract, in the context of a micro-controller purchase issue, using generic chips is a viable option; see Yan, Liu, and Hsu [8]). It is important for the buyer to know when a purchase contract is attractive. Yan, Liu, and Hsu [8] solve the problem of two supply modes with demand-information updates, where the second-stage decision is the quantity of the generic components that should be used. The expected cost functions at stage 2 and stage 1 are

$$\begin{aligned}\widetilde{\Pi}_2(q_1, q_2, i) &= \widetilde{c}_2 \cdot q_2 + h\int_{i-\varepsilon a/2}^{q_1+q_2} [q_1 + q_2 - z] \cdot \psi(z|i)\mathrm{d}z \\ &\quad +p\int_{q_1+q_2}^{i+\varepsilon a/2} [z - q_1 - q_2] \cdot \psi(z|i)\mathrm{d}z, \\ \widetilde{\Pi}_1(q_1) &= c_1q_1 + \mathsf{E}\left[\widetilde{\Pi}_2(q_1, q_2, I)\right],\end{aligned}$$

where $\widetilde{c}_2$ is the per unit ordering cost for generic component. Comparing with the explicit solution derived in previous subsection, it is possible for us to have the following results.

THEOREM 7.6 *As for the contract-cost function $\Pi_1(q_1)$ and the substitution cost function $\widetilde{\Pi}_1(q_1)$, we have*

(i) $\min_{q_1 \geq 0}\left\{\Pi_1(q_1)\right\}$ *is a monotone nondecreasing and continuous function with respect to the contract-exercise cost K;*

(ii) *there exists a K_0 such that*

$$\begin{aligned} \min_{q_1\geq 0}\left\{\Pi_1(q_1)|_{K=0}\right\} &\leq \min_{q_1\geq 0}\left\{\widetilde{\Pi}_1(q_1)\right\} \\ &\leq \min_{q_1\geq 0}\left\{\Pi_1(q_1)\Big|_{K\geq K_0}\right\}; \end{aligned}$$

(iii) *there exists a unique K_1 such that*

$$\min_{q_1\geq 0}\left\{\Pi_1(q_1)|K=K_1\right\} = \min_{q_1\geq 0}\left\{\widetilde{\Pi}_1(q_1)\right\}.$$

Proof (i) Recall that the optimal reorder point $\sigma_1(i) = \Sigma_1(i) - \mu(K)$, the reduction point $\sigma_2(i) = \Sigma_2(i) + \mu(K)$, and $\mu(K) = \sqrt{\frac{2\varepsilon a K}{p+h}}$. Therefore, $\sigma_1(i)$ decreases and $\sigma_2(i)$ increases as K increases. The feasible set of q_2 shrinks when K increases. Therefore, $\min_{q_1\geq 0}\{\Pi_1(q_1)\}$ is a monotone nondecreasing function of K.

(ii) The validity of the left-hand side inequality lies in the fact that the feasible set of q_2 becomes larger when negative q_2 is allowed. In addition, a sufficiently large contract cost K forces the initial order quantity to remain unchanged. This fact results in the validity of the right-hand side inequality.

(iii) Properties (i) and (ii) ensure the existence of K_1. □

Theorem 7.6 reveals a rule for hedging strategy selection. When $K \geq K_1$, it is unwise for the buyer to sign the purchase contract. By noting that a purchase contract is a real option, K can be considered as the option price in the case when $c_{21} = c_{22} = c_1$. It is interesting for the buyer to know the value of a purchase contract. In particular, the buyer needs to know what is the best strategy against demand uncertainty.

7.6. Concluding Remarks

In this chapter, we study a purchase contract with a demand-forecast update. We formulate the problem as a two-stage dynamic programming problem. We obtain an optimal solution for the contract management for a class of demand distributions. In particular, we obtain an explicit optimal solution for a uniformly distributed updated demand. The explicit nature of the optimal solution enables us to gain managerial insights into better supply chain management. More specifically, we find a critical value of the contract-exercise cost, which determines the direction of further improvement in demand forecasts. In comparison with other uncertainty hedging approaches such as substitution, we obtain another critical value of the contract-exercise cost, below (resp. above) which the buyer would (resp. would not) sign the contract.

7.7. Notes

This chapter is based on Huang, Sethi, and Yan [5].

Traditional inventory models assume a simple buyer-supplier arrangement. The buyer places an order at any time for any amount at a fixed cost and a given unit price, and the supplier provides the product. However, this results in a great deal of uncertainty for both parties, since very little is known about the eventual demand at the time of order. In many industries, complicated forms of arrangements, known as contracts, exist to strike a balance between flexibility and uncertainty. Barnes-Schuster, Bassok, and Anupindi [1] study the role of a supply contract between the buyer and the supplier in a two-stage model. In their model, the option of volume flexibility applies to the second stage. Structural properties of the objective functions for both the buyer and the supplier are explored. They show that to achieve channel coordination, the contract-exercise cost must be in the form of a piecewise linear function. The contract-option price is also evaluated numerically. Donohue [3] considers a supply contract as a risk-sharing mechanism between the buyer and the supplier. She focuses on the channel coordination by determining the wholesale price and the return policy. Eppen and Iyer [4] discuss a so-called backup agreement in the fashion industry for a catalog company. It entails that the supplier holds back a constant fraction of the commitment and delivers the remaining units to the catalog company before the start of the fashion season. It allows the company to make decisions after observing the early demand. That is, the company may order up to the backup quantity at the original cost, along with a penalty cost for any backup units that are not ordered. They find that a backup agreement has an impact on expected profits. Cachon [2] reviews and extends the literature of supply chain coordination for a class of contracts. In particular, he addresses the coordination issues of the two-stage newsvendor. The newsvendor is allowed only to increase, at the second stage, the original order at a higher unit cost and no fixed cost. It is found that it is possible to coordinate the supply chain with a buy-back contract.

In contrast with Eppen and Iyer [4], we allow the buyer to adjust the initial order at the second stage. As in Chapter 6, we allow in stage 2 for cancellation of a part of the initial order issued in stage 1. This means that our cost function is no longer K-convex. Consequently, we specialize our distribution to be PF_2 or uniform to ensure a unimodal cost function. Our work differs from Barnes-Schuster, Bassok, and Anupindi [1] in a couple of ways. We update both the mean and the spread of the the demand forecast, whereas Barnes-Schuster, Bassok, and Anupindi [1] update the minimum demand. In addition, we consider a fixed contract-exercise cost and use the contract to hedge the demand uncertainty. In comparison with the models of Donohue [3], we consider a fixed contract-exercise cost and obtain an explicit solution for any degree of the demand-information update. The worthless and perfect information updates are therefore special cases of our model.

References

[1] D. Barnes-Schuster, Y. Bassok, and R. Anupindi. Coordination and flexibility in supply contracts with options. *Manufacturing and Service Operations Management*, 4:171–207, 2002.

[2] G.P. Cachon. Supply chain coordination with contracts. To appear in *Handbook of Operations Management*, S. Graves and Ton de Kok (editors), North-Holland, Amsterdam, The Netherlands.

[3] K.L. Donohue. Efficient supply contracts for fashion goods with forecast updating and two production modes. *Management Science*, 46:1397–1411, 2000.

[4] G.D. Eppen and A.V. Iyer. Improved fashion buying with Bayesian updates. *Operations Research*, 45:805–819, 1997.

[5] Y. Huang, S. Sethi, and H. Yan. Purchase contract management with demand forecast updates. *IIE Transactions*, to appear.

[6] S. Karlin. *Total Positivity*, Vol. 1. Stanford University Press, Stanford, CA, 1968.

[7] E.L. Porteus. On the optimality of generalized (s, S) policies. *Management Science*, 17:411–426, 1971.

[8] H. Yan, K. Liu, and A. Hsu. Optimal ordering in a dual-supplier system with demand-forecast updates. *Production and Operations Management*, 12:30–45, 2003.

Chapter 8

PURCHASE CONTRACT MANAGEMENT: TWO-PLAYER GAMES

8.1. Introduction

In this chapter, we consider contract pricing and information dynamics in the problem of purchase-contract management. We investigate the competitive behavior by using noncooperative game models: a static game and a two-step dynamic game, where the Nash equilibrium and the subgame-perfect Nash equilibrium are proved to exist. While it has been widely reported that information sharing benefits the supplier, our focus is on its impact on the buyer. We demonstrate that information sharing benefits the buyer only when the supplier overestimates the true demand.

This chapter is organized as follows. In Section 8.2, we formulate the problem as a static game between a supplier and a buyer. For a general demand process, we study the structure properties of the cost or payoff functions, which lead to the existence of a Nash equilibrium. In Section 8.3, based on the uniformly distributed demand forecasts, we characterize the best reaction strategy for each player. In Section 8.4, we explore the Nash equilibria of the static game for cases with and without information sharing. With Nash equilibria, it is possible to discuss the impacts of information sharing on the supplier and the buyer. In Section 8.5, we obtain a subgame-perfect Nash equilibrium for the dynamic game. The issue of information sharing is treated as in Section 8.4. We conclude in Sections 8.6 and 8.7 with a discussion of incentive design issues for channel coordination and some endnotes.

8.2. Problem Formulation

In this chapter, we consider the competitive behaviors of the buyer and the supplier in a special purchase-contract problem where the buyer is allowed to increase the initial order quantity q_1 at a later stage—that is, the reorder quantity $q_2 \geq 0$ at a cost $c_2 \geq c_1$, where c_1 denotes the unit cost of the initial order. The supplier faces a production cost w_1 per unit at stage 1 and a more expensive production cost $w_2 \geq w_1$ at stage 2. The demand process possesses the same structure as described in Chapter 7.

Differing from Chapter 7, we are here interested in the contract pricing and the value of information sharing in the supplier-buyer game framework. There are two ways to formulate this problem: a static game and a two-step dynamic game. In the static game, the supplier and the buyer figure out their own reaction functions independently. Specifically, the buyer's reaction function provides a mapping rule from the contract-exercise cost K to the initial order quantity q_1. Likewise, the supplier's reaction function defines a mapping from the initial order q_1 to the contract-exercise cost K. The simultaneous solution of these two reaction functions, if it exists, reveals a Nash equilibrium where no party is willing to deviate.

To the empty threat that exists in the static game, we propose a two-step dynamic game where the players move sequentially. Without loss of generality, we assume that the supplier is a leader and the buyer is a follower. The backward induction procedure provides a way to derive the subgame-perfect Nash equilibrium.

To make the above decision process clear, we insert one more argument K and q_1 in the buyer's cost or value functions and the supplier's payoff function defined in Chapter 7, respectively. Specifically the related notation includes

$$
\begin{array}{rl}
\Pi_2(q_1, q_2, K, i) & \text{instead of } \Pi_2(q_1, q_2, i), \\
\Pi_1(q_1, K) & \text{instead of } \Pi_1(q_1), \\
\pi_2^{*b}(q_1, K, i) & \text{instead of } \pi_2^*(q_1, i), \\
\pi_1^{*b}(K) & \text{instead of } \pi_1^*, \\
J_1(q_1, K) & \text{the supplier's payoff function, and} \\
\pi_1^{*s}(q_1) & \text{the supplier's optimal payoff function,}
\end{array}
$$

where the superscripts b and s represent the buyer and the supplier, respectively. Other notations are the same as those introduced in Chapter 7. Because there is no cancellation at stage 2, for notation simplicity in this chapter we write

$$(\sigma_1(i), \Sigma_1(i)),$$

which in Chapter 7 is given by

$$(\sigma(i, K), \Sigma(i, K)).$$

In the remainder of this section, we investigate the existence of a Nash equilibrium for a general demand process. Since the buyer's problem has been formulated in a more general purchase-contract context in Chapter 7, we directly study the structural properties of the buyer's cost functions in the following theorem.

THEOREM 8.1 *The buyer's cost functions under optimal decisions at two stages, $\pi_1^{*b}(K)$ and $\pi_2^{*b}(q_1, K, i)$, are monotone nondecreasing in K.*

Proof The proof can be developed similarly as in Theorem 7.1. Here we omit it. □

Let $q_2^*(q_1, K, i)$ be the optimal order quantity at stage 2. Using Theorem 7.1,

$$q_2^*(q_1, K, i) = \begin{cases} \Sigma(i, K) - q_1, & \text{if } q_1 \le \sigma(i, K), \\ 0, & \text{otherwise.} \end{cases} \tag{8.1}$$

By Theorem 7.1 again, we know that

$$\pi_2^{*b}(q_1, K, i) = \begin{cases} K + c_2 \cdot (\Sigma(i, K) - q_1) \\ \quad + \mathsf{E}\left[h \cdot (\Sigma(i, K) - D)^+ + p \cdot (D - \Sigma(i, K))^+ | \, i\right], \\ \qquad\qquad\qquad\qquad\qquad \text{if } q_1 \le \sigma(i, K), \\ \mathsf{E}\left[h \cdot (q_1 - D)^+ + p \cdot (D - q_1)^+ | \, i\right], \quad \text{if } q_1 > \sigma(i, K). \end{cases} \tag{8.2}$$

Furthermore, for any observed information i, if $K \le \bar{K}$, then

$$\Sigma(i, K) = \Sigma(i, \bar{K}) \quad \text{and} \quad \sigma(i, K) \ge \sigma(i, \bar{K}). \tag{8.3}$$

In the following, we introduce some terminology from game theory (for more details, see Milgrom and Roberts [10], Topkis [11], and Yao [12]). The buyer's strategy space is $[0, \infty)$, and the supplier's strategy space is also $[0, \infty)$. Refer to (q_1, K) as a *strategy profile.* The strategy space for the buyer and supplier is defined as

$$\mathcal{R}^2 = [0, \infty) \times [0, \infty).$$

DEFINITION 8.1 *A function $L(x_1, x_2)$ defined on $\mathcal{R}^2$ is* supermodular (submodular) *if for all $x = (x_1, x_2)$, and $y = (y_1, y_2) \in \mathcal{R}^2$,*

$$L(x \wedge y) + L(x \vee y) \ge (\le) L(x) + L(y),$$

where $\wedge$ and $\vee$ denote, respectively, the "min" and the "max" operators (both in the componentwise sense).

Using (8.2) and (8.3), we can get the following theorem.

THEOREM 8.2 *The buyer's cost function* $\Pi_2(q_1, q_2, K, i)$ *at stage* 2 *is supermodular in* (q_2, K) *given* q_1, *and the buyer's cost function* $\Pi_1(q_1, K)$ *at stage* 1 *is submodular in* (q_1, K).

Proof It is directly verifiable that for any given q_1, $\Pi_2(q_1, q_2, K, i)$ is supermodular in (q_2, K). We show the submodularity for (q_1, K). For any (q_1, K), $(\bar{q}_1, \bar{K}) \in \mathcal{R}^2$, if

$$q_1 \leq \bar{q}_1, \quad \text{and} \quad K \leq \bar{K},$$

then

$$(q_1, K) \wedge (\bar{q}_1, \bar{K}) = (q_1, K) \quad \text{and} \quad (q_1, K) \vee (\bar{q}_1, \bar{K}) = (\bar{q}_1, \bar{K}).$$

Thus,

$$\begin{aligned} &\Pi_1\left((q_1, K) \wedge (\bar{q}_1, \bar{K})\right) + \Pi_1\left((q_1, K) \vee (\bar{q}_1, \bar{K})\right) \\ &= \Pi_1(q_1, K) + \Pi_1(\bar{q}_1, \bar{K}). \end{aligned}$$

Therefore, to get the submodularity for (q_1, K), without loss of generality we can assume that

$$q_1 < \bar{q}_1 \quad \text{and} \quad K > \bar{K}. \tag{8.4}$$

In the light of (8.3), the proof is divided into several cases. For any given i, we have the following cases.

Case 1: $[q_1 \leq \sigma(i, K)$ and $\bar{q}_1 \leq \sigma(i, K)]$

For Case 1, it follows from (8.1) that

$$\pi_2^{*b}(q_1, \bar{K}, i) + \pi_2^{*b}(\bar{q}_1, K, i) = \pi_2^{*b}(q_1, K, i) + \pi_2^{*b}(\bar{q}_1, \bar{K}, i). \tag{8.5}$$

Case 2: $[q_1 \leq \sigma(i, K)$ and $\sigma(i, K) < \bar{q}_1 \leq \sigma(i, \bar{K})]$

For Case 2, it follows from (8.1) and (8.2) that

$$\begin{aligned} \pi_2^{*b}(q_1, \bar{K}, i) &= \bar{K} + c_2 \cdot (\Sigma(i, \bar{K}) - q_1) \\ &\quad + \mathsf{E}\left[h \cdot (\Sigma(i, \bar{K}) - D)^+ + p \cdot (D - \Sigma(i, \bar{K}))^+ \middle| I = i\right], \\ \pi_2^{*b}(\bar{q}_1, K, i) &= \mathsf{E}\left[h \cdot (\bar{q}_1 - D)^+ + p \cdot (D - \bar{q}_1)^+ \middle| I = i\right], \\ \pi_2^{*b}(q_1, K, i) &= K + c_2 \cdot (\Sigma(i, K) - q_1) \\ &\quad + \mathsf{E}\left[h \cdot (\Sigma(i, K) - D)^+ + p \cdot (D - \Sigma(i, K))^+ \middle| I = i\right], \\ \pi_2^{*b}(\bar{q}_1, \bar{K}, i) &= \bar{K} + c_2 \cdot (\Sigma(i, \bar{K}) - \bar{q}_1) \\ &\quad + \mathsf{E}\left[h \cdot (\Sigma(i, \bar{K}) - D)^+ + p \cdot (D - \Sigma(i, \bar{K}))^+ \middle| I = i\right]. \end{aligned}$$

Note that

$$
\begin{aligned}
&\mathsf{E}\left[h \cdot (\bar{q}_1 - D)^+ + p \cdot (D - \bar{q}_1)^+ \middle| I = i\right] \\
&\quad \le \bar{K} + c_2 \cdot (\Sigma(i, \bar{K}) - \bar{q}_1) \\
&\quad\quad + \mathsf{E}\left[h \cdot (\Sigma(i, \bar{K}) - D)^+ + p \cdot (D - \Sigma(i, \bar{K}))^+ \middle| I = i\right].
\end{aligned}
$$

Therefore,

$$
\pi_2^{*b}(q_1, \bar{K}, i) + \pi_2^{*b}(\bar{q}_1, K, i) \le \pi_2^{*b}(q_1, K, i) + \pi_2^{*b}(\bar{q}_1, \bar{K}, i). \tag{8.6}
$$

Case 3: $[q_1 \le \sigma(i, K)$ and $\bar{q}_1 > \sigma(i, \bar{K})]$
For Case 3, it follows from (8.1) and (8.2) that

$$
\begin{aligned}
\pi_2^{*b}(q_1, \bar{K}, i) &= \bar{K} + c_2 \cdot (\Sigma(i, \bar{K}) - q_1) \\
&\quad + \mathsf{E}\left[h \cdot (\Sigma(i, \bar{K}) - D)^+ + p \cdot (D - \Sigma(i, \bar{K}))^+ \middle| I = i\right], \\
\pi_2^{*b}(\bar{q}_1, K, i) &= \mathsf{E}\left[h \cdot (\bar{q}_1 - D)^+ + p \cdot (D - \bar{q}_1)^+ \middle| I = i\right], \\
\pi_2^{*b}(q_1, K, i) &= K + c_2 \cdot (\Sigma(i, K) - q_1) \\
&\quad + \mathsf{E}\left[h \cdot (\Sigma(i, K) - D)^+ + p \cdot (D - \Sigma(i, K))^+ \middle| I = i\right], \\
\pi_2^{*b}(\bar{q}_1, \bar{K}, i) &= \mathsf{E}\left[h \cdot (\bar{q}_1 - D)^+ + p \cdot (D - \bar{q}_1)^+ \middle| I = i\right].
\end{aligned}
$$

Noting $K > \bar{K}$, hence, we have

$$
\pi_2^{*b}(q_1, \bar{K}, i) + \pi_2^{*b}(\bar{q}_1, K, i) < \pi_2^{*b}(q_1, K, i) + \pi_2^{*b}(\bar{q}_1, \bar{K}, i). \tag{8.7}
$$

Case 4: $[\sigma(i, K) < q_1 \le \sigma(i, \bar{K})$ and $\sigma(i, K) < \bar{q}_1 \le \sigma(i, \bar{K})]$
For Case 4, similarly, we have

$$
\begin{aligned}
\pi_2^{*b}(q_1, \bar{K}, i) &= \bar{K} + c_2 \cdot (\Sigma(i, \bar{K}) - q_1) \\
&\quad + \mathsf{E}\left[h \cdot (\Sigma(i, \bar{K}) - D)^+ + p \cdot (D - \Sigma_1(i, \bar{K}))^+ \middle| I = i\right], \\
\pi_2^{*b}(\bar{q}_1, K, i) &= \mathsf{E}\left[h \cdot (\bar{q}_1 - D)^+ + p \cdot (D - \bar{q}_1)^+ \middle| I = i\right], \\
\pi_2^{*b}(q_1, K, i) &= \mathsf{E}\left[h \cdot (q_1 - D)^+ + p \cdot (D - q_1)^+ \middle| I = i\right], \\
\pi_2^{*b}(\bar{q}_1, \bar{K}, i) &= K + c_2 \cdot (\Sigma(i, K) - \bar{q}_1) \\
&\quad + \mathsf{E}\left[h \cdot (\Sigma(i, K) - D)^+ + p \cdot (D - \Sigma(i, K))^+ \middle| I = i\right].
\end{aligned}
$$

By the convexity of $\Pi_2(q_1, 0, K, i)$ and $q_1 < \bar{q}_1 < \Sigma(i, K)$, we have

$$
\begin{aligned}
&\mathsf{E}\left[h \cdot (q_1 - D)^+ + p \cdot (D - q_1)^+ \middle| I = i\right] \\
&\quad > \mathsf{E}\left[h \cdot (\bar{q}_1 - D)^+ + p \cdot (D - \bar{q}_1)^+ \middle| I = i\right].
\end{aligned}
$$

Consequently, (8.7) still holds for Case 4.

Case 5: $[\sigma(i, K) < q_1 \le \sigma(i, \bar{K})$ and $\bar{q}_1 > \sigma(i, \bar{K})]$

Case 6: $[q_1 > \sigma(i, \bar{K})$ and $\bar{q}_1 > \sigma(i, \bar{K})]$

Similarly, we can show that (8.7) holds for Case 5, and (8.5) holds for Case 6. Hence, the proof is completed. □

The direct implication of these properties is that the buyer's optimal decision q_1^* is increasing in K, while q_2^* is decreasing in K. These properties are consistent with our intuition: if the supplier's contract-exercise cost is high, then the buyer will tend to order more at stage 1 and order less at stage 2.

The supplier's objective is to find an optimal contract-exercise cost K that maximizes the profit function for given q_1:

$$\begin{aligned}
&\max_K \{J_1(q_1, K)\} \\
&= \max_K \Big\{(c_1 - w_1) \cdot q_1 \\
&\quad + \mathsf{E}\Big[K \cdot \delta(q_2^*(q_1, K, I)) + (c_2 - w_2) \cdot q_2^*(q_1, K, I)\Big]\Big\} \\
&= \max_K \Big\{(c_1 - w_1) \cdot q_1 \\
&\quad + \mathsf{E}\Big\{\Big[K + (c_2 - w_2) \cdot (\Sigma(I, K) - q_1)\Big] \cdot \delta(\sigma(I, K) - q_1)\Big\}\Big\},
\end{aligned}$$

where $(\sigma(i, K), \Sigma(i, K))$ is the buyer's optimal policy at stage 2 when $I = i$

Similar to Theorem 8.2, the following theorem implies that the supplier's optimal decision K^* is decreasing in q_1.

THEOREM 8.3 *The supplier's payoff function $J_1(q_1, K)$ is supermodular in (q_1, K).*

Based on the monotone properties of the optimal decisions described by Theorem 8.2 and Theorem 8.3, it is possible to consider the Nash equilibrium of the static game.

THEOREM 8.4 *There exists a Nash equilibrium in the supplier-buyer game.*

8.3. Reaction Strategies Under Uniformly Distributed Demand

In this section, we try to reveal more insights from the behaviors of the supplier and the buyer under the assumption that the information and the conditional demand given that information follow uniform distributions. Formally, I is uniformly distributed over the interval $\left[\gamma - \frac{a}{2},\ \gamma + \frac{a}{2}\right]$ and, given $I = i$, D follows the uniform distribution over the interval $\left[i - \frac{\varepsilon a}{2},\ i + \frac{\varepsilon a}{2}\right]$.

8.3.1 The Buyer's Reaction Strategy

As a special case of the purchase contract studied in Chapter 7, we sketch the buyer's optimal decisions in the following lemma.

LEMMA 8.1 (i) *At stage 2, the buyer's optimal policy is* $(\sigma(i,K),\Sigma(i,K))$ *policy, where* $\Sigma(i,K) = i + \varepsilon a(\beta - \frac{1}{2})$ *and* $\sigma(i,K) = \Sigma(i,K) - \mu(K)$ *with* $\mu(K) = \sqrt{\frac{2\varepsilon a K}{p+h}}$ *and* $\beta = \frac{p-c_2}{p+h}$.

(ii) *The buyer's total cost* $\Pi_1(q_1, K)$ *is a unimodal function of* q_1 *and attains its minimum at*

$$q_1^* = \begin{cases} \gamma - \frac{a}{2} + \varepsilon a(\beta - \frac{1}{2}) + \mu(K + c_2 a - c_1 a), \\ \qquad\qquad\qquad \textit{if } \ \mu(K) \le \frac{a}{2} - \varepsilon a \frac{c_2 - c_1}{p+h}, \\ \gamma + \varepsilon a(\frac{p-c_1}{p+h} - \frac{1}{2}), \\ \qquad\qquad\qquad \textit{if } \ \mu(K) > \frac{a}{2} - \varepsilon a \frac{c_2 - c_1}{p+h}. \end{cases}$$

Proof Recall the discussion before Theorem 7.4. We know that the cost function is a piecewise continuous function, and it is nonincreasing in the interval

$$\left[\gamma - \frac{a}{2} - \frac{\varepsilon a}{2},\ \gamma - \frac{a}{2} + \varepsilon a\left(\beta - \frac{1}{2}\right) - \mu(K)\right],$$

if $c_1 \le c_2$. Therefore, the possible candidate for the optimal order quantity is either in the right-open interval

$$\left[\gamma - \frac{a}{2} + \varepsilon a\left(\beta - \frac{1}{2}\right) - \mu(K),\ \gamma + \frac{a}{2} + \varepsilon a\left(\beta - \frac{1}{2}\right) - \mu(K)\right)$$

or in the interval

$$\left[\gamma + \frac{a}{2} + \varepsilon a\left(\beta - \frac{1}{2}\right) - \mu(K),\ \gamma + \frac{a}{2} + \frac{\varepsilon a}{2}\right].$$

If the minimum is obtained in the interval

$$\left[\gamma - \frac{a}{2} + \varepsilon a\left(\beta - \frac{1}{2}\right) - \mu(K),\ \gamma + \frac{a}{2} + \varepsilon a\left(\beta - \frac{1}{2}\right) - \mu(K)\right),$$

the optimal order quantity is

$$z_1 = \gamma - \frac{a}{2} + \varepsilon a\left(\beta - \frac{1}{2}\right) + \mu(K + c_2 a - c_1 a). \tag{8.8}$$

Otherwise, the optimal order quantity is

$$z_2 = \gamma + \varepsilon a\left(\frac{p - c_1}{p+h} - \frac{1}{2}\right). \tag{8.9}$$

To get the lemma, we show that

$$\begin{aligned} & z_1 < \gamma + \frac{a}{2} + \varepsilon a \left(\beta - \frac{1}{2}\right) - \mu(K) \\ \Longleftrightarrow\ & z_2 < \gamma + \frac{a}{2} + \varepsilon a \left(\beta - \frac{1}{2}\right) - \mu(K) \\ \Longleftrightarrow\ & \mu(K) < \frac{a}{2} - \varepsilon a \frac{c_2 - c_1}{p + h}, \end{aligned} \tag{8.10}$$

and

$$\begin{aligned} & z_1 = \gamma + \frac{a}{2} + \varepsilon a \left(\beta - \frac{1}{2}\right) - \mu(K) \\ \Longleftrightarrow\ & z_2 = \gamma + \frac{a}{2} + \varepsilon a \left(\beta - \frac{1}{2}\right) - \mu(K) \\ \Longleftrightarrow\ & \mu(K) = \frac{a}{2} - \varepsilon a \frac{c_2 - c_1}{p + h}. \end{aligned} \tag{8.11}$$

By some simple algebraic calculations, we have

$$\begin{aligned} & z_1 < \gamma + \frac{a}{2} + \varepsilon a \left(\beta - \frac{1}{2}\right) - \mu(K) \\ \Longleftrightarrow\ & \mu(K + c_2 a - c_1 a) < a - \mu(K) \\ \Longleftrightarrow\ & \mu(k) < \frac{a}{2} + \varepsilon a \cdot \left(\beta - \frac{p - c_1}{p + h}\right) \\ \Longleftrightarrow\ & \varepsilon a \frac{p - c_1}{p + h} < \frac{a}{2} + \varepsilon a \beta - \mu(K) \\ \Longleftrightarrow\ & z_2 < \gamma + \frac{a}{2} + \varepsilon a \left(\beta - \frac{1}{2}\right) - \mu(K). \end{aligned}$$

So we have (8.10). Similarly, we can show (8.11).

Based on these equivalent relationships, in the case of $\mu(K) \leq \frac{a}{2} - \varepsilon a \frac{c_2 - c_1}{p+h}$, $\Pi_1(q_1, K)$ is nondecreasing in

$$\left[\gamma + \frac{a}{2} + \varepsilon a \left(\beta - \frac{1}{2}\right) - \mu(K),\ \ \gamma + \frac{a}{2} + \frac{\varepsilon a}{2}\right],$$

since $z_2 \leq \gamma + \frac{a}{2} + \varepsilon a(\beta - \frac{1}{2}) - \mu(K)$. In addition, $\Pi_1(q_1, K)$ is a nonincreasing function in

$$\left[\gamma - \frac{a}{2} - \frac{\varepsilon a}{2},\ \ \gamma - \frac{a}{2} + \varepsilon a \left(\beta - \frac{1}{2}\right) - \mu(K)\right].$$

Further, it is a convex function on the interval

$$\left[\gamma - \frac{a}{2} + \varepsilon a\left(\beta - \frac{1}{2}\right) - \mu(K),\quad \gamma + \frac{a}{2} + \varepsilon a\left(\beta - \frac{1}{2}\right) - \mu(K)\right).$$

Therefore, $\Pi_1(q_1, K)$ is a nonincreasing function on the left of z_1; it is a nondecreasing function on the right of z_1. Hence, $\Pi_1(q_1, K)$ is unimodal and attains its minimum at z_1. Similar proof can be developed for the case of $\frac{a}{2} - \varepsilon a\frac{c_2-c_1}{p+h} < \mu(K)$ by using the following relationship, which can be considered as a different (but equivalent) version of (8.10) and (8.11):

$$\begin{aligned}
& z_1 > \gamma + \frac{a}{2} + \varepsilon a\left(\beta - \frac{1}{2}\right) - \mu(K) \\
\Longleftrightarrow\ & z_2 > \gamma + \frac{a}{2} + \varepsilon a\left(\beta - \frac{1}{2}\right) - \mu(K) \\
\Longleftrightarrow\ & \mu(K) > \frac{a}{2} - \varepsilon a\frac{c_2 - c_1}{p + h}.
\end{aligned} \tag{8.12}$$

□

The corresponding optimal cost function $\pi_1^{*b}(K)$ is given by

$$\begin{cases}
c_1\gamma + \frac{\varepsilon a}{2(p+h)}\left[(h+c_1)(p-c_1) + (c_2-c_1)^2\right] + K + \frac{a}{2}(c_2 - c_1) \\
\quad -\frac{2}{3}\sqrt{\frac{2\varepsilon}{a(p+h)}}\left[K^{\frac{3}{2}} + (K + c_2 a - c_1 a)^{\frac{3}{2}}\right], \\
\qquad\qquad \text{if } \mu(K) \leq \frac{a}{2} - \varepsilon a\frac{c_2-c_1}{p+h}, \\
c_1\gamma + \frac{a(p+h)}{24\varepsilon} + \frac{\varepsilon a}{2(p+h)}(h+c_1)(p-c_1), \\
\qquad\qquad \text{if } \mu(K) > \frac{a}{2} - \varepsilon a\frac{c_2-c_1}{p+h}.
\end{cases}$$

Taking derivative with respect to K yields

$$\frac{\partial \pi_1^{*b}(K)}{\partial K} = \begin{cases}
1 - \frac{1}{a}[\mu(K) + \mu(K + c_2 a - c_1 a)], \\
\qquad\qquad \text{if } \mu(K) \leq \frac{a}{2} - \varepsilon a\frac{c_2-c_1}{p+h}, \\
0, \qquad\qquad \text{if } \mu(K) > \frac{a}{2} - \varepsilon a\frac{c_2-c_1}{p+h}.
\end{cases}$$

The following corollary is straightforward. In fact, it is a special case of Theorem 8.1.

COROLLARY 8.1 *The buyer's optimal cost $\pi_1^{*b}(K)$ is a monotone nondecreasing function of K.*

REMARK 8.1 By Lemma 8.1, when $\mu(K) > \frac{a}{2} - \varepsilon a \frac{c_2 - c_1}{p+h}$, the best strategy is to take no action at stage 2. In this case, the purchase-contract model reduces to a simple action problem—that is,

$$\min_{q_1} \left\{ c_1 q_1 + \mathsf{E}\left[\mathsf{E}\left[h \cdot (q_1 - D)^+ + p \cdot (D - q_1)^+ \middle| I\right]\right] \right\}. \tag{8.13}$$

Therefore,

$$\frac{p+h}{2\varepsilon a}\left(\frac{a}{2} - \varepsilon a \frac{c_2 - c_1}{p+h}\right)^2$$

is the maximum contract-exercise cost K below which the buyer would exercise the purchase contract.

8.3.2 The Supplier's Reaction Strategy

Following the $(\sigma(i, K), \Sigma(i, K))$ policy, the buyer will exercise the purchase contract if the initial order quantity q_1 is less than the reordering point

$$\sigma(i, K) = i + \varepsilon a \left(\beta - \frac{1}{2}\right) - \sqrt{\frac{2\varepsilon a K}{p+h}},$$

and the reorder quantity q_2 is equal to

$$q_2^* = \Sigma(i, K) - q_1 = i + \varepsilon a \left(\beta - \frac{1}{2}\right) - q_1. \tag{8.14}$$

Therefore, the supplier's payoff function is

$$\begin{aligned}
& J_1(q_1, K) \\
&= (c_1 - w_1) q_1 + \int_{q_1 - \varepsilon a(\beta - \frac{1}{2}) + \mu(K)}^{\gamma + \frac{a}{2}} \Big\{ K + (c_2 - w_2) \cdot \\
&\qquad \cdot \Big[i + \varepsilon a \Big(\beta - \frac{1}{2}\Big) - q_1 \Big] \Big\} \mathrm{d}\Lambda(i) \\
&= (c_1 - w_1) \cdot q_1 \\
&\quad + \Big[\gamma + \frac{a}{2} - q_1 + \varepsilon a \Big(\beta - \frac{1}{2} - \frac{c_2 - w_2}{p+h}\Big) - \mu(K)\Big] \frac{K}{a} \\
&\quad + \frac{c_2 - w_2}{2a} \Big[\gamma + \frac{a}{2} - q_1 + \varepsilon a \Big(\beta - \frac{1}{2}\Big)\Big]^2.
\end{aligned} \tag{8.15}$$

Taking the first- and second-order partial derivatives with respect to the contract-exercise cost K gives

$$\begin{aligned}\frac{\partial J_1(q_1, K)}{\partial K} &= \frac{1}{a}\Big[\gamma + \frac{a}{2} - q_1 \\ &\quad + \varepsilon a\Big(\beta - \frac{1}{2} - \frac{c_2 - w_2}{p+h}\Big) - \frac{3}{2}\mu(K)\Big], \\ \frac{\partial^2 J_1(q_1, K)}{\partial K^2} &= -\frac{3}{4a}\sqrt{\frac{2\varepsilon a}{K(p+h)}} \le 0.\end{aligned}$$

LEMMA 8.2 *The supplier's payoff $J_1(q_1, K)$ is a concave function with respect to the contract-exercise cost K and attains its maximum at*

$$K^* = \frac{2(p+h)}{9\varepsilon a}\Big[\gamma + \frac{a}{2} - q_1 + \varepsilon a\Big(\beta - \frac{1}{2} - \frac{c_2 - w_2}{p+h}\Big)\Big]^2. \tag{8.16}$$

Note that the supplier's optimal decision K^* is a parabola with a vertex $\Big(\gamma + \frac{a}{2} + \varepsilon a(\beta - \frac{1}{2} - \frac{c_2 - w_2}{p+h}), 0\Big)$. By Theorem 8.3, K^* is a decreasing function of q_1. Hence, only the left branch of the parabola is the supplier's reaction-strategy curve. The supplier's optimal payoff function is

$$\begin{aligned}\pi_1^{*s}(q_1) &= \max_K \{J_1(q_1, K)\} \\ &= (c_1 - w_1) \cdot q_1 + \frac{2(p+h)}{27\varepsilon a^2}\Big[\gamma + \frac{a}{2} - q_1 \\ &\quad + \varepsilon a\Big(\beta - \frac{1}{2} - \frac{c_2 - w_2}{p+h}\Big)\Big]^3 \\ &\quad + \frac{c_2 - w_2}{2a}\Big[\gamma + \frac{a}{2} - q_1 + \varepsilon a\Big(\beta - \frac{1}{2}\Big)\Big]^2.\end{aligned} \tag{8.17}$$

8.4. A Static Noncooperative Game

It has been widely documented that a channel coordination would improve the efficiency of a supply chain. One of the most popular mechanisms in achieving channel coordination is the information-sharing scheme—that is, each party provides its private information to the other party. In this purchase-contract setting, information sharing can take place in many forms, such as the cost or payoff structure, inventory replenishment policies, and demand (the form and parameters of the demand distribution). In this chapter, we assume that both the buyer and the supplier know each other's cost or payoff structure and inventory replenishment policies. The information-sharing scheme means that the buyer provides the demand information to the supplier.

In this section, a static noncooperative game is used to analyze the decisions of the supplier and the buyer. Two cases, with and without information sharing, are considered.

8.4.1 The Static Game with Information Sharing

In this section, we explore the Nash equilibrium in the case with information sharing. Denote $q_1 = r_b(K)$ and $K = r_s(q_1)$ as the reaction functions for the buyer and the supplier, respectively—that is,

$$\begin{aligned} q_1 &= r_b(K) = \arg\min_{q_1}\{\Pi_1(q_1, K)\}, \quad \text{for all } K \geq 0; \\ K &= r_s(q_1) = \arg\max_{K}\{J_1(q_1, K)\}, \quad \text{for all } q_1 \geq 0. \end{aligned}$$

By Lemma 8.1, the buyer's reaction function is

$$\begin{aligned} q_1 &= r_b(K) \\ &= \begin{cases} \gamma - \frac{a}{2} + \varepsilon a\left(\beta - \frac{1}{2}\right) + \mu(K + c_2 a - c_1 a), \\ \qquad\qquad \text{if } \mu(K) \leq \frac{a}{2} - \varepsilon a \frac{c_2 - c_1}{p+h}, \\ \gamma + \varepsilon a\left(\frac{p-c_1}{p+h} - \frac{1}{2}\right), \\ \qquad\qquad \text{if } \mu(K) > \frac{a}{2} - \varepsilon a \frac{c_2 - c_1}{p+h}. \end{cases} \end{aligned}$$

It is worth noting that the buyer's reaction function $r_b(K)$ is an increasing function of K.

By Lemma 8.2, the supplier's reaction function is the left branch of the parabola (8.16), when $0 \leq q_1 \leq \gamma + \frac{a}{2} + \varepsilon a(\beta - \frac{1}{2} - \frac{c_2 - w_2}{p+h})$. For the case of $q_1 > \gamma + \frac{a}{2} + \varepsilon a(\beta - \frac{1}{2} - \frac{c_2 - w_2}{p+h})$, we define the contract-exercise cost as zero. Therefore, the supplier's reaction function can be written as follows:

$$\begin{aligned} K &= r_s(q_1) \\ &= \begin{cases} \frac{2(p+h)}{9\varepsilon a}\left[\gamma + \frac{a}{2} - q_1 + \varepsilon a(\beta - \frac{1}{2} - \frac{c_2 - w_2}{p+h})\right]^2, \\ \qquad\qquad \text{if } q_1 \leq \gamma + \frac{a}{2} + \varepsilon a\left(\beta - \frac{1}{2} - \frac{c_2 - w_2}{p+h}\right), \\ 0, \qquad \text{if } q_1 > \gamma + \frac{a}{2} + \varepsilon a\left(\beta - \frac{1}{2} - \frac{c_2 - w_2}{p+h}\right). \end{cases} \end{aligned} \tag{8.18}$$

Since this reaction function is strictly decreasing in q_1, when

$$0 \leq q_1 \leq \gamma + \frac{a}{2} + \varepsilon a\left(\beta - \frac{1}{2} - \frac{c_2 - w_2}{p + h}\right),$$

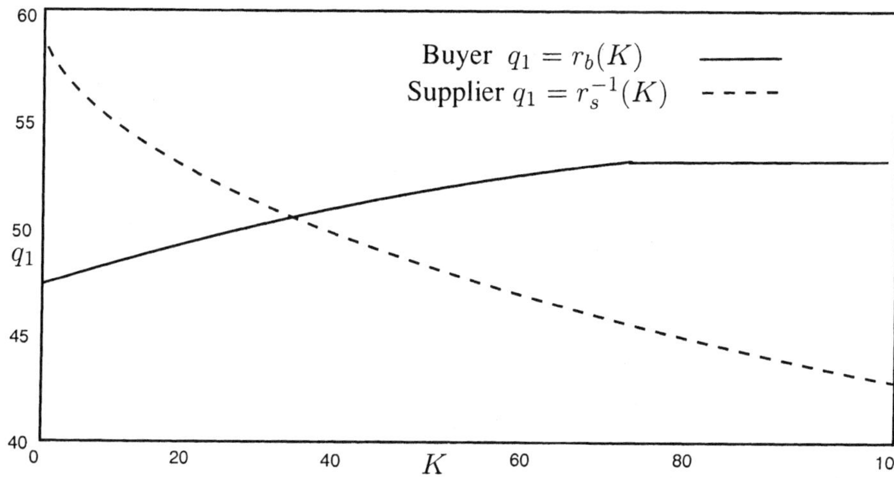

Figure 8.1. Reaction functions of the buyer and the supplier

the inverse function $q_1 = r_s^{-1}(K)$ exists, and $r_s^{-1}(K)$ is strictly decreasing in K as well. To demonstrate the competitive behaviors of the buyer and the supplier, we depict the reaction functions $r_b(K)$ and $r_s^{-1}(K)$ in Figure 8.1.

THEOREM 8.5 *There exists a unique equilibrium (q_1^e, K^e) that is the intersection of the two reaction curves $r_b(K)$ and $r_s^{-1}(K)$. Formally,*

(i) *if*

$$\gamma + \frac{a}{2} + \varepsilon a\left(\beta - \frac{1}{2} - \frac{c_2 - w_2}{p+h}\right) = \gamma - \frac{a}{2} + \varepsilon a\left(\beta - \frac{1}{2}\right) + \mu(c_2 a - c_1 a)$$

—that is, $a - \varepsilon a\frac{c_2 - w_2}{p+h} - \mu(c_2 a - c_1 a) = 0$—then the equilibrium is

$$\begin{cases} q_1^e &= \gamma - \frac{a}{2} + \varepsilon a\left(\beta - \frac{1}{2}\right) + \mu(c_2 a - c_1 a), \\ K^e &= 0; \end{cases} \tag{8.19}$$

(ii) *if*

$$\gamma + \frac{a}{2} + \varepsilon a\left(\beta - \frac{1}{2} - \frac{c_2 - w_2}{p+h}\right) > \gamma - \frac{a}{2} + \varepsilon a\left(\beta - \frac{1}{2}\right) + \mu(c_2 a - c_1 a)$$

—that is, $a - \varepsilon a\frac{c_2-w_2}{p+h} - \mu(c_2 a - c_1 a) > 0$ *and* $\mu(K) \le \frac{a}{2} - \varepsilon a\frac{c_2-c_1}{p+h}$*—then the equilibrium is*

$$\begin{cases} q_1^e &= \gamma - \frac{13}{10}a + \varepsilon a[\beta - \frac{1}{2} + \frac{4(c_2-w_2)}{5(p+h)}] + \frac{3a}{5(p+h)}\cdot \\ & \quad \cdot\sqrt{10(c_2-c_1)\varepsilon(p+h) + 4(p+h-c_2\varepsilon+w_2\varepsilon)^2}, \\ K^e &= \frac{2a}{25\varepsilon(p+h)}\Big[3(p+h-c_2\varepsilon+w_2\varepsilon) \\ & \quad -\sqrt{10(c_2-c_1)\varepsilon(p+h)+4(p+h-c_2\varepsilon+w_2\varepsilon)^2}\Big]^2; \end{cases} \tag{8.20}$$

(iii) *if*

$$a - \varepsilon a\frac{c_2 - w_2}{p+h} - \mu(c_2 a - c_1 a) > 0,$$

and

$$\frac{a}{2} - \varepsilon a\frac{c_2 - c_1}{p+h} < \mu(K),$$

then the equilibrium is

$$\begin{cases} q_1^e &= \gamma + \varepsilon a\left(\frac{p-c_1}{p+h} - \frac{1}{2}\right), \\ K^e &= \frac{2(p+h)}{9\varepsilon a}\left[\frac{a}{2} - \varepsilon a\frac{2c_2-c_1-w_2}{p+h}\right]^2. \end{cases} \tag{8.21}$$

REMARK 8.2 Note that the buyer's reaction function $r_b(K)$ is an increasing function of K. By (8.17) and (8.18), to make the reaction functions $r_b(K)$ and $r_s^{-1}(K)$ have an intersection, the inequality

$$\gamma - \frac{a}{2} + \varepsilon a\left(\beta - \frac{1}{2}\right) + \mu(K + c_2 a - c_1 a)\Big|_{K=0}$$
$$\le \gamma + \frac{a}{2} + \varepsilon a\left(\beta - \frac{1}{2} - \frac{c_2 - w_2}{p+h}\right)$$

should hold. In view of this fact, the theorem gives a complete description for the Nash equilibrium between the supplier and the buyer.

Proof of Theorem 8.5 (i) If

$$\gamma + \frac{a}{2} + \varepsilon a\left(\beta - \frac{1}{2} - \frac{c_2 - w_2}{p+h}\right) = \gamma - \frac{a}{2} + \varepsilon a\left(\beta - \frac{1}{2}\right) + \mu(c_2 a - c_1 a),$$

then for any $K \geq 0$, the buyer's reaction function is

$$\begin{aligned} q_1 = r_b(K) \geq r_b(0) &= \gamma - \frac{a}{2} + \varepsilon a\left(\beta - \frac{1}{2}\right) + \mu(c_2 a - c_1 a) \\ &= \gamma + \frac{a}{2} + \varepsilon a\left(\beta - \frac{1}{2} - \frac{c_2 - w_2}{p+h}\right). \end{aligned}$$

By (8.18), for any q_1, the supplier's reaction function is $K = 0$. Therefore, the intersection (8.19) of two players' reaction functions is a Nash equilibrium.

(ii) If

$$\begin{aligned} &\gamma + \frac{a}{2} + \varepsilon a\left(\beta - \frac{1}{2} - \frac{c_2 - w_2}{p+h}\right) \\ &> \gamma - \frac{a}{2} + \varepsilon a\left(\beta - \frac{1}{2}\right) + \mu(c_2 a - c_1 a), \end{aligned}$$

then for any $q_1 \geq \gamma + \frac{a}{2} + \varepsilon a(\beta - \frac{1}{2} - \frac{c_2 - w_2}{p+h})$, the supplier's reaction function is $K = 0$. However, the buyer's corresponding reaction strategy is

$$\begin{aligned} q_1 &= r_b(0) \\ &= \gamma - \frac{a}{2} + \varepsilon a\left(\beta - \frac{1}{2}\right) + \mu(c_2 a - c_1 a) \\ &< \gamma + \frac{a}{2} + \varepsilon a\left(\beta - \frac{1}{2} - \frac{c_2 - w_2}{p+h}\right). \end{aligned} \tag{8.22}$$

As a result, it is impossible for the two reaction curves to intersect at $K = 0$.

For any $q_1 < \gamma + \frac{a}{2} + \varepsilon a(\beta - \frac{1}{2} - \frac{c_2 - w_2}{p+h})$, by (8.18), the supplier's reaction function is

$$K = \frac{2(p+h)}{9\varepsilon a}\left[\gamma + \frac{a}{2} - q_1 + \varepsilon a\left(\beta - \frac{1}{2} - \frac{c_2 - w_2}{p+h}\right)\right]^2. \tag{8.23}$$

When $\mu(K) \leq \frac{a}{2} - \varepsilon a \frac{c_2 - c_1}{p+h}$, the buyer's reaction function is

$$q_1 = \gamma - \frac{a}{2} + \varepsilon a\left(\beta - \frac{1}{2}\right) + \mu(K + c_2 a - c_1 a). \tag{8.24}$$

The simultaneous solution of (8.23) and (8.24) gives the Nash equilibrium (8.20).

(iii) The proof is similar to (ii). □

The equilibrium provides the optimal strategy pair for both parties. Any action to deviate from it would make at least one party worse off. Therefore, K^e is the competitive purchase-contract exercise cost.

8.4.2 The Static Game Without Information Sharing

In this subsection, we investigate the competitive behaviors of the buyer and the supplier where the information-sharing scheme does not exist.

Since the buyer keeps its demand information private, the supplier has to rely on its own estimation. To fully understand the impacts of information sharing, we concentrate on a specific and simple scenario. Specifically, the supplier knows the form of the distribution function but does not know the parameter of the distribution function. We take the location parameter of the information as an example.

Let $\hat{\gamma}$ be the supplier's estimate of the location parameter γ of the information I, and let $\hat{\Lambda}(i)$ be the corresponding distribution when $I = i$. Then the supplier's estimated payoff function is

$$\begin{aligned}
&\hat{J}_1(q_1, K)\\
&= (c_1 - w_1) \cdot q_1 + \mathsf{E}[\hat{J}_2(q_1, K)]\\
&= (c_1 - w_1) \cdot q_1 + \mathsf{E}\left[K \cdot \delta(\hat{q}_2^*) + (c_2 - w_2) \cdot \hat{q}_2^*\right]\\
&= (c_1 - w_1) \cdot q_1\\
&\quad + \int_{q_1 - \varepsilon a(\beta - \frac{1}{2}) + \mu(K)}^{\hat{\gamma} + \frac{a}{2}} \Big\{K + (c_2 - w_2) \cdot\\
&\qquad\qquad \cdot \left[i + \varepsilon a\left(\beta - \frac{1}{2}\right) - q_1\right]\Big\} \mathrm{d}\hat{\Lambda}(i)\\
&= (c_1 - w_1) \cdot q_1 + \Big[\hat{\gamma} + \frac{a}{2} - q_1\\
&\qquad + \varepsilon a \cdot \left(\beta - \frac{1}{2} - \frac{c_2 - w_2}{p + h}\right) - \mu(K)\Big] \cdot \frac{K}{a}\\
&\quad + \frac{c_2 - w_2}{2a}\left[\hat{\gamma} + \frac{a}{2} - q_1 + \varepsilon a\left(\beta - \frac{1}{2}\right)\right]^2.
\end{aligned}$$

Note that the difference between the estimated payoff function and the true payoff function of the supplier lies in the location parameter of the information. Similar to Lemma 8.2, we develop the following result.

LEMMA 8.3 *The supplier's estimated payoff $\hat{J}_1(q_1, K)$ is a concave function of K and attains its maximum at*

$$\hat{K}^* = \frac{2(p + h)}{9\varepsilon a} \cdot \left[\hat{\gamma} + \frac{a}{2} - q_1 + \varepsilon a \cdot \left(\beta - \frac{1}{2} - \frac{c_2 - w_2}{p + h}\right)\right]^2. \qquad (8.25)$$

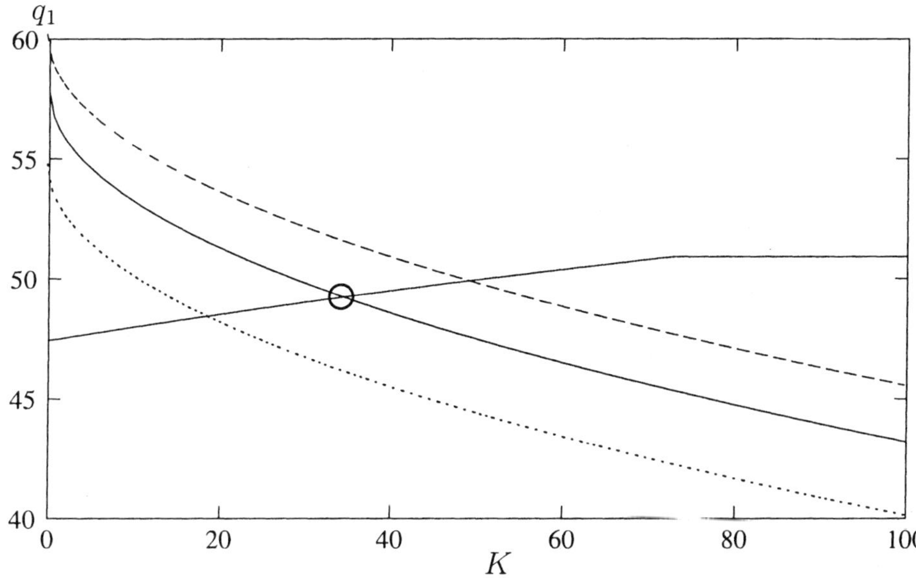

Figure 8.2. Reaction functions of both parties with and without information sharing

Let $K = \hat{r}_s(q_1)$ be the supplier's estimated reaction function, then for any $q_1 \geq 0$,

$$K = \hat{r}_s(q_1) = \arg\max_K \{\hat{J}_1(q_1, K)\} = \hat{K}^*. \tag{8.26}$$

It is worth noting that the supplier's estimated reaction function has a form similar to the true reaction function $r_s(q_1)$ of (8.18) except that the vertex is shifted. In what follows, we say that the supplier underestimates (resp. overestimates) the demand when $\hat{\gamma} < \gamma$ (resp. $\hat{\gamma} > \gamma$). If the supplier underestimates (resp. overestimates) the demand, the vertex of the reaction curve $\hat{r}_s(q_1)$ moves downward (resp. upward). We depict the reaction curves of both parties in the cases with and without information sharing in Figure 8.2, where the intersection of reaction curves $r_b(K)$ and $\hat{r}_s(q_1)$ gives the equilibrium $(\hat{q}_1^e, \hat{K}^e)$.

8.4.3 Impact of Information Sharing

With competitive solutions for cases with or without information sharing, it is interesting to explore the impacts of information sharing on both parties and the channel. A significant body of literature has reported on the benefits of information sharing, both analytically and empirically. It has been widely reported that the supplier and the channel are better off under an information-sharing scheme. However, the benefits to the buyer have not been clearly addressed. With the results obtained in the previous sections, we find that it is necessary for the supplier to provide an incentive to entice the buyer to practice information sharing. We summarize our findings in the following theorem.

THEOREM 8.6 *If the supplier underestimates (resp. overestimates) the demand in the case without information sharing, then*

(i) *the equilibrium contract-exercise cost and the initial order quantity are less (resp. larger) than those in the case with information sharing—that is,* $\hat{K}^e \leq K^e$ *and* $\hat{q}_1^e \leq q_1^e$ *if and only if* $\hat{\gamma} \leq \gamma$;

(ii) *the buyer's equilibrium cost is less (resp. larger) than that in the case with information sharing—that is,* $\pi_1^{*b}(\hat{K}^e) \leq \pi_1^{*b}(K^e)$ *if and only if* $\hat{\gamma} \leq \gamma$.

Proof (i) Let $\hat{r}_s^{-1}(K)$ denote the inverse function of the supplier's reaction function $\hat{r}_s(q_1)$. Lemma 8.1 implies that the buyer's reaction curve $r_b(K)$ is nondecreasing in K, while Lemmas 8.2 and 8.3 ensure that parabola $\hat{r}_s^{-1}(K)$ moves downward relatively to parabola $r_s^{-1}(K)$ when $\hat{\gamma} \leq \gamma$. Thus, $\hat{K}^e \leq K^e$ and $\hat{q}_1^e \leq q_1^e$.

(ii) $\pi_1^{*b}(\hat{K}^e) \leq \pi_1^{*b}(K^e)$ follows from Corollary 8.1 and $\hat{K}^e \leq K^e$. □

Theorem 8.6 leads to an interesting and intuitively appealing competitive behavior. The supplier observes the buyer's initial order and makes decisions on the possibility of exercising a contract. Since the buyer makes decisions based on the true demand distribution, the supplier compares the buyer's initial order and its own estimation. If the supplier believes that the buyer has ordered a sufficient quantity, the supplier will reduce the exercise cost to entice the buyer to change its initial order. On the other hand, if the supplier believes that the buyer ordered too little and expects additional orders at stage 2, the supplier has the tendency to increase the contract-exercise cost. As the results show, the buyer would be better off for not sharing demand information with the supplier if the supplier underestimates the demand.

To this end, it is natural to evaluate the impact of information sharing on the supplier and further to evaluate the impact on the channel. Although there is no doubt that the supplier is always worse off when the supplier underestimates the demand, we are not able to demonstrate that the supplier is always worse off when the supplier overestimates the demand. Actually, we have found examples where the buyer and the supplier do better without information sharing, respectively.

EXAMPLE 8.1 The information I is uniformly distributed with $a = 20$ and $\gamma = 50$ as its spread and center parameters, respectively. Inventory holding and shortage penalty costs are 0.3 and 10, respectively. Both the stage 1 and stage 2 ordering costs are 3 per unit. Production costs of the supplier are $w_1 = 1$ and $w_2 = 2$ for stage 1 and stage 2, respectively. The forecast-improvement factor is 0.75.

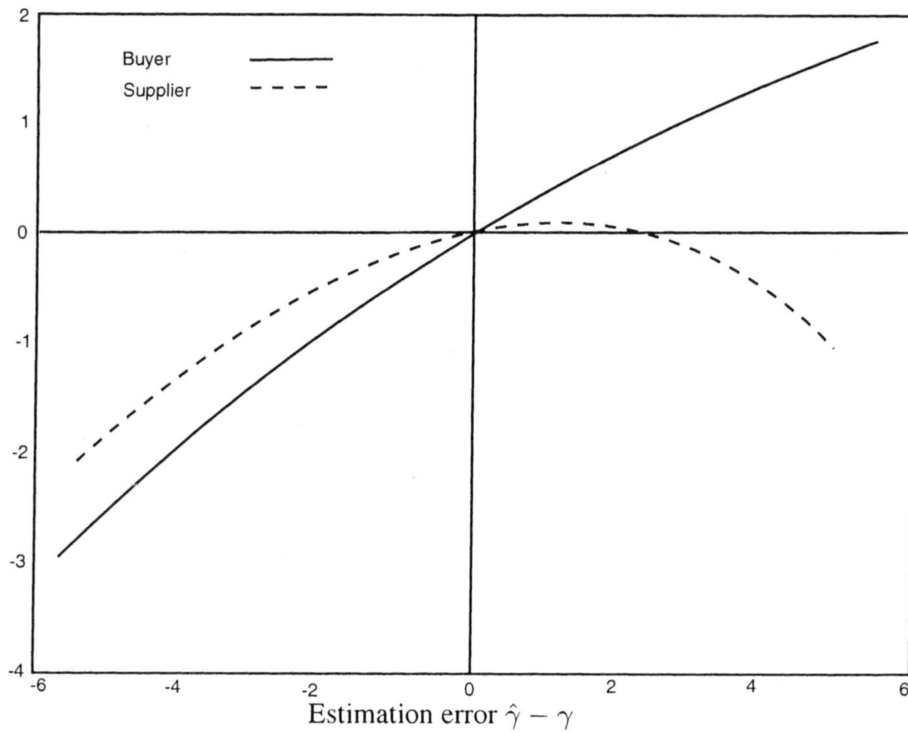

Figure 8.3. Objective functions as functions of the estimation error in the static game

In the case with information sharing, the equilibrium contract-exercise cost and initial order size are 18.89 and 50.11, respectively. The cost and the payoff for the buyer and the supplier are 176.37 and 107.68, respectively.

If the supplier underestimates the demand by one unit—that is, $\hat{\gamma} = 49$—then the equilibrium contract-exercise cost and the initial order size are 16.91 and 49.71, respectively. The buyer's cost is reduced by 0.55, and the supplier's payoff is reduced by 0.24 as well. This example cooperates our earlier findings in Theorem 8.6—that is, the buyer is better off and the supplier is worse off without information sharing.

If the supplier overestimates the demand by one unit—that is, $\hat{\gamma} = 51$—that is then the equilibrium contract-exercise cost and the initial order size are 20.98 and 50.51. The buyer's cost is increased by 0.50, and the supplier's payoff is increased by 0.10. From this example, it is worth noting that without information sharing, the buyer is worse off and the supplier is better off.

The relationship of estimation error and changes in the buyer's cost and the supplier's payoff, are depicted in Figure 8.3. We observe that the buyer is always better off if the supplier underestimates its demand, as concluded in Theorem 8.6. On the other hand, if the supplier overestimates the buyer's demand, the supplier can be better off without information sharing, especially for the cases where the estimation errors are small.

Example 8.3 presents an interesting result. It is expected that the supplier would do better with the true demand information. We believe that the phenomenon results from the known pitfalls of static game, such that a simultaneous move leads to an empty threat and rival cheating. Since the supplier knows that the buyer would do better if the supplier underestimates the demand, it is necessary for the supplier to overestimate the demand. To prevent empty threats and rival cheating, dynamic game theory ensures that both parties make decisions based on each other's true information. In the next section, we explore the same issues in a dynamic game setting.

8.5. A Dynamic Noncooperative Game

We consider a two-step dynamic game where two players move in sequence. The game is played as follows: In Step 1, the supplier provides a contract-exercise cost K. In Step 2, the buyer chooses the optimal initial order quantity q_1 for the given contract-exercise cost K. The process terminates until two players reach an equilibrium from which no party is willing to deviate. The subgame-perfect Nash equilibrium is the optimal solution of the dynamic game, and it can be obtained by the following backward-induction procedure:

(i) For the given contract-exercise cost K, find the buyer's reaction function $q_1 = r_b(K) = \arg\min_{q_1}\{\Pi_1(q_1, K)\}, \forall K$, which is the same as in the static game.

(ii) Substitute $q_1 = r_b(K)$ into the supplier's payoff function $J_1(q_1, K)$, and find K such that $K^d = \arg\max_{K}\{J_1(r_b(K), K)\}$.

(iii) The subgame-perfect Nash equilibrium is K^d and $q_1^d = r_b(K^d)$.

The backward induction scheme significantly increases the difficulties that are inherent in exploring analytical equilibrium solutions. It seems to be very difficult to find an explicit form of equilibrium for the general case. However, it is possible to find solutions for some special cases—for example, when the purchase costs for the two stages are the same—(that is, $c_1 = c_2$). In what follows, we concentrate on the case of $c_1 = c_2$. It is our goal to investigate the explicit subgame-perfect Nash equilibrium in the cases with and without information sharing and further explore the impacts of an information-sharing scheme on both players and the channel.

8.5.1 The Subgame-Perfect Nash Equilibrium with Information Sharing

Since the buyer moves after the supplier announces the contract-exercise cost, the reaction of the buyer is the same as in the static game. Therefore, by (8.17), the buyer's reaction function is

$$q_1 = r_b(K) = \gamma - \frac{a}{2} + \varepsilon a \left(\beta - \frac{1}{2}\right) + \mu(K). \tag{8.27}$$

LEMMA 8.4 *The supplier's payoff function $J_1(r_b(K), K)$ is a concave function of K and attains its maximum at*

$$\begin{aligned} K^d \;=\; & \frac{a}{36\varepsilon}\Big[p + h + 6\varepsilon\cdot(w_2 - w_1) \\ & + \sqrt{(p+h)[p+h+12\varepsilon\cdot(w_2-w_1)]}\,\Big]. \end{aligned} \tag{8.28}$$

Proof Substitute (8.27) into $J_1(q_1, K)$. The first-order and second-order derivatives of $J_1(r_b(K), K)$ are

$$\begin{aligned} & \frac{\partial J_1(r_b(K), K)}{\partial K} \\ & = \frac{1}{\mu(K)\cdot(p+h)}\Big[\mu(K)\cdot(p+h) + \varepsilon a\cdot(w_2 - w_1) - 6\varepsilon K\Big], \\ & \frac{\partial^2 J_1(r_b(K), K)}{\partial K^2} \\ & = -\frac{\varepsilon}{2}\,\frac{6K + (w_2 - w_1)a}{K(p+h)\mu(K)} \le 0. \end{aligned}$$

Therefore, the supplier's payoff function $J_1(r_b(K), K)$ is concave in the contract-exercise cost K, and the first-order condition yields the optimal solution K^d.

□

Following the backward-induction procedure, we have the following theorem.

THEOREM 8.7 *The subgame-perfect Nash equilibrium is (q_1^d, K^d), where the equilibrium contract-exercise cost K^d is characterized by* (8.28), *and the equilibrium initial order quantity $q_1^d = r_b(K^d)$ is determined by* (8.27).

8.5.2 The Subgame-Perfect Nash Equilibrium Without Information Sharing

Similar to the information structure of the static game without information sharing in Section 8.4.2, we assume that the buyer keeps its private information of demand distribution and that the supplier relies on its own estimation. Therefore, the equilibrium contract-exercise cost can be found by

$$\max_K \left\{ \hat{J}_1(r_b(K), K) \right\}. \tag{8.29}$$

For simplicity, we denote Δ as the supplier's estimation error $\hat{\gamma} - \gamma$.

LEMMA 8.5 *For the supplier's payoff function $\hat{J}_1(r_b(K), K)$ considered as a function K, there exists an inflection point at $\frac{1}{6}[(c_2 - w_2)\Delta - (w_2 - w_1)a]$. If the contract-exercise cost K is greater than or equal to the inflection point, the payoff function is concave, and its local maximum is obtained at*

$$\begin{aligned} \hat{K}^d &= \frac{1}{36\varepsilon a}\Big[(a+\Delta)^2(p+h) + 6\varepsilon a[(w_2 - w_1)a - (c_2 - w_2)\Delta] \\ &\quad +(a+\Delta)\sqrt{p+h}\,\cdot \\ &\quad \cdot\sqrt{(a+\Delta)^2(p+h) + 12\varepsilon a[(w_2 - w_1)a - (c_2 - w_2)\Delta]}\Big]. \end{aligned} \tag{8.30}$$

Otherwise, the payoff function is convex, and the local maximum is obtained at $K = 0$.

Proof The lemma is the immediate results of the following derivatives of $\hat{J}_1(r_b(K), K)$:

$$\begin{aligned} &\frac{\partial \hat{J}_1(r_b(K), K)}{\partial K} \\ &= \frac{1}{\mu(K)(p+h)a}\Big\{ -3(p+h)\mu^2(K) + \mu(K)(p+h)(a+\Delta) \\ &\qquad + \varepsilon a\left[(w_2 - w_1)a - (c_2 - w_2)\Delta\right]\Big\}, \end{aligned} \tag{8.31}$$

$$\begin{aligned} &\frac{\partial^2 \hat{J}_1(r_b(K), K)}{\partial K^2} \\ &= -\frac{\varepsilon}{2}\,\frac{6K + (w_2 - w_1)a - (c_2 - w_2)\Delta}{K(p+h)\mu(K)}. \end{aligned} \tag{8.32}$$

□

Specifically, if the inflection point is negative, then $\hat{J}_1(r_b(K), K)$ is concave function of K, and $\hat{K}^d$ is the global maximum. If the inflection point is nonnegative, then $\hat{J}_1(r_b(K), K)$ is concave if the contract-exercise cost K is greater than or equal to the inflection point, while it is convex if the contract-exercise cost K is less than the inflection point. Similar to Theorem 8.7, we have the following theorem.

THEOREM 8.8 *The subgame-perfect Nash equilibrium is*

$$(\hat{q}_1^d, \hat{K}^d) \text{ or } \left(\gamma - \frac{a}{2} + \varepsilon a \left(\beta - \frac{1}{2}\right), 0\right),$$

where the equilibrium contract-exercise cost $\hat{K}^d$ is characterized by (8.30), *and the equilibrium initial order quantity $\hat{q}_1^d = r_b(\hat{K}^d)$ is determined by* (8.27).

8.5.3 Effects of Information Sharing on the Decisions

Parallel to our analysis for the static game in Section 8.4.3, we are able to explore the impacts of an information-sharing scheme on both parties in the dynamic game. Recall that we were not able to make a conclusive statement for the supplier in the static game setting. However, for the dynamic game setting, we are able to prove that the supplier is always better off with information sharing. We present the main conclusion in the following theorem.

THEOREM 8.9 *In the dynamic game, the supplier is always better off in the case with an information-sharing scheme—that is, $J_1(q_1^d, K^d) \geq J_1(\hat{q}_1^d, \hat{K}^d)$.*

Proof By Lemma 8.4, with an information-sharing scheme, the equilibrium contract-exercise cost K^d maximizes the payoff function $J_1(r_b(K), K)$. Therefore, without information sharing, the estimation error is not zero in general—that is, $\hat{\gamma} \neq \gamma$—and as a result, $\hat{K}^d \neq K^d$. Hence, $J_1(q_1^d, K^d) \geq J_1(\hat{q}_1^d, \hat{K}^d)$. □

Recall that the buyer's equilibrium cost $\pi_1^{*b}(K) = \Pi_1(r_b(K), K)$ is an increasing function of the contract-exercise cost K (Corollary 8.1). In the static game, by showing that the contract-exercise cost K is an increasing function of the estimation error (Theorem 8.6), the impact of information sharing is identified. Although we conjecture that the monotone property of the contract-exercise cost preserves in the dynamic game, we are able to prove the property only in the following two cases.

LEMMA 8.6 *Assume that the supplier's production cost remains the same. If $(a+\Delta)(p+h) \geq 6\varepsilon a(c_2 - w_2)$ or $p+h \geq 3\varepsilon(c_2 - w_2)$, then $\hat{K}^d$ is increasing with respect to the supplier's estimation error Δ.*

Proof When $c_1 = c_2$, $w_1 = w_2$, the first-order condition $\frac{\partial \hat{J}_1(r_b(K),K)}{\partial K} = 0$ is simplified as

$$\frac{(c_2 - w_2)\varepsilon\Delta}{\mu(K)(p+h)} + \frac{3\mu(K)}{a} - \frac{\Delta}{a} - 1 = 0. \tag{8.33}$$

Denote the left-hand side function as M. Then

$$\begin{aligned}\frac{\partial K}{\partial \Delta} &= -\frac{\partial M}{\partial \Delta} \Big/ \frac{\partial M}{\partial K} \\ &= \frac{2K}{\varepsilon a} \cdot \frac{\mu(K)(p+h) - \varepsilon a \cdot (c_2 - w_2)}{6K - (c_2 - w_2)\Delta}. \end{aligned} \tag{8.34}$$

Solve (8.33) to obtain

$$\mu(K) = \frac{\varepsilon a}{(a+\Delta)(p+h)}[6K + (c_2 - w_2)\Delta]. \tag{8.35}$$

With (8.35), (8.34) can be further simplified as

$$\frac{\partial K}{\partial \Delta} = \frac{2K}{a+\Delta} \cdot \frac{6K - (c_2 - w_2)a}{6K - (c_2 - w_2)\Delta}, \tag{8.36}$$

where

$$K = \frac{1}{36\varepsilon a}[(a+\Delta)^2(p+h) - 6\varepsilon a(c_2 - w_2)\Delta + (a+\Delta) \cdot \Delta_1 \tag{8.37}$$

and

$$\Delta_1 = \sqrt{(a+\Delta)^2(p+h)^2 - 12\varepsilon a\Delta(c_2 - w_2)(p+h)}.$$

Substitute (8.37) into the right-hand side of (8.36):

$$\frac{\partial K}{\partial \Delta} = 2K\frac{(a+\Delta)(p+h) - 6\varepsilon a(c_2 - w_2) + \Delta_1}{(a+\Delta)^2 \cdot (p+h) - 12\varepsilon a\Delta \cdot (c_2 - w_2) + (a+\Delta) \cdot \Delta_1}. \tag{8.38}$$

It follows from the nonnegativity of

$$(a+\Delta)^2(p+h)^2 - 12\varepsilon a\Delta(c_2 - w_2)(p+h)$$

that

$$(a+\Delta)^2 \cdot (p+h) - 12\varepsilon a\Delta \cdot (c_2 - w_2)$$

is nonnegative. Therefore, the denominator of the right-hand side of (8.38) is nonnegative. Consequently, if $(a+\Delta)(p+h) \geq 6\varepsilon a(c_2 - w_2)$, then $\partial K/\partial\Delta \geq 0$. This implies that $\hat{K}^d$ is monotone increasing with respect to Δ.

If $(a+\Delta)(p+h) < 6\varepsilon a(c_2 - w_2)$, then the numerator of the right-hand side of (8.38) is rewritten as

$$\begin{aligned}&\Delta_1 - [6\varepsilon a \cdot (c_2 - w_2) - (a+\Delta)(p+h)] \\ &= \frac{12\varepsilon a^2(c_2 - w_2)[p+h-3\varepsilon(c_2 - w_2)]}{\Delta_1 + [6\varepsilon a(c_2 - w_2) - (a+\Delta)(p+h)]}. \end{aligned} \tag{8.39}$$

Note that the fraction in (8.39) is nonnegative. Then the results are straightforward. □

REMARK 8.3 Conditions in the above lemma can be interpreted intuitively. For example, let $p + h \geq 3\varepsilon(c_2 - w_2)$ and $c_2 \leq 2w_2$. In other words, if the supplier's profit margin is less than 100%, the contract-exercise cost is increasing with respect to the supplier's estimation error.

THEOREM 8.10 *If conditions in Lemma* 8.6 *hold, then the buyer is better off when the supplier underestimates the demand—that is, if* $\hat{\gamma} \leq \gamma$, *then* $\pi_1^{*b}(\hat{K}^d) \leq \pi_1^{*b}(K^d)$. *Otherwise, the buyer is worse off—that is, if* $\hat{\gamma} \geq \gamma$, *then* $\pi_1^{*b}(\hat{K}^d) \geq \pi_1^{*b}(K^d)$.

Proof If $\hat{\gamma} \leq \gamma$, using Lemma 8.6, we have $\hat{K}^d \leq K^d$. Then $\pi_1^{*b}(\hat{K}^d) \leq \pi_1^{*b}(K^d)$ directly follows from Corollary 8.1. Similarly, we can prove the other result of the theorem. □

EXAMPLE 8.2 Continuing from Example 8.1, find the subgame-perfect Nash equilibria. With information sharing, the equilibrium contract-exercise cost is 21.41, and the equilibrium initial order quantity is 50.59. The cost and the payoff for the buyer and the supplier are 176.96 and 107.79, respectively.

Unlike in the static setting, without information sharing the supplier can underestimate or overestimate the true demand and still always be worse off, as claimed in Theorem 8.9. We depict the supplier's payoff and the buyer's cost curves with respect to the estimation error in Figure 8.4.

8.6. Concluding Remarks

In this chapter, we develop equilibrium solutions for the purchase-contract problem. With equilibria for the cases with and without information sharing, it is possible to evaluate the impacts of an information-sharing scheme on both parties in the dynamic game setting. We conclude that (1) information sharing is always beneficial to the party that lacks true information (the supplier in this problem) and that (2) information sharing may hurt the party with the true information (the buyer in this problem). We further demonstrate that the outcome depends on how well the less-informed party estimates the information.

It is clear that an incentive mechanism is necessary to entice the well-informed party to practice information sharing. The incentive should be no less than the gain for the well-informed party and should be no more than the loss for the less-informed party. If this incentive-design criterion is acceptable to both parties, then the issue becomes whether the information-sharing mechanism benefits the channel.

As is demonstrated in Sections 8.4 and 8.5, the benefit of information sharing depends on both parties' cost or payoff structures and the quality of the supplier's

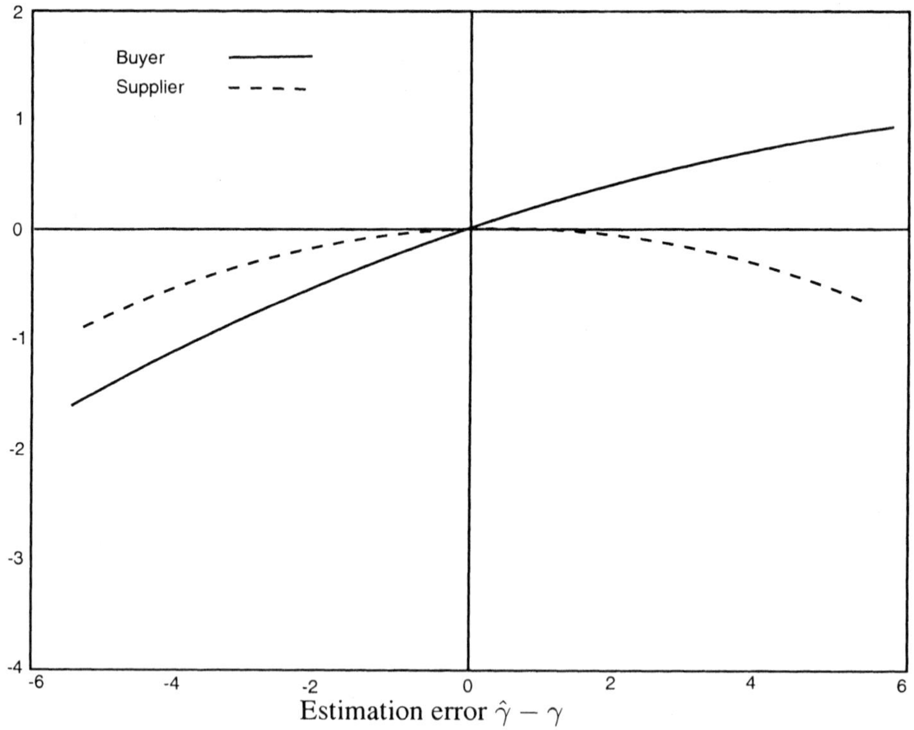

Figure 8.4. Objective functions as functions of the estimation error in the dynamic game

estimation of the demand. There is no doubt that information sharing results in a significant benefit when the supplier's estimation is poor. Further, the estimation quality also affects the benefit of information sharing for the channel.

Based on Example 8.3, we explore the benefit of information sharing to the buyer, the supplier, and the channel. Suppose that the supplier's estimation is unbiased with errors of 4 and -4 and probability of 0.5 each. By calculation, the buyer's cost function increases by 0.90 and -1.17 for the estimation error 4 and -4, respectively. The supplier's payoff function decreases by 0.35 and 0.48 for the estimation error 4 and -4, respectively. Therefore, the buyer's average cost increase is $0.5 \times 0.9 - 0.5 \times 1.17 = -0.135$, and the supplier's average payoff decrease is $0.5 \times 0.35 + 0.5 \times 0.48 = 0.415$. As the result, the channel is better off by $0.415 - 0.135 = 0.28$. It is possible for the supplier to provide an incentive that is larger than 0.135 to make the information sharing work.

Next, suppose that the supplier's estimation is biased with errors of 1 and -4 and probability 0.3 and 0.7, respectively. In this scenario, the average cost or payoff increase is -0.455 and -0.190 for the buyer and the supplier, respectively. Actually, the information sharing reduces the channel efficiency. Note that such a biased estimation could happen, especially when the product is in the ramp-up period.

From the above discussion, we would like to point out that in the noncooperative game setting, it is possible to find cases where an information-sharing scheme and an incentive program are difficult to construct. We believe that cooperation between the buyer and the supplier and mechanism of profit sharing such as the Shapley formula might be the solution.

8.7. Notes

This chapter is based on Huang and Yan [7]

Competitive supply chain management has attracted much attention recently. Research covers topics such as characterization of the competitive behavior, coordination mechanism, and incentives design. Cachon and Zipkin [3] study competitive inventory policies in a two-level inventory system constructed by base-stock policies. They demonstrate that each player chooses a competitive policy that is featured by a Nash equilibrium and further that the optimal solution can be established from the Nash equilibrium by a linear transfer payment. Lippman and McCardle [9] study the competitive newsvendor problem, where newsvendors are allowed to switch firms to secure inventory. Chen, Fedegruen, and Zheng [4] investigate a pricing (accounting) scheme in a distribution system where the supplier announces the wholesale price and the retailer determines its own retail price. They argue that the retailer should share some of profits to reward the supplier's participation. For a complete review in competitive models in a supply chain, we refer to a recent survey paper by Cachon [1] and the references therein.

Information sharing, the value of information, and using shared information to enhance performance in a supply chain are areas of importance. In a serial inventory system, Lee, So, and Tang [8] investigate the value of information sharing in assisting ordering functions. Cheung and Lee [5] study the benefit of shipment coordination with information sharing. For the Vendor Managed Inventory (VMI) program, Cheung and Lee [5] find that shared information allows suppliers to consolidate replenishment and enables retailers to balance inventories. Cachon and Fisher [2] compare ordering policies with and without shared information. Their findings reveal that policies with shared information reduce supply chain cost. In their study, the shared information is the retailer's inventory position.

References

[1] G.P. Cachon. Competitive supply chain inventory management. In *Quantittative Models for Supply Chain Management*, S. Tayur, R. Ganeshan, and M. Magazine (editors), pp. 113–146. Kluwer Academic Publishers, Boston, 1999.

[2] G.P. Cachon and M. Fisher. Supply chain inventory management and the value of shared information. *Management Science*, 46:1032–1048, 2000.

[3] G.P. Cachon and P. Zipkin. Competitive and cooperative inventory policies in a two-stage supply chain. *Management Science*, 45:936–953, 1999.

[4] F. Chen, A. Federgruen, and Y. Zheng. Coordination mechanisms for a distribution system with one supplier and multiple retailers. *Management Science*, 47:693–708, 2001.

[5] K.L. Cheung and H.L. Lee. Coordinated replenishments in a supply chain with vendor-managed inventory programs. Working Paper, The University of Science and Technology of Hong Kong, Hong Kong, 1998.

[6] D. Fudenberg and J. Tirole. *Game Theory*. MIT Press, Cambridge, MA, 1993.

[7] H.Y. Huang and H. Yan. Information sharing and updating in supply chain management—A non-corporative game approach. Working Paper, the Chinese University of Hong Kong, Hong Kong, 2002.

[8] H.L. Lee, K.C. So, and C.S. Tang. The value of information sharing in a two-level supply chain. *Management Science*, 46:626–643, 2000.

[9] S. Lippman and K. McCardle. The competitive newsboy. *Operations Research*, 45:54–65, 1997.

[10] P. Milgrom and J. Roberts. The economics of modern manufacturing: Technology, strategy, and organization. *American Business Review*, 80:511–528, 1990.

[11] D.M. Topkis. Equilibrium points in nonzero-sum n-person submodular games. *SIAM Journal of Control and Optimization*, 17:773–787, 1979.

[12] D.D. Yao. S-modular games with queueing applications. *Queueing Systems: Theory and Applications*, 21:449–475, 1995.

Copyright Permissions

Selected portions of the publications below have been reprinted with permissions as indicated.

"Peeling layers of an onion: Inventory model with multiple delivery modes and forecast updates" by Sethi, S.P., Yan, H. and Zhang, H., *Journal of Optimization Theory and Applications*, **108**, 253–281. Copyright ©2001 by Kluwer/Plenum Publishers, 233 Spring Street, 7th Floor, New York, NY 10013-1578, USA.

"Inventory models with fixed costs, forecast updates, and two delivery modes" by Sethi, S.P., Yan, H. and Zhang, H., *Operations Research*, **51**, 321–328. Copyright ©2003 by the INFORMS, 901 Elkridge Landing Road, Suite 400, Linthicum, MD 21090-2909, USA.

"Information revision and decision making in supply chain management" by Yan, H. and Zhang, H., in *Stochastic Modelings and Optimization*, D.D. Yao, H. Zhang, and X. Zhou (editors), 429–457. Copyright ©2003 by Springer-Verlag New York, Inc.

"Optimal ordering in a dual-supplier system with demand updates" by Yan, H., Liu, K. and Hsu, A., *Production and Operations Management*, **12**, 30–45. Copyright ©2003 by POMS Executive Office, College of Engineering, Florida International University, EAS 2460, 10555 West Flagler Street, Miami, FL 33174, USA.

"Quantity-flexibility contracts: Optimal decisions with information updates" by Sethi, S.P., Yan, H. and Zhang, H., *Decision Sciences*, **35**, 691–712. Copyright ©2004 by Blackwell Publishing Ltd., 9600 Garsington Road, Oxford, UK.

Index

Early Titles in the
INTERNATIONAL SERIES IN OPERATIONS RESEARCH & MANAGEMENT SCIENCE
Frederick S. Hillier, Series Editor, *Stanford University*

Saigal/ *A MODERN APPROACH TO LINEAR PROGRAMMING*
Nagurney/ *PROJECTED DYNAMICAL SYSTEMS & VARIATIONAL INEQUALITIES WITH APPLICATIONS*
Padberg & Rijal/ *LOCATION, SCHEDULING, DESIGN AND INTEGER PROGRAMMING*
Vanderbei/ *LINEAR PROGRAMMING*
Jaiswal/ *MILITARY OPERATIONS RESEARCH*
Gal & Greenberg/ *ADVANCES IN SENSITIVITY ANALYSIS & PARAMETRIC PROGRAMMING*
Prabhu/ *FOUNDATIONS OF QUEUEING THEORY*
Fang, Rajasekera & Tsao/ *ENTROPY OPTIMIZATION & MATHEMATICAL PROGRAMMING*
Yu/ *OR IN THE AIRLINE INDUSTRY*
Ho & Tang/ *PRODUCT VARIETY MANAGEMENT*
El-Taha & Stidham/ *SAMPLE-PATH ANALYSIS OF QUEUEING SYSTEMS*
Miettinen/ *NONLINEAR MULTIOBJECTIVE OPTIMIZATION*
Chao & Huntington/ *DESIGNING COMPETITIVE ELECTRICITY MARKETS*
Weglarz/ *PROJECT SCHEDULING: RECENT TRENDS & RESULTS*
Sahin & Polatoglu/ *QUALITY, WARRANTY AND PREVENTIVE MAINTENANCE*
Tavares/ *ADVANCES MODELS FOR PROJECT MANAGEMENT*
Tayur, Ganeshan & Magazine/ *QUANTITATIVE MODELS FOR SUPPLY CHAIN MANAGEMENT*
Weyant, J./ *ENERGY AND ENVIRONMENTAL POLICY MODELING*
Shanthikumar, J.G. & Sumita, U./ *APPLIED PROBABILITY AND STOCHASTIC PROCESSES*
Liu, B. & Esogbue, A.O./ *DECISION CRITERIA AND OPTIMAL INVENTORY PROCESSES*
Gal, T., Stewart, T.J., Hanne, T. / *MULTICRITERIA DECISION MAKING: Advances in MCDM Models, Algorithms, Theory, and Applications*
Fox, B.L. / *STRATEGIES FOR QUASI-MONTE CARLO*
Hall, R.W. / *HANDBOOK OF TRANSPORTATION SCIENCE*
Grassman, W.K. / *COMPUTATIONAL PROBABILITY*
Pomerol, J-C. & Barba-Romero, S. / *MULTICRITERION DECISION IN MANAGEMENT*
Axsäter, S. / *INVENTORY CONTROL*
Wolkowicz, H., Saigal, R., & Vandenberghe, L. / *HANDBOOK OF SEMI-DEFINITE PROGRAMMING: Theory, Algorithms, and Applications*
Hobbs, B.F. & Meier, P. / *ENERGY DECISIONS AND THE ENVIRONMENT: A Guide to the Use of Multicriteria Methods*
Dar-El, E. / *HUMAN LEARNING: From Learning Curves to Learning Organizations*
Armstrong, J.S. / *PRINCIPLES OF FORECASTING: A Handbook for Researchers and Practitioners*
Balsamo, S., Personé, V., & Onvural, R./ *ANALYSIS OF QUEUEING NETWORKS WITH BLOCKING*
Bouyssou, D. et al. / *EVALUATION AND DECISION MODELS: A Critical Perspective*
Hanne, T. / *INTELLIGENT STRATEGIES FOR META MULTIPLE CRITERIA DECISION MAKING*
Saaty, T. & Vargas, L. / *MODELS, METHODS, CONCEPTS and APPLICATIONS OF THE ANALYTIC HIERARCHY PROCESS*
Chatterjee, K. & Samuelson, W. / *GAME THEORY AND BUSINESS APPLICATIONS*
Hobbs, B. et al. / *THE NEXT GENERATION OF ELECTRIC POWER UNIT COMMITMENT MODELS*
Vanderbei, R.J. / *LINEAR PROGRAMMING: Foundations and Extensions, 2nd Ed.*
Kimms, A. / *MATHEMATICAL PROGRAMMING AND FINANCIAL OBJECTIVES FOR SCHEDULING PROJECTS*
Baptiste, P., Le Pape, C. & Nuijten, W. / *CONSTRAINT-BASED SCHEDULING*
Feinberg, E. & Shwartz, A. / *HANDBOOK OF MARKOV DECISION PROCESSES: Methods and Applications*
Ramík, J. & Vlach, M. / *GENERALIZED CONCAVITY IN FUZZY OPTIMIZATION AND DECISION ANALYSIS*
Song, J. & Yao, D. / *SUPPLY CHAIN STRUCTURES: Coordination, Information and Optimization*
Kozan, E. & Ohuchi, A. / *OPERATIONS RESEARCH/ MANAGEMENT SCIENCE AT WORK*
Bouyssou et al. / *AIDING DECISIONS WITH MULTIPLE CRITERIA: Essays in Honor of Bernard Roy*

Early Titles in the
INTERNATIONAL SERIES IN
OPERATIONS RESEARCH & MANAGEMENT SCIENCE
(Continued)

Cox, Louis Anthony, Jr. / *RISK ANALYSIS: Foundations, Models and Methods*
Dror, M., L'Ecuyer, P. & Szidarovszky, F. / *MODELING UNCERTAINTY: An Examination of Stochastic Theory, Methods, and Applications*
Dokuchaev, N. / *DYNAMIC PORTFOLIO STRATEGIES: Quantitative Methods and Empirical Rules for Incomplete Information*
Sarker, R., Mohammadian, M. & Yao, X. / *EVOLUTIONARY OPTIMIZATION*
Demeulemeester, R. & Herroelen, W. / *PROJECT SCHEDULING: A Research Handbook*
Gazis, D.C. / *TRAFFIC THEORY*

A list of the more recent publications in the series is at the front of the book